CONTROLLING
CHEMICAL HAZARDS

CONTROLLING CHEMICAL HAZARDS

Fundamentals of the
management of toxic chemicals

Edited by

RAYMOND P. CÔTÉ

*School for Resource and Environmental Studies,
Dalhousie University, Halifax, Nova Scotia*

&

PETER G. WELLS

*Environment Canada, Dartmouth, Nova Scotia, and
School for Resource and Environmental Studies;
Dalhousie University, Halifax, Nova Scotia*

London
UNWIN HYMAN
Boston   Sydney   Wellington

Published by the Academic Division of
Unwin Hyman Ltd
15/17 Broadwick Street, London W1V 1FP, UK

Unwin Hyman Inc.
955 Massachusetts Avenue, Cambridge, MA 02139, USA

Allen & Unwin (Australia) Ltd
8 Napier Street, North Sydney, NSW 2060, Australia

Allen & Unwin (New Zealand) Ltd
in association with the Port Nicholson Press Ltd
Compusales Building, 75 Ghuznee Street, Wellington 1, New Zealand

First published in 1991

British Library Cataloguing in Publication Data
Controlling Chemical Hazards
Fundamentals of the management of toxic chemicals. –
(The risks & hazards series, ISSN 0261-0507).
1. Toxic chemicals. Regulation
I. Côté, R.   II. Wells, Peter G.   III. Series
363.179

ISBN-13: 978-94-011-6851-9

Library of Congress Cataloging-in-Publication Data
Controlling Chemical Hazards
Fundamentals of the management of toxic chemicals / [edited by]
Raymond Côté and Peter G. Wells.
    p.   cm.
Includes bibliographical references (p. 305) and index.
ISBN 978-94-011-6851-9   ISBN 978-94-011-6849-6 (eBook)
DOI 10.1007/978-94-011-6849-6
1. Hazardous substances – Management.
I. Côté, Raymond, 1947–   II. Wells, P. G.
T55.3.H3F86   1990
660′.2804—dc20   90-12716
                                                  CIP

Typeset in 10 on 12 point Bembo by Computape (Pickering) Ltd,
North Yorkshire

Contents

List of figures

List of tables

Preface

This book presents environmental protection managers and advanced students in environmental studies programs with an overview of the principles, facts, multidisciplinary approaches, and some of the complexities of the management of toxic substances.

The text explores critical issues facing managers responsible for preventing and controlling problems associated with the manufacture, transport, storage, use and disposal of chemicals. It does this from two perspectives. The first is a disciplinary perspective, that is environmental chemistry, toxicology, engineering, economics, sociology and political science, all of which play a role in implementing comprehensive programs to manage chemicals. The second perspective is from the view of industry, government, academia and non-government organizations. For example, Chapter 5 is authored by technical managers of a major chemical company, Chapters 6 and 12 by government scientists and managers respectively, and Chapter 10 by a respected member of the environmental lobby. An appreciation of these perspectives is very important for developing and running effective chemical management programs.

This is not a book for scientific or legal experts unless someone wants to become more familiar with the different disciplines, considerations and components of a toxic substances management program. Nor is it a toxicology, or an environmental engineering text. Instead, the book includes a judicious combination of topics, disciplines and contributors, to provide managers with guidance on the issues, management strategies and tools from a 1990s perspective. It should be useful to professionals and students in the rapidly advancing environmental protection and management fields. Comments and criticism from all readers are welcomed.

R. P. Côté
P. G. Wells
Halifax, Nova Scotia
1990

Acknowledgements

This text is based in part on a course taught at Dalhousie University but it would not have been initiated without the encouragement of Professor Ian Burton of the University of Toronto and the International Federation of Institutes for Advanced Studies. Dr Burton also reviewed one of the chapters. A number of other colleagues were kind enough to review draft chapters and these are gratefully acknowledged: Dr Donald Chant, Ontario Waste Management Corporation, Toronto, Ontario; Dr Guido Persoone, State University of Ghent, Belgium; Dr John B. Sprague, J. B. Sprague Associates Ltd, Guelph, Ontario; Dr Donald Mackay, Department of Chemical Engineering, University of Toronto, Toronto, Ontario; Dr James Butler, Harvard University, Boston, Massachusetts: Dr Henry Melcer and Hugh Monteith, P.Eng., Department of Environment, Burlington, Ontario; Dr Melvin Cross, Department of Economics, Dalhousie University, Halifax, Nova Scotia; Dr Leslie Whitby, Department of Environment, Ottawa, Ontario; and Professor David VanderZwaag, Faculty of Law, Dalhousie University, Halifax, Nova Scotia. The expert typing and editorial assistance of Gina MacPhail, Debbie Chiasson and especially Linda Mercer is also gratefully acknowledged. The index was compiled by Elaine Toms of the School of Library and Information Studies, Dalhousie University.

Contributors

Dollof F. Bishop is the Chief of the Biosystems Branch in the Risk Reduction Engineering Laboratory at the Andrew W. Breidenbach Environmental Research Center in Cincinnati, Ohio. Mr Bishop has 27 years of experience in treatment of wastewater and hazardous wastes with the United States Environmental Protection Agency and its predecessor agencies. His current responsibilities involve research on control of toxics in wastewater and hazardous waste specializing in toxics management using biological treatment approaches. He has both a Bachelor of Science and a Master of Science degree in Chemical Engineering and is an Adjunct Professor for Research in Environmental Engineering with the University of Cincinnati.

Raymond P. Côté is Associate Director of the School for Resource and Environmental Studies and Co-Director of the Marine Affairs Program at Dalhousie University where he is also working toward the establishment of a Collaborative Program in Toxicology. He was Chairman of the Pest Control Products Advisory Committee for the province of Nova Scotia from 1987–1990. He is also a member of the Associate Committee on Toxicology of the National Research Council of Canada. Prior to joining Dalhousie, he held a number of senior positions with the Canadian Department of Environment including responsibilities for laboratories, field services and the toxic chemicals management program for the Atlantic Regional Office of that department. He is the author of numerous papers and conference presentations.

Norman O. Crossland graduated in entomology at Imperial College, London University. He spent 10 years in Africa and in Europe studying snail and parasite biology in relation to control of trematode diseases. In environmental research he has designed and developed the use of outdoor ponds and streams to provide a bridge between the laboratory and the natural environment. He has published book chapters on 'Integrated Control of Trematode Diseases' and 'Use of Outdoor Ponds in Experimental Ecotoxicology'. He is employed by Shell Research at the Sittingbourne Research Centre.

Stacy L. Daniels received a B.S.E. degree in Chemical Engineering in 1960, M.S.E. degrees in Environmental and Chemical Engineering, respectively, in 1961 and 1963, and a PhD in 1967, all from the University of Michigan. He is currently a Senior Environmental Specialist in the Environmental Services Department of the Michigan Division of the Dow Chemical Company in Midland, Michigan. He is responsible for regulatory and technical issues as they pertain to reporting of environmental data. His special interests involve assessment of hazardous waste management technologies, particularly incineration. He has more than 200 presentations and publications on a variety of environmental subjects.

Bruce Doern is Professor of Public Policy in the School of Public Administration at Carleton University. A former Director of the School, he is a graduate of the University of Manitoba, Carleton University, and Queen's University. He is the author of over 20 books on various aspects of public policy and public management in Canada. In the field of Canadian environmental policy and social regulation, Professor Doern has authored or co-authored several books and articles, including: *The Environmental Imperative* (C. D. Howe Institute, 1990), *Living with Contradictions: Health and Safety Regulation in Ontario* (1982), *The Politics of Risk: the Identification of Hazardous Substances in Canada (1982)*, *Regulating Herbicides in the Canadian Forestry Industry* (1986), and *Government Intervention and the Canadian Nuclear Industry* (1980).

Susan Holtz received a Bachelor of Arts degree, Magna cum Laude, from Clark University in Worcester, Massachusetts in 1967. Her professional life has been in the field of resource and environmental policy, including 8 years on the staff of Ecology Action Centre, a Nova Scotia-based environmental citizens' organization. Since 1983, she has worked as a private consultant. She has served on many environmental group boards and government advisory bodies, including the recently established National Round Table on Environment and Economy, and is the author of numerous articles, studies and reports on topics ranging from public consultation to energy planning.

Huib M. A. Jansen graduated in econometrics at the University of Amsterdam in 1971. Since then he specialized in environmental economics. He works at the Institute for Environment Studies, Free University in Amsterdam. He spent two years working at UNEP, Nairobi, and carried out research as a consultant for UNEP, IFAD, EC and OECD. Most of his publications are in Dutch, but he co-authored James, Jansen and Opschoor, *Economic Approaches to Environmental Problems* (Elsevier, 1978), a book that recently also appeared in Chinese. He is the Executive Managing Editor of *Environmental and Resource Economics*, a new journal.

Robert Paehlke teaches in the Political Studies Department and the Environmental and Resource Studies Program at Trent University, Peterborough, Ontario. He is the founding editor of *Alternatives: Perspectives of Society, Technology and Environment*, currently published at the University of Waterloo. Recent publications include *Environmentalism and the Future of Progressive Politics* (Yale, 1989), *Managing Leviathan: Environmental Politics and the Administrative State* (Broadview, 1990), and articles in *International Political Science Review*, *Society*, *Canadian Public Administration*, *Environment*, *Environmental Ethics*, and *Administration and Society*.

Colin N. Park received a B.S. degree in mathematics in 1965 from the University of British Columbia. He earned an M.S. and PhD in applied statistics from Purdue University in 1970. He is an environmental manager of the Dow Chemical Company, Midland, Michigan. He is involved in the technical and management side of issues as they impact the health and environmental sciences function of Dow. He served on the Science Advisory Board of the National Council for Toxicology Research, and is currently a member of the risk assessment group of the Council of Environmental Quality for the State of Michigan.

Gary M. Rand received his PhD degree from Texas A & M University. Dr Rand is presently Vice President and Director of Toxikon Environmental Sciences, a comprehensive environmental toxicology testing and consulting company. His research interests are on the effects of chemicals on aquatic and avian species. He is a member of the Society of Environmental Toxicology and Chemistry, the Society of Toxicology, and AAAS. Dr Rand has written and presented over 50 papers in ecotoxicology at scientific meetings and national symposia. He edited the aquatic toxicology textbook entitled *Fundamentals of Aquatic Toxicology* (Hemisphere Publications, 1985).

Murray Rankin is a professor of Law at the University of Victoria, in British Columbia, Canada. He holds law degrees from the University of Toronto and Harvard University. His academic and teaching interests include environmental regulations and administrative law and Professor Rankin has written extensively in these fields. He has spent some time at the OECD and is co-author of a chapter entitled 'Persuasion, Penalties and Prosecution: Administrative v. Criminal Sanctions' in M. Friedland (ed.)., *Securing Compliance: Seven Case Studies* (University of Toronto Press, 1990).

Patrick Sheehan received his PhD in Aquatic Toxicology from the University of California at Davis, and has over ten years of experience in assessing the effects of chemicals on human health and ecosystems. Dr Sheehan is currently a Supervising Toxicologist at ChemRisk, a division of McLaren, where he manages risk assessment and toxicology consulting services for Fortune 100 clients. Current projects include human health and ecological risk assessments at some of the most high profile hazardous waste sites in the United States. Dr Sheehan also teaches courses on the environmental fate of pollutants and risk assessment at the University of California at Santa Cruz. He has published over 40 papers in books and peer-reviewed journals on the environmental fate, toxicology, ecotoxicology, and human health effects of chemicals released into the environment.

Emmanuel Somers was educated in England where he received a B.Sc (Leeds), M.Sc (Leeds), PhD (Bristol) and D.Sc (Leeds). He is a Fellow of both the Royal Society of Chemistry (U.K.) and the Chemical Institute of Canada. He was Director General, Environmental Health Directorate, Health and Welfare Canada from 1974 to 1987. Currently he is Director General, Drugs Directorate. Dr Somers was the first manager of the International Programme on Chemical Safety (IPCS), at the World Health Organization (WHO), Geneva and served as Chairman of the IPCS Programme Advisory Committee from 1984 to 1986. Dr Somers has been a member and chairman of a range of national and international committees concerned with environmental health and environmental monitoring. Dr Somers is the author of more than 140 research and review publications on toxicology, health risk assessment and decision-making.

Peter G. Wells is the Senior Advisor, Marine Environmental Quality, Conservation and Protection, at Environment Canada, Nova Scotia. He was educated in England, Calgary and Montreal, and received degrees at McGill University, University of Toronto (MSc., Marine Zoology) and the University of Guelph (Ph.D., Zoology). He co-teaches a graduate course in environmental toxicology at the School for Resource and Environmental Studies, Dalhousie University, as an Adjunct Professor. His primary interests are aquatic and marine toxicology, hydrocarbon and dispersant ecotoxicology in marine systems and the role of marine ecotoxicology in strategies for identifying, assessing, managing and monitoring marine pollution problems. He has produced or contributed to over one hundred primary and technical publications, including book chapters in these and other areas of aquatic science and environmental management. Most recently he edited the Proceedings of the Canadian Conference on Marine Environmental Quality (1988), and was a contributor to the US National Academy of Science review 'Using Oil Spill Dispersants on the Sea' (1989).

C. J. M. Wolff obtained his original qualifications in physical chemistry. Since 1970 he has been involved in environmental research including the environmental fate and behaviour of chemicals in the laboratory, natural systems, ponds and lakes. Currently, he is an advisor for the Shell Group on environmental matters in the Health, Safety and Environment Division. He has produced computer models on evaporation of chemicals from water and on chemical spills at sea and is co-author of a book chapter on 'Outdoor Ponds: Their Construction, Management and Use in Environmental Toxicology'.

1 The nature and scope of the toxic chemicals issue

R. P. CÔTÉ

This chapter describes the nature and scope of the management complexities facing chemical producers, regulators and those at risk due to exposure to toxic chemicals. It also provides insights into the components of a comprehensive chemical management program although these are explored in greater detail in subsequent chapters.

The scope of the issue

Some reports indicate that there are now approximately 100,000 distinct substances in commercial use with 1,000 new chemicals introduced each year (Connell, 1987). Of these, there are more than 1,000 basic pesticide ingredients currently available in more than 30,000 different commercial formulations. A single pesticide, such as toxaphene, may consist of dozens of separate compounds, isomers and congeners. (Ware, 1978). In addition, there are also several thousand food additives and drugs available on the market. The production, importation, transportation and use of these products result in tens of thousands of accidents each year involving the release of more than one billion liters of toxic chemicals. Almost 2,000 chemicals, for example, are noted in the chemical hazards index of the International Maritime Organization (GESAMP, 1989), representing chemicals carried worldwide by ships. To expand the scope of the problem, at least one billion tonnes of hazardous wastes are generated, the majority of which are flushed into sewers, stored or buried inappropriately in countries around the globe.

Trade figures provide another perspective on the scope of the issue. Schweitzer (1983) compiled statistics for 1977 which indicated that total world trade in bulk chemicals exceeded $88 billion (US). United States trade (imports and exports) was $16 billion while that of the Federal Republic of Germany was $18 billion, France $11 billion, United Kingdom $10 billion, and Japan's trade was $7 billion. In 1979 trade involving OECD countries amounted to $181 billion (Lemerle, 1981). These figures, now exceeding $200 billion, belie the actual values because a large amount of the national production is used within each country.

On the other side of the ledger, the measurable economic costs involving remediation and compensation of some toxic chemical incidents have also given governments and industries reason to consider alternative management approaches to the traditional strategies. The remedial costs of Love Canal in New York State have been estimated at $100 million (US); law suits totalling $15 billion were initiated for compensation of victims of the Bhopal disaster though the former government of India settled for $470 million in compensation in 1989; the award from a suit brought by residents against a town in New Jersey for contamination of their water supply totalled $16 million of which $8.2 million was set aside for a medical surveillance program; the town of Times Beach, Missouri, was purchased for $30 million by the government; the remediation program to remove and incinerate 500,000 metric tonnes of tarry matter including polynuclear aromatic hydrocarbons discharged by the coke ovens at Sydney, Nova Scotia is estimated at $35 million (Can) (Travers and Leydon, 1986); the costs to all parties in the explosion of the Icmesa factory at Seveso, Italy, has been estimated at more than $150 million (US). Liability claims for damages to asbestos workers have been estimated at more than $10 billion. These costs are probably one of the better examples of externalities in environmental issues often discussed by economists.

Other impacts have been identified but there has been reluctance to place a dollar value on them. Ecological and human health effects cannot be measured directly and the indirect measures used provide only a proportion of the actual costs. The impacts of chlorinated hydrocarbons on bird populations through eggshell thinning or the formation of neoplastic tumors in marine fish due to exposure to mutagens and carcinogens are not easily amenable to economic measurement. Partial measurement may be possible in the latter case if the tumors lead to the closure of a recreational or commercial fishery. Nor is it possible to quantify in economic terms, the full cost of human losses caused by pesticide poisonings. For example, the International Labour Organization (1982) reported 677 cases of intoxication in one state of Brazil during a one-year period beginning in December 1978. Pearson (1987) has suggested that 100,000 people die each year in developing countries from causes directly related to pesticide production and use. Damage claims, in some instances, may be paid in partial compensation but these occur in the more widely publicized cases or in those where cause–effect relationships have been established.

The most insidious of the accidents such as Bhopal (Table 1.1) and the waste disposal sites, e.g. Love Canal, have become the *causes célèbres* of the movement by governments and non-government organizations to reduce the risks associated with chemicals. While toxic chemical problems have been identified in the past, e.g. speculation about lead and the decline of the Roman Empire, or chimney soot and scrotal cancer in Britain in 1976, and more recently chemicals which have been linked to a number of diseases in the workplace in the past fifty years, none has galvanized the public's

Table 1.1   Anatomy of a toxic chemical problem – Bhopal, India

---

Cause: Release of high pressure methyl isocyanate from a Union Carbide India Ltd chemical factory, 1984

Casualties: 3,400 dead, 17,000 seriously injured, 300,000 people affected

Cost of clean-up and treatment: $40 million by Indian government

Legal implications: 130 lawsuits filed in the United States and 2,700 filed in India

Other economic implications: Union Carbide's stock fell 16 points and contributed to a restructuring in which 4,500 employees lost their jobs

---

attention as have Minimata, Love Canal, Valley of Drums, Seveso, Bhopal, and Basel in precipitating action by governments.

Toxic chemical problems were introduced by Rachel Carson in 1962 to a post Second World War world that believed the chemical substances being produced and used in increasing amounts were safe. The facts that billions of dollars were being expended in manufacturing facilities and that governments did not disapprove of chemical use were assumed by an unsuspecting public to mean that these products presented no risk either to human health or the natural environment. This assumption might have been due to several factors: misunderstanding, misrepresentation of the facts, ignoring information and perhaps covering up facts. The biomagnification of organochlorine insecticides described by Carson (1962) and their distribution has now been confirmed for other chlorinated hydrocarbons. Polybrominated biphenyls (PBBs) mistakenly mixed into cattle feed at a Michigan feed mill in 1979 had spread to 97 per cent of that state's residents within five years of the incident, according to a report of the American Medical Association (Department of Public Health, 1982). Polychlorinated biphenyls (PCBs) have now been found in samples in locations far removed from industrial activitiy and human habitation. It has recently been shown that PCBs reach the far north primarily through the atmosphere (Muir *et al.*, 1988, Society for Environmental Toxicology and Chemistry annual meeting, October 1989). Chlorofluorocarbons have now been implicated in the destruction of the ozone layer of the upper atmosphere.

The nature of the issue

Since Carson, the identification of real, potential and perceived problems has increased exponentially. This increase is due to a combination of events. Larger numbers of chemicals are now produced and some of these are produced in very large quantities. Methods of chemical analysis have become quite sophisticated and sensitive; chemists are now able to measure some chemical substances at the part per quadrillion level. Many new chemical methods have been developed, thus we can expect to find a number of anthropogenic chemicals whenever samples are taken of air, water or biota. Unfortunately, our capacity to measure these extremely low levels of chemicals has outpaced our ability, in most cases, to interpret these levels toxicologically and ecologically. In 1984, an expert group convened by the US National Academy of Sciences found that there was no toxicity information available on 38 per cent of the pesticides and inert ingredients of pesticide formulations, 56 per cent of cosmetic ingredients, 25 per cent of the drugs and excipients in drug formulations, 45 per cent of food additives, or 77 per cent of chemicals in commerce. For a large percentage of the remaining products and chemicals in these groups, there was minimal toxicity information available (NAS, 1984). Standardized laboratory tests may determine that a chemical substance causes lethal or sublethal effects on test animals under highly controlled conditions and at relatively high concentrations in most cases. When publicized, these findings have and will continue to raise the fears of the public. Unfortunately, public health and environmental managers have little or no solid evidence upon which to base any attempts to reduce those fears. To complicate matters, debates regarding the safety of a chemical are often generated in the media by interviewing different scientists, some of whom may be expressing subjective opinions. Research has been underway for many years to improve our ability to extrapolate laboratory data generated from tests using specially bred animals to field, domestic or workplace situations. Toxicological and ecotoxicological science has progressed from the days of 30-day lethal dose tests with the white rat or mouse and 96-hour lethal concentration tests with the rainbow trout but not to the same degree of sensitivity and interpretation as analytical advances in environmental chemistry.

Many of today's commercial chemicals can be found in diverse products with different uses. It is easy to argue that the problem will only be managed when a comprehensive approach is taken because the same

Table 1.2 Categories and sources of toxic chemicals

Categories of toxic chemical issues	Sources of toxic chemicals
Ecotoxicological	Air emissions
− contamination of air, water, soil	Liquid effluents
− contamination of biota	Solid and hazardous wastes
Public health	Spills
− food contamination	− transportation
− water contamination	− operations
− indoor air quality	Pesticide application
− ambient air quality	Drug use
Occupational health	Food and water additives
− contamination of air	− intentional
− hazardous chemicals	− unintentional

chemical may be found in the workplace, in food, drinking water, air and consumer products (Table 1.2). But acutely toxic effects attributed to the use of a chemical or product demand immediate, even if piecemeal, action to protect the exposed population or to control the activity or discharge. For example, during the past twenty-five years, governmental action has been focused on DDT, PCBs, CFCs and dioxins as individual groups or classes requiring attention at different times. While the immediate problem cannot be ignored, a reactive control program may not result in the most effective allocation of resources. This is especially serious when, as stated earlier, a substance or similar substances, may be present in dozens of products involving many uses resulting in widespread distribution (Table 1.2). Where does an agency place its priority? Paehlke (1981) has argued that 'careful control of workplace toxins may be the only way to protect the larger public'. Others would place more emphasis on training, education and protective equipment to reduce exposures to chemicals viewed as absolutely necessary in a variety of situations. Many would undoubtedly agree that more hazard and risk assessment is required before a chemical is produced and released into the environment, but how much assessment is desirable or necessary? The toxic chemical issue can be categorized into several sub-issues, e.g. environmental, occupational, public health, food contamination, etc. and it is for these reasons that integrated management approaches are necessary. One of these is described as 'life-cycle' or 'cradle-to-grave' (Department of the Environment, Canada, 1986) which is supported by environmental sensing (Wells and Côté, 1988).

Every chemical is toxic at some dose or concentration. But what are the characteristics of toxic chemicals? A review of selected definitions indicates that they are more or less comprehensive and the reasons behind the adoption of those definitions in specific jurisdictions may reflect a combination of political, institutional, legal or scientific factors. For example, the definition of a toxic substance from an occupational perspective will usually emphasize different criteria from the definition used by an environmental protection agency. Definitions are important because they determine the scope of a management system that may attempt to assess and control the effects, exposures and risks of hundreds of thousands of products. This could entail a very large expenditure of human and financial resources which are not available in most countries. What is an appropriate definition to facilitate management of the issue? Some examples of definitions are:

Department of Environment (Canada)
A toxic substance is one which is capable of being dispersed in the environment or of being transformed into matter capable of being dispersed, any element or free radical or combination of elements, in manufactured items or in effluents, emissions or wastes which can have an immediate or long term effect on the environment or danger to

human life. (Canadian Environmental Protection Act 35, 36, 37 Elizabeth II June 28, 1988)

National Institute of Occupational Safety and Health, United States
A toxic substance is any chemical or biological agent which has the potential to result in short or long term disease, bodily injury, affect health adversely or endanger the life of men. (Federal Register, Vol. 37 No. 215, November 7, 1972)

Environmental Protection Board (Sweden)
The term 'products hazardous to health and to the environment' is used in Swedish legislation and the products are further defined as any substances or preparations which by reason of their chemical or physico-chemical properties or handling are liable to cause harm to human beings or to the environment. These substances are further classified as very toxic, toxic, harmful, corrosive, irritant, allergenic and carcinogenic. (National Swedish Environmental Protection Board, 1984)

Ministry of Health, Tanzania
A hazardous chemical substance is one which (a) does not lend itself easily to chemical changes caused by natural effects and is also easily accumulated in biological organisms, (b) is suspected of harming human health when ingested continuously, (c) through chemical changes caused by natural effects, produces substances corresponding to, (a) or (b) (Madati, 1981).

While still very broad, these definitions do emphasize certain effects. The definition used by NIOSH with its concern for the workplace makes general comments about short and long term effects or bodily injury but specifically refer to carcinogens, mutagens and teratogens. The Canadian Department of the Environment with a general responsibility for protecting the natural environment emphasizes persistence, bioconcentration, and bioaccumulation. The Swedish Board, on the other hand, is much less specific but this may reflect its broader responsibilities for health and environment and its authority over environmental contaminants and hazardous consumer products. The Tanzanian definition is in some ways more general (a) and in others more specific (b). But none of these definitions covers the breadth of the toxic chemicals issue. Rather, they reflect the mandate of their respective agencies. In some jurisdictions, e.g. Canada and the United States, pesticides are not legally considered toxic substances because they are controlled by other legislation and in Canada by another government agency.

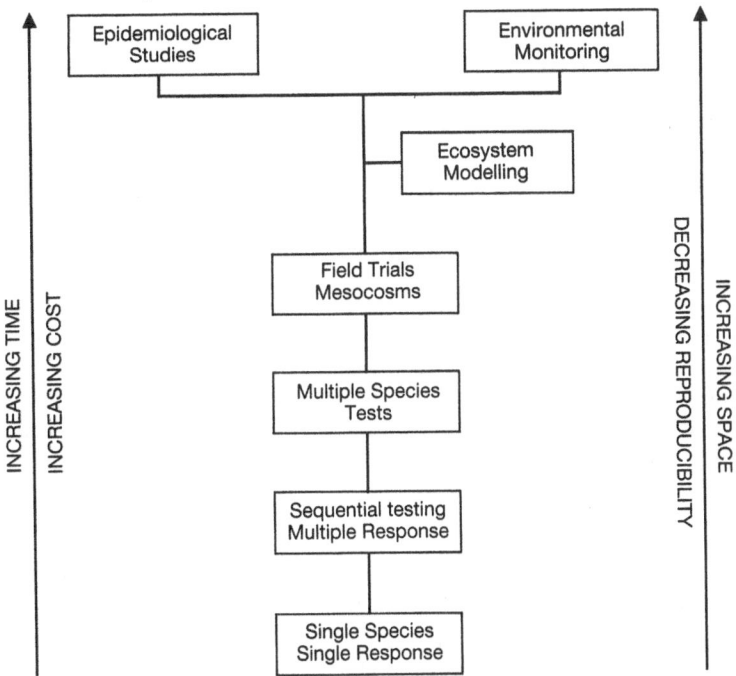

Figure 1.1   The hierarchy of toxicological and ecotoxicological tests

The management of toxic chemicals must address existing as well as new chemicals. With the multitude of products in use today, and the limited financial and trained human resources in most jurisdictions, approaches must be developed to set priorities for assessment of the risk to human health and the environment (Konemann and Visser, 1983). These risk ... necessarily involve analysis of the effects, and the likely exposure. Critical physical and chemical properties can be determined leading to predictions of transport and fate of the chemicals. The status and limitations of these tests are described by Wolff and Crossland in Chapter 2 of this book. The toxicological component of the management framework is more complex. A suite of tests ranging from single species, single response bioassays to field trials and epidemiological studies is presented in Figure 1.1 (Suter, 1983). Though epidemiological studies, monitoring programs and field studies reflect actual conditions more closely, the tests at the lower end of the biological scale are the ones in which toxicologists have the most confidence in terms of predictability and reproducibility. As indicated in Figure 1.1, these tests are also cheaper and less time consuming. The nature of these tests is explored by Rand in Chapter 3. Unfortunately, extrapolating from small groups of mice, trout, cladocerans or bacterial cultures to humans or to natural field populations is a task fraught with danger. In courts of law, such data have often proved less than satisfactory because of the difficulty of establishing cause–effect relationships and controlling all the variables which can influence the toxicity of a substance. However, research is continuing to improve the reliability and usefulness of tests at all levels in the biological hierarchy as well as understanding the ecotoxicological implications of the chemicals. Sheehan discusses many of these approaches in Chapter 4. As the physical, chemical and biological interrelationships are being decyphered, a variety of computer based models are being developed to improve the predictive process (MacKay, 1981; Lyman and Potts, 1987). Researchers emphasize that none of the models, utilized unilaterally, will provide results conclusive enough to be the sole basis of predictions or management decisions. Some argue that laboratory experiments which quantify the biological effects of pollutants for a controlled set of conditions and knowledge of ambient concentrations combined with a stochastic model, will permit a reasonable extrapolation for management purposes.

Having collated the physical, chemical and toxicological data for the chemicals in question, a hazard ranking can be prepared. Several schemes have been developed by industries, e.g. 3M, Dow Chemical and Ciba-Geigy, as well as governments, e.g. the State of Michigan Critical Materials Register, the province of Ontario's Chemical Evaluation Search and Retrieval System (CESARS), and the Dutch government's scoring system and those of international agencies, e.g. the International Maritime Organization. For example, the Dutch government's system developed by the Directorate-General for Environmental Protection scores the following parameters: general mammalian toxicity, mutagenicity, carcinogenicity, aquatic toxicity, environmental exposure based on use volume and percentage release to the environment, degradation in soil, water and air, occurrence in soil, water and/or air, bioconcentration, and finally exposure through products based on use patterns, frequency and intensity (Konemann and Visser, 1983).

The potential exposure of ecosystem components, either designated by legislation or valued by society (e.g. humans, endangered species, commercial fish species), can then be analyzed. That analysis also can be quite complex. This is explained, in part, by the fact that many commercial products contain the same substance. For example, lead is a constituent of many products, resulting in multiple exposures to humans in the domestic, occupational and ambient environments (Figure 1.2). Thus, while the substance might be measured in air, water or foods, the actual source of the chemical may not be easily discernible nor can the effect of one source be segregated because several sources may exist within a geographical area. In the latter case, the cumulative effects of exposure from multiple sources through various routes (dermal, inhalation, ingestion) may be more serious in the longer term than a single exposure through one route. Furthermore,

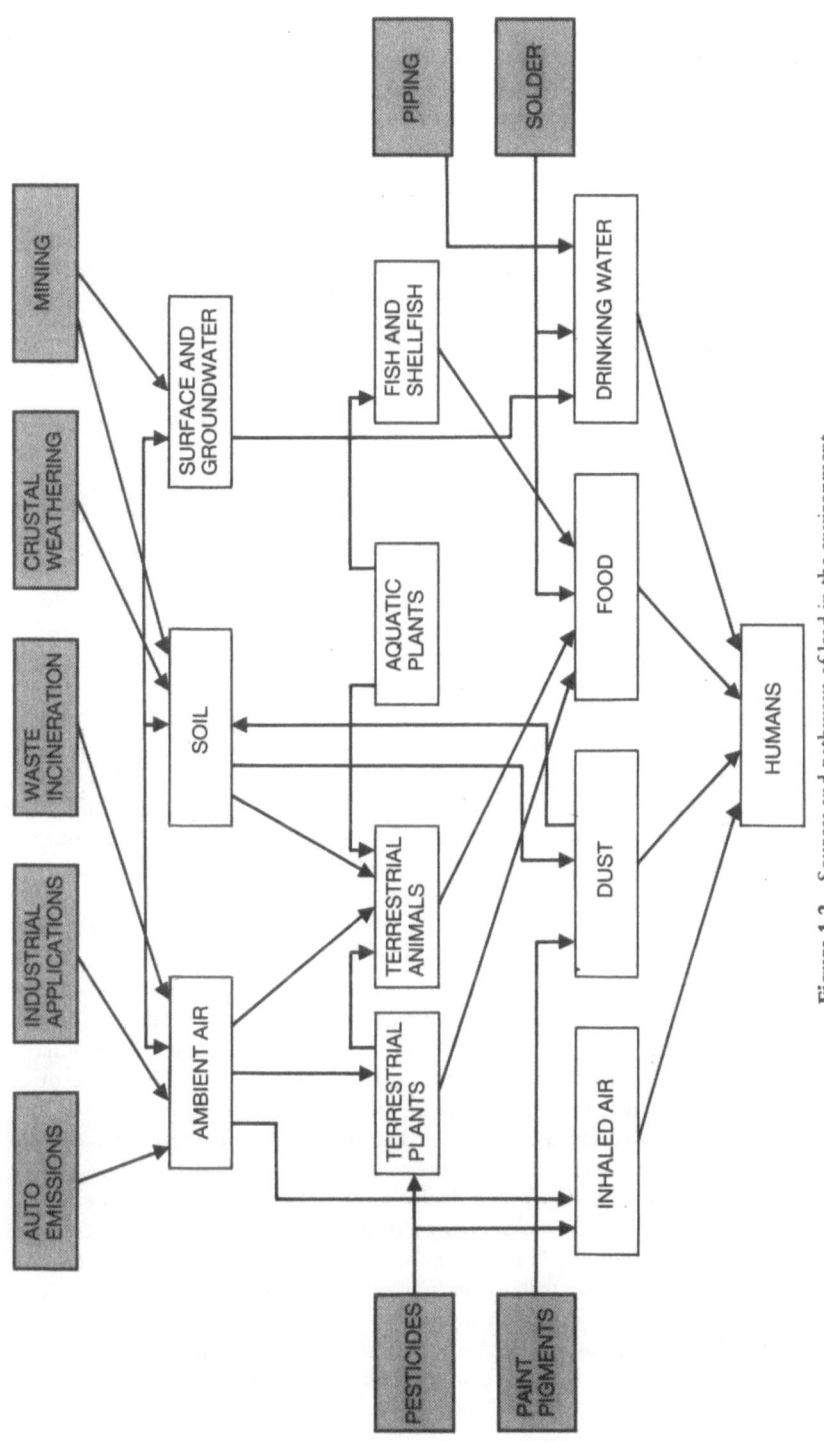

Figure 1.2 Sources and pathways of lead in the environment

transformation can occur to reduce or alter the chemical in some cases, producing more toxic byproducts, and synergistic effects are also possible.

It is argued that exposure analysis should be an integral part of the assessment of existing and new chemicals if management action is to be directed at the most significant sources. Exposure to the majority of chemicals in use, as process intermediates or as final formulations, cannot be eliminated, though in many instances, it can be reduced by a combination of management measures. More often than not, this is easier to implement for new chemicals than for those substances which have been in widespread use or discharged into the environment for some time. The United States Environmental Protection Agency issued proposed Guidelines for Exposure Assessment in 1984 which provide useful guidance to any government agency or industry establishing a management framework for a chemical substance (EPA, 1984). Although the degree of analysis in these guidelines is extensive, a fully integrated exposure analysis might only be required in selected cases. It is worth reminding the reader at this point that requirements imposed on manufacturers, users or importers by government to undertake an exposure analysis or detailed toxicological evaluations also entails a responsibility on the part of regulators to evaluate that analysis and pass judgement on it.

The effect and exposure analyses are important contributors to the decision-making process regarding the level and type of control in the workplace, the external environment, product transportation, use and waste disposal. This aspect is explored further by Daniels and Park in Chapter 5. By developing hierarchical systems for analysis of effect and exposure, which permit users to select levels of complexity depending on the expected multiplicity of products and distribution of the substance, priorities can be established for the scope and degree of risk assessment.

Until recently, waste management has been equated with waste treatment, and emission controls continue to be the main form of treatment in many countries. A large range of treatment technologies is available. These technologies are generally categorized as primary, e.g. separation and sedimentation; secondary, e.g. aerated lagoons and trickling filters; and tertiary, e.g. reverse osmosis, activated carbon adsorption. Primary and secondary technologies represent the largest percentage of treatment systems employed by industries. Not surprisingly, tertiary technologies are currently used in limited numbers as costs escalate quickly upon proceeding from primary to tertiary. However, emission control is only one of many strategies which can be employed by commercial establishments, industries and other users of chemicals. An effective control program will undoubtedly require a combination of strategies. Some of the control strategies and technologies listed in Table 1.3 are explored in greater detail by Bishop in Chapter 6.

The first four strategies, reduction, recovery, re-use and recycling are also known as 'the 4 Rs' and form a group of control technologies termed

Table 1.3   The range of control strategies available for the management of chemicals in an industrial facility

---

1. Recycle process solutions and chemicals
2. Re-use waste water for secondary uses
3. Recovery for sale or exchange
4. Reduction through waste concentration
5. Waste segregation
6. Increase purity of raw materials
7. End of pipe treatment
8. Land disposal

---

waste minimization. Several established firms are employing these alternatives very successfully to reduce wastes, thereby protecting the environment and saving money. The 3M manufacturing company based in the United States has been promoted as a model since Royston (1979) described its waste minimization program, *Pollution Prevention Pays*. Many other cases have now been described in the technical literature (Curi, 1985). In some cases, these practices have the added advantage of reducing energy requirements. Retrofitting an existing facility is costly but as energy costs increase, the price of raw materials escalates and the social and economic costs of disposal increase, the relative cost of waste minimization methods will decrease.

Many of these strategies rely heavily on technology. Events such as the Seveso explosion, the Mississauga train derailment, the Bhopal gas release, and Chernobyl in the USSR, among many others, indicate that man-made technologies will fail. Causes of such failures include inadequate design, inappropriate funding, poor maintenance, lack of training, and inadequate regulation. Although all events and consequences can probably not be identified as was demonstrated following the study by the US Nuclear Regulatory Commission (1975), efforts must be made to do so to some reasonable level. Where some probability exists that all initiating events have not been determined, fail-safe systems and exposure reduction methods are put into place.

As was highlighted by the Bhopal disaster, human and financial resources must also be devoted to the maintenance of such systems so that they are in good working order. In that unfortunate incident, several safety measures failed. The refrigeration system constructed to keep the methyl isocyanate (MIC) under 15°C was shut down; the vent gas scrubber designed to spray the gas with caustic soda solution was not operational; a water curtain designed to neutralize MIC at a height of 12–15 m before it spread beyond the immediate production area was not effective because the gas was released at 33 m; the flare tower which was expected to burn toxic gases was partially corroded and in any event the unit was under maintenance; finally, a valve leading to a spare tank could not open automatically (Shrivastava, 1987). While this may be a worst-case scenario, Kumar

(1986) writes that this disaster shows the need for better laws and inspections, more effective siting requirements, risk analysis and public awareness of the potential dangers of chemical processes and products. From a public and environmental health perspective, siting is an important consideration in a strategy to reduce exposure. However, siting as a preventive strategy is generally only effective when a new industrial facility is proposed. It is highly unlikely that hazardous industries, even those located in sensitive locations, would be shut down or moved elsewhere though a combination of factors may dictate such a move. In cases where industries using or producing hazardous chemicals cannot be moved, more emphasis is placed on back-up and alarm systems, training and maintenance programs.

This brings workers into focus as critical players in toxic chemical management strategies. Workers in a number of industries, as described earlier, have been the victims of chemical poisoning. The occupational environment has provided many examples of morbidity associated with chemical exposures beginning with the findings of Sir Percival Pott of scrotal cancer among chimney sweeps in the United Kingdom (Doull *et al.*, 1980). Other examples are listed in Table 1.4. Such incidents have provoked debates in many countries on the worker's right to know, the right to refuse unsafe work, screening programs for high risk groups and discrimination against female workers in some industries (Lavine, 1982).

In the workplace, there are a number of preventive measures which can be employed to reduce exposures. The traditional methods have been protective clothing, e.g. masks and gloves, and safety-conscious designs, e.g. safety caps and berms in the floor. Several technological measures have been employed in various industries to prevent or reduce release of chemicals: increased mechanization, remote control procedures, continuous instead of batch processing, sealing of equipment, increased maintenance and substitution of chemicals. Other methods act as early warning systems for potential problems particularly among workers: ambient monitoring, periodic medical examinations and personal hygiene programs.

Thus, there are a variety of technologies and strategies available to reduce discharges and exposures. Generally speaking, the industrial processes are not fail safe and therefore spills of products will continue to occur,

Table 1.4 Some examples of human health effects caused by chemicals in the workplace

Chemical	Effect
Chromium	dermatoses
Benzene	effects on blood, leukemia
Dibromo chloropropane	low sperm counts and sterility
Lead	sterility, miscarriages, stillbirths, nervous disorders
Vinyl chloride	angiosarcoma of the liver
Asbestos	lung cancer
Polychlorinated byphenyls	chloracne

unwanted contaminants, e.g. dioxins may be synthesized, and hazardous wastes will be produced. Contingency planning by industries and government then becomes an essential part of the management framework. Technological 'fixes' are as much a part of the problem as they are of the solution. It appears that a balance is required between preventive action, protective measures and response systems when technology is involved. Because the public is increasingly distrustful of technological solutions, informing the public that a comprehensive approach is being taken to the management of toxic chemicals is becoming a necessity.

These issues are intensively covered by the media and information is circulated widely often in a disjointed and inflammatory manner. This information may lead to some very real social, psychological and medical problems for people exposed to toxic chemicals through accidents, contamination of drinking water, food, and exposure in the workplace. These should not be downplayed. A study by Madisso (1985) listed the following impacts: helplessness, loss of trust in authority, depression, behavioral changes, insecurity in children and adolescents, effects on marital relations and loss of property values.

The social, cultural and perceptual aspects of the issue are becoming more complex and in the case of the siting of hazardous facilities, e.g. pesticide manufacturing industries and hazardous waste disposal sites, increasingly intractable. This should not surprise industrial and governmental officials given the catastrophic and near catastrophic incidents which have occurred in the past ten years. The misery for all concerned associated with the Love Canal, the explosion at a 2,4,5-trichlorophenol manufacturing plant in Seveso, Italy resulting in the release of dioxins into the air and the consequent dislocation of residents and destruction of farm animals, the natural gas explosion and fires in Mexico City, the temporary dislocation of 250,000 people in Mississauga, Ontario due to the derailment and fire of a train carrying chlorine, the Bhopal disaster described in Table 1.1, and the Chernobyl nuclear accident and evacuation of thousands of people in the Ukraine, have contributed to public mistrust.

The nature of the issue is, in large measure, the genesis of the 'not in my backyard' syndrome which has frustrated efforts to site necessary hazardous waste storage, treatment and disposal facilities. Increasingly, that attitude is being displayed in public hearings regarding the siting of industries which use or produce large quantities of chemicals. To residents, the costs appear to be borne by the community and the environment while the benefits will accrue to a select few, namely the owners and shareholders of the companies involved. However, there are benefits for the local industry, the workers and community. The difficulty exists of apportioning benefits and costs when considering the safety of transportation, storage, use and disposal of chemicals. Without properly designed and constructed management facilities, wastes will continue to be flushed down sewers or buried as they are today. At worst, communities will be at the mercy of the 'midnight dumpers', companies which indiscriminately dispose of wastes to make a profit. From a larger perspective, this situation raises the concern that continued opposition may well result in the siting of industries or waste management sites in areas or countries with less stringent standards, resulting in contamination that may haunt society in the future.

On the legislative and regulatory fronts, a wide range of control policies and statutes have been enacted, especially in industrialized countries, to protect human health, the workplace and the environment from toxic chemical contamination. Some statutes, e.g. the US Toxic Substances Control Act, are extremely detailed with very elaborate rule-making procedures and guidelines for hazard and exposure analysis (EPA, 1984). The Swedish Act on Products Hazardous to Health and the Environment is not as detailed as some other statutes but covers a wider range of hazardous substances: poisons, pesticides, PCBs, gasoline, cadmium, medical supplies and some consumer products (National Swedish Environmental Protection Board, 1984). In contrast, some European countries, e.g. the UK, France and West Germany, have enacted broad enabling legislation with simplified administrative procedures. Japanese legislation is also much more straight-forward and appears to set out an efficient assessment system.

Another facet of the toxic chemicals issue is its increasingly international nature. In the past three years, there has been exclusive media coverage of efforts to dispose of hazardous waste generated in industrialized countries in less developed countries especially in Africa, the Caribbean and South America. These events spurred international efforts under the auspices of the United Nations Environment Program (UNEP) to adopt the Cairo Guidelines and Principles for the Environmentally-Sound Management of Hazardous Wastes proposed in 1987. The international community has now gone further in drafting and signing the Global Convention on the Control of the Transboundary Movements of Hazardous Wastes (the Basel Convention). But implementation of these guidelines and conventions is difficult. Schweitzer (1983) identified a number of factors which have influenced the success and failure of such international mechanisms:

1 the scientific aspects have dominated the history of international co-operation in toxic chemical management;
2 there have been continuing concerns over proprietary information;
3 responsibilities are fragmented within national governments and this fragmentation differs from one country to another;
4 there has been a reluctance to abrogate national prerogatives regardless of international lobbying;
5 inconsistencies exist in the level of environmental concerns among countries.

One aspect is highlighted here, that is, the concern of industrialized countries that internationally inconsistent regulatory frameworks will affect the trade of commercial products. Trade in commercial chemicals occurs on a very large scale. The OECD has recognized the potential impact of regulatory systems which are not co-ordinated and harmonious at the international level. Following several high-level meetings, the OECD in 1984 identified six priority areas for international co-ordination (OECD, 1984):

1 the development of consistent data requirements and test methods;
2 the development of standards for good laboratory practice including quality assurance;
3 facilitation of information exchange on test data and administrative actions;
4 provision of adequate protection for trade secrets;
5 methodology for analyzing the economic and trade impacts of national legislation;
6 publication of an international glossary of key terms.

The OECD is continuing its work of harmonizing approaches to chemicals management with an emphasis on the impact of control policies on the public and private sector. Other international agencies such as the International Maritime Organization, the United Nations Environment Program and the World Health Organization are also implicated in worldwide efforts to harmonize approaches and information dealing with chemicals.

A further aspect is that of trade in chemicals such as DDT banned in some or all developed countries, which also has an ethical side to it. The export of these chemicals to developing countries has moral connotations. If the risk associated with a particular chemical is considered unacceptable in one country, should that chemical be exported and sold in another country with less stringent standards? Should one country impose its standards on another? Whose needs should take precedence, eggshell thinning in North America or malaria control in Africa? What is the responsibility of the exporting country for ensuring that the importing country understands the risks? If the chemical is banned, should agricultural or manufactured products from another country contaminated by the chemical also be banned? All these questions have implications for world trade and clearly require careful analysis as they infringe on the sovereignty of nations. The United Nations has produced a consolidated list of banned chemicals which provides some of the information on which these questions can be debated (United Nations Secretariat, 1984).

Strategic elements

The nature and scope of toxic chemicals issues are such that specific issues cannot be resolved by one discipline, by one group or at one institutional level. Each issue is multi-faceted and requires a multi-disciplinary effort to put an effective management program in place. Because managers and society are dealing with many uncertainties, a balancing of risks and benefits will be required. And as new data are generated, as the capability to measure and understand impacts both ecologically and economically increases, as social values and legal standards change, the balancing of risks and benefits will continue to evolve. Though no country can claim to have put a truly effective management program in place, progress is being made internationally, nationally and locally. A recent Canadian exercise is a demonstration of that progress. A group of stakeholders representing industry, government, labor, environmental groups and consumers came to an agreement on a set of principles for a comprehensive approach to managing chemicals in a 'cradle to grave' approach (Department of the Environment, Canada, 1986).

All of these dimensions are addressed in the chapters which follow. The chapters explore in greater depth the different perspectives and disciplines which must be brought to bear on individual and global toxic chemical problems in order to reduce risks and resolve conflicts. Chapters 2, 3 and 4 provide an overview of the physical, chemical and toxicological concepts and approaches which create the primary information upon which management programs must be based. In Chapter 2, Wolff and Crossland discuss physical and chemical properties and their relationship to the transport, transformation and fate of organic chemicals in particular. Rand in Chapter 3 describes the science of toxicology and identifies factors which are important to the assessment of the toxicity of aquatic toxicants. In Chapter 4, Sheehan focuses on ecotoxicology, a subset of toxicology which addresses the fate and effects of chemicals in natural ecosystems. Daniels and Park provide industry's perspective on the integration of physical, chemical and toxicological information and on the need to set priorities for effects and exposure analyses in Chapter 5. Examples are given of several systems for setting priorities for chemical hazard assessment. Chapter 6 by Bishop describes some of the control strategies and technologies available to reduce the environmental impacts of chemicals and wastes.

But clearly while comprehensive scientific and technical information is an essential and vital part of the management process, other factors influence the selection of policies, the priorities for action, the choice of control strategies, the degree of public involvement and the legal sanctions that might be used. The degrees of hazard and risk can be estimated scientifically with more or less accuracy depending on the comprehensiveness of the information, but the acceptability of particular risks involves economic, social, political and legal considerations.

In Chapter 7, Paehlke links occupational and environmental health assessments pointing out that an integrated approach is necessary to effectively manage widely used chemicals. Jansen in Chapter 8 describes the costs and benefits associated with the manufacture and use of toxic chemicals and further presents information on economic analysis techniques that can assist decision makers. The manufacture, transport, use and disposal of chemicals present many legal challenges to managers and these are discussed by Rankin in Chapter 9. The author also describes some of the new legislation which is being promulgated to manage chemicals.

As anyone who has been involved with the management of toxic chemicals is aware, the social and cultural concerns and the political response to them are often the most difficult to resolve. Holtz, an environmental advocate, reviews the social and ethical aspects of the problem in Chapter 10. In Chapter 11, Doern considers the interaction of the various interest groups in the political process leading to the selection of acceptable policies. The resulting risk management decisions which can be taken by government are then described in Chapter 12 by Somers.

The final chapter synthesizes the foregoing chapters by identifying factors and components essential to the management of toxic chemicals. Wells and Côté also suggest some initiatives that might be taken to improve the effectiveness of the chemical management process.

References

Baram, M. S. (1986), 'Chemical industry accident hazards and the emerging legal framework for risk communication and community right to know', In *Proceedings of an International Conference: Avoiding and Managing Environmental Damage from Major Industrial Accidents* (Vancouver: Air Pollution Control Association), pp. 365–76.

Campbell, M. (1982), 'Industrial waste reduction and recovery', *Alternatives* 10(2)(3), pp. 59–64.

Carson, R. (1962), *Silent Spring* (Boston: Houghton Mifflin Co.).

Connell, D. W. (1987), 'Ecotoxicology – A framework for investigations of hazardous chemicals in the environment', *Ambio* 16(1), pp. 47–50.

Curi, K. (1985), *Appropriate Waste Management for Developing Countries* (New York: Plenum Press).

Department of the Environment, Canada (1986), *From Cradle to Grave: A Management Approach to Chemicals* (Ottawa, Canada), September.

Department of Public Health, Canada (1982), 'Extrapolation shows 97% of Michigan residents have PBB in their bodies', *Toxic Material News*, Jan 6.

Doull, J., Klassen, C., and Amdur, M. (eds) (1980), *Toxicology: The Basic Science of Poisons* (Cassarett and Doull's 2nd edn) (New York: Macmillan Publishing).

EPA (Environmental Protection Agency) (1984), *Proposed Guidelines for Exposure Assessment*, Federal Register Vol. 49, No. 227, November (Washington), pp. 46304–12.

GESAMP (Joint Group of Experts on the Scientific Aspects of Marine Pollution) (1989), *The Evaluation of the Hazards of Harmful Substances Carried by Ships: Revision of GESAMP Reports and Studies No. 17*, Reports and Studies, No. 35 (London: International Maritime Organization).

International Labour Organization (1982), *Occupational safety and health on plantations, with special reference to mechanisation and the use of chemicals and to labour inspection*, Report III (Geneva: ILO).

Konemann, H. and Visser, R. (1983), *Netherlands approach for setting environmental priorities for giving attention to existing chemicals: WMS scoring system* (The Netherlands: Ministry of Housing, Physical Planning and Environment).

Kumar, S. (1986), 'The three legacies of Bhopal', *Alternatives* Vol. 13, No. 4, pp. 3–10.

Lavine, M. P. (1982), 'Industrial screening programs for workers', *Environment* 24(5), pp. 26–38.

Lemerle, P., (1981), Opening address in *Proceedings of the Workshop on the Control of Existing Chemicals Under the Patronage of the Organisation for Economic Cooperation and Development* (West Berlin: OECD), June 10–12.

Lyman, W. J. and Potts, R. G. (1987), CHEMEST *User's Guide* (Cambridge, Mass: Arthur D. Little, Inc.).

Mackay, D. (1981), 'Calculating fugacity', *Environ. Sci. Technol.* 15(9), pp. 1006–14.

Madati, P. J. (1981), 'Procedures and plans for controlling toxic chemicals in Tanzania', *Industry and Environment* (UNEP), Oct–Nov–Dec, Vol. 4, No. 4.

Madisso, U. (1985), *A synthesis of social and psychological effects of exposure to hazardous substances*, A study for Environment Canada. Contract No. TOB 84–106311.

Muir, D. C. G., Norstrom, R. J. and Simon, M. (1988), 'Organochlorine contaminants in arctic marine food chains: accumulation of specific polychlorinated biphenyls and chlordane-related compounds', *Envir. Sci. Tech.* 22: pp. 1071–8.

NAS (National Academy of Sciences) (1984), *Toxicity Testing: Strategies to Determine Needs and Priorities* (Washington, DC: National Academy Press).

National Swedish Environmental Protection Board (1984), An Act on Products Hazardous to Health and the Environment. Act 1973:329. Solna.

OECD (1984) *The OECD Chemicals Programme* (Paris: OECD).

Paehlke, R. C. (1981), 'Much to be done: Exploring the interface between occupational and environmental health', *Alternatives*, pp. 13–20.

Pearson, C. S. (ed.) (1987), *Multinational Corporations, Environment and the Third World.* (Durham USA: Duke University Press).

Royston, M. (1979), *Pollution Prevention Pays* (Toronto: Pergamon Press).

Schrivastava, P. (1987), *Bhopal: Anatomy of a Crisis* (Cambridge, Mass: Ballinger Publishing).

Schweitzer, G. E. (1983), 'Toxic chemicals, steps toward their evaluation and control', in Kay, D. A. and Jacobsen, H. K. (eds) *Environmental Protection: The International Dimension* (New Jersey: Allenheld, Osmun).

Suter, G. W. (1983), 'Multiple species tests for environmental toxicology', *Environ. International* 9:157–60.

Stokes, P.M. (ed.) (1986), *Pathways, cycling, transformation of lead in the environment. The Commission on Lead in the Environment* (Ottawa: The Royal Society of Canada).

Travers, I. and Leydon, J. (1986), *The Sydney Tar Ponds Remediation Program.* Eighth Canadian Waste Management Conference, Halifax, Environment Canada, Ottawa, Canada, pp. 377–85.

US Nuclear Regulatory Commission (1975), *Reactor Safety Study: An Assessment of Accident Risks in US Commercial Nuclear Power Plants* (Wash: US NRC).

United Nations Secretariat (1984), *Consolidated List of Products Whose Consumption and/or Sale Have Been Banned, Withdrawn, Severely Restricted or Not Approved by Governments*, DIESA/WP/1/Rev.1. (New York: United Nations).

Ware, G. W. (1978), *The Pesticide Book* (San Francisco: W. H. Freeman and Company).

Wells, P. G. and Côté, R. P. (1988), 'Protecting marine environmental quality from land-based pollutants: The strategic role of ecotoxicology', *Marine Policy*, Jan, 12(1), pp. 9–21.

2 The environmental fate of organic chemicals

C. J. M. WOLFF & N. O. CROSSLAND

Introduction

The environmental fate of chemicals is a very broad subject. In the context of this book, the present chapter on organic chemicals will concentrate on the basic relationships describing their environmental fate, the incorporation of these relationships into mathematical models and the use of these models for management purposes. In the environment, chemicals are subjected to a variety of transport and transformation processes. The rates of these processes can have a profound influence on the exposure of organisms to these chemicals. Therefore the study of rates is of paramount importance when evaluating chemical hazards. The rates depend on (1) physico-chemical properties, (2) environmental parameters and (3) for some processes, release characteristics.

To evaluate the environmental exposure to a chemical, data for transport and transformation processes can be combined with data on release rates and characteristics in mathematical models that can be used to estimate the distribution of chemicals. Such mathematical models can be used for a variety of management purposes. The choice of an appropriate model depends on the purpose, e.g. to estimate the global distribution and persistence of a synthetic organic chemical, to assess the distribution of several emissions of a contaminant within a defined area, or to assess the distribution of a chemical or an effluent from a point-source discharge.

For modelling purposes it is conventional to divide the environment into the compartments of air, water, soil or sediment and biomass. These compartments may be further divided into sub-compartments or boxes which are considered to be homogeneous. Chemicals may be transported within compartments by homogeneous advection and dispersion processes which do not involve a change of chemical phase, i.e. gaseous, dissolved or sorbed. Alternatively, they may be transported within or between compartments by heterogeneous transport processes, such as evaporation and sorption, that involve phase changes. In addition, chemical transformations may occur within each compartment. Figure 2.1 shows a schematic overview of the main transport and transformation reactions occurring in the environment.

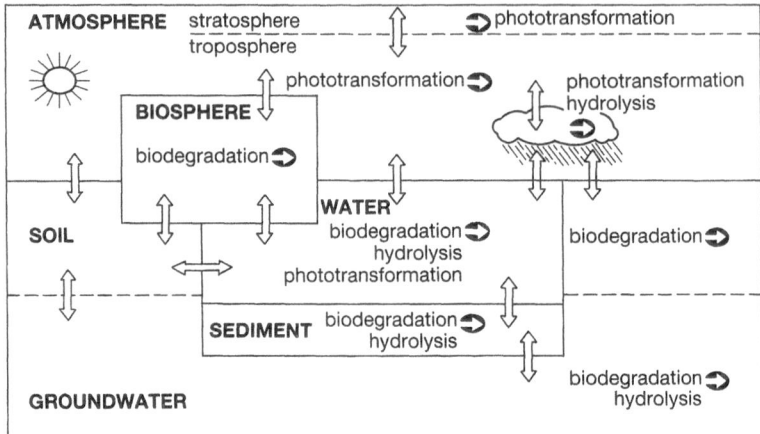

Figure 2.1   Schematic overview of the main transport (<⟹>) and transformation (⊃) processes in the environment

Data on the rate of the transport and transformation processes in the environment are needed as inputs for exposure analysis modelling systems. Alternatively, appropriate subroutines may be incorporated in some of the models to estimate these rates, using physico-chemical properties. However, for many chemicals essential physico-chemical properties have not been measured. Actual measurements for all required data would be very expensive and would require many more laboratory facilities than are currently available. Much research is therefore aimed at developing methods for estimating physico-chemical properties on the basis of chemical structure.

Transport and transformation in the environment

Conventionally, the environment is subdivided into three abiotic compartments: the air, water, and soil or sediment compartments. These compartments are characterized by their most important component, i.e. air, water or solid material, respectively. There are, however, heterogeneous systems containing smaller or larger amounts of the other environmental components. Soil and sediment, especially, may contain substantial amounts of air or water.

In addition to the three abiotic environmental compartments a fourth compartment is generally defined, namely biomass. Because of its relatively small volume, this compartment is less important when considering the overall distribution of chemicals in the environment in terms of the mass balance. But, obviously, transport of chemicals by and into biomass is of prime importance when considering exposure of the biota to toxic chemicals and the resulting biotic transport and effects. A fifth compartment may also exist, namely the condensed liquid or solid phase of the chemical, for chemicals such as polymers that have a low affinity for any of the environmental components.

Chemicals will be transported between the different compartments and between different components within the same compartment. At the same time, chemicals may be transformed. The rate of each transformation process will vary depending on the compartment in which the chemical is present and on the component with which it is associated.

Transport

Transport of chemicals in the environment (see Figure 2.1) may be homogeneous or heterogeneous. Homogeneous transport involves movement of a chemical without any change of chemical phase, e.g. by movement in air or water currents or by diffusion. Heterogeneous transport, on the other hand, always involves a change of chemical phase. For example, when a chemical moves across the air–water boundary layer it changes from the gaseous to the dissolved phase. Transfer of chemicals between environmental compartments takes place during heterogeneous transport but does not occur during homogeneous transport. Both homogeneous and heterogeneous transport may occur between and within the different environmental compartments but heterogeneous transport of chemicals will predominate. However, transport of environmental components between compartments may occur and in this case chemicals associated with these components will be simultaneously transported. In Figure 2.2 a schematic overview is given of the important homogeneous and heterogeneous transport processes in and into the atmosphere.

Within the air and water compartments the main process is homogeneous transport of gaseous and dissolved chemicals. When a chemical has a high affinity for other environmental components that are present in small amounts in the form of small particles, heterogeneous transport to and subsequent homogeneous transport associated with these particles may also be important.

The soil and sediment compartments may consist of up to 50 to 60 per cent of air and water. These environmental components are transported, although usually slowly, within the soil compartment, and therefore both homogeneous and heterogeneous transport of chemicals will be important within the soil and sediment.

HOMOGENEOUS TRANSPORT

There are two types of homogeneous transport, namely advection and dispersion. Through advection chemicals are displaced along with a parcel or a particle of an environmental component without any change in

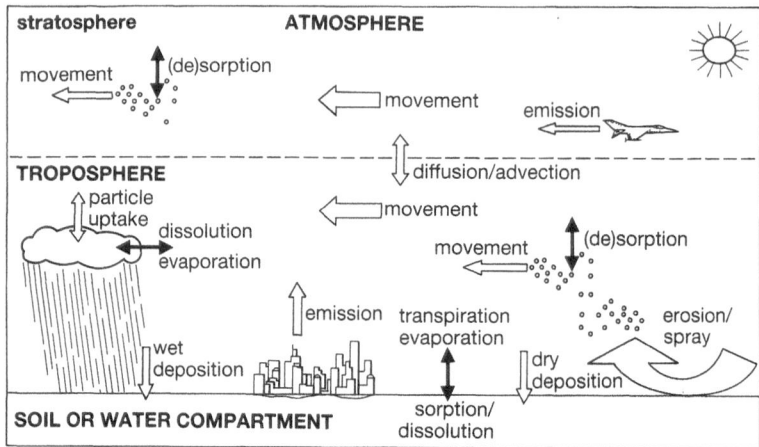

Figure 2.2   Schematic overview of important homogeneous (⇔/⇐) and heterogeneous (↔) transport processes in/into the atmosphere

chemical concentrations. The importance of this process is that chemicals may be transported to areas that are remote from the area where they are released, e.g. long-range transport of airborne pollutants or transport of chemicals associated with suspended solids in rivers and subsequent sedimentation in estuaries. Dispersive transport always leads to dilution of chemical concentrations. The dilution rate can be calculated using equations as given by Geankoplis (1972); Beek and Mutzall (1975); Fischer *et al.* (1979); and Genuchten and Alves (1982).

Let us look first at homogeneous transport in the air and water compartments. *Advection* is caused by gravitational or stress forces. Through these forces environmental components, i.e. air, water or solid particles, are displaced taking along chemicals that are associated with them. Important parameters for this process are the wind speed for air and the current speed for water. For solid particles the rate of advection depends on the density difference between particles and the surrounding air or water, particle sizes and shapes and system turbulence. For dense gases (Colenbrander, 1980; Puttock, 1986) density differences between gas and air can also cause advective transport to occur. The direction of transport may be constant as in rivers, or alternating through the tidal movement as in seas and estuaries, or highly variable as in the atmosphere.

The mass transfer coefficient for *dispersive* transport of gaseous chemicals in air (Chiang and Sill, 1985) or dissolved chemicals in water (List, 1982) depends on the dispersion coefficient and the path length for transport. In stagnant water bodies, for example, the system turbulence is very small or even nonexistent. The dispersion coefficient is then determined only by the molecular diffusion coefficient of the chemical and thus dispersive transport is slow. When the system turbulence is increased by environmental processes such as currents, wind and temperature differences, these will overwhelm diffusive transport, and then the process will depend on environmental processes only.

A special type of dispersive transport is the so-called initial dilution process where system turbulence is generated by jets or buoyant plumes, such as is the case for most aqueous and gaseous effluents. For the mass flux and process kinetics during initial dilution, empirical relationships have been developed (List, 1982; Chiang and Sill, 1985; Muellenhoff *et al.*, 1985).

Homogeneous transport of chemicals associated with small particles of other environmental components within the air or water compartments can be advective and dispersive. Its rate depends mainly on gravitational or stress forces (erosion). Wind or water currents may also be important depending on the magnitude of these parameters and on particle size and weight. Descriptions of such processes have been published by various authors, e.g. for wet and dry deposition (Mackay *et al.*, 1986), sedimentation and resuspension (Harris *et al.*, 1984) and wind erosion of particles loaded with pesticides (Hartley and Graham-Bryce, 1980). This type of transport will be important only for chemicals having a high affinity for these particles. These are typically chemicals that have a high water solubility, for example showing a high affinity to rain droplets, or chemicals that have a low vapour pressure and a log octanol/water partition coefficient greater than 5 to 6, indicating a high affinity of the chemical for the organic fraction of solid particles. However, only a small fraction of such chemicals will usually be present in the air or water compartments. Therefore, this process may be neglected in most cases when calculating mass balances.

Secondly, let us consider homogeneous transport in or into the soil or sediment compartment. Soil can be divided into two regions, namely, the unsaturated and the saturated zones. The boundary between the two zones is the so-called water table. The water table often fluctuates, creating an intermediate zone which is alternately saturated and unsaturated. In the unsaturated zone the space between the soil particles is occupied by air and water and in the saturated zone by water only. In aquatic sediments only a saturated zone exists.

In the unsaturated zone hardly any advection of air occurs but advective transport of water will occur due to rainfall and evaporation/evapotranspiration. The rate of this advective transport depends mainly on the permeability of the soil and on gravitational forces. Its direction is vertically downwards. In the saturated zone advective transport of water depends on the permeability of the soil or sediment and on the hydraulic gradient (American Petroleum Institute, 1982). It is usually very slow, in the order of several tens of metres or less per year, and its direction is more or less horizontal.

Hardly any turbulence occurs in the soil and sediment. Therefore, dispersive transport via the air and water in the soil or sediment is usually slow, the mass transfer coefficient depending on the diffusion coefficient of the chemical in air and water, the porosity of the soil and the path length for transport (Thibodeaux and Scott, 1985). At present it is not possible to estimate mass transfer coefficients from empirical relationships between these parameters. Therefore, mass transfer coefficients are usually determined experimentally. For example, Crossland *et al.* (1986, 1987) determined rates of transport from water into the sediment of outdoor experimental ponds using a PCB homologue (2,5,4'-trichlorobiphenyl). Data for residues of this compound in the water, sediment and vegetation were fitted to a three-compartment model from which the mass transfer coefficient was calculated.

HETEROGENEOUS TRANSPORT

Heterogeneous transport involves the transfer of the chemical between different environmental compartments. Thus, unlike homogeneous transport, the concentration difference is also dependent on the partition coefficient of the chemical between the environmental components. Mass flux equations for heterogeneous transport are given by Geankoplis (1972) and Beek and Mutzall (1975).

For heterogeneous transport *between* the air and the water compartments the partition coefficient is expressed as Henry's law constant. The generally accepted kinetic description for this process is the two-resistance layer model originally developed by Lewis and Whitman (1924) and applied to heterogeneous transport between the air and water compartments by Liss and Slater (1974). In the two-resistance layer model it is assumed that the bulk of the two compartments is homogeneously mixed. The interfacial transport is hindered only by two laminar boundary layers across which transport occurs through chemical diffusion only. Mass flux equations for this process have been derived by Wolff and van der Heijde (1982) and Mackay and Yeun (1983).

For heterogeneous transport *between* the air or water and the soil or sediment compartment similar mass flux equations could be developed. However, because of the obvious lack of turbulence in the soil or sediment compartment, the resistance to heterogeneous interfacial transport in the boundary layer of this compartment is very great. Therefore, heterogeneous interfacial transport between these compartments can be neglected. Exchange of chemicals between the air or water and the soil or sediment compartments occurs mainly through homogeneous transport, transport via the interstitial air or water in the soil or sediment compartment.

Heterogeneous transport will occur *within* all environmental compartments. In the air compartment it may occur between air and solid material or water droplets and in the water compartment between water and suspended solids. Within the soil or sediment compartment, heterogeneous transport will occur between interstitial air or water and the solid material.

Partition coefficients between water and suspended solids or sediment depend on hydrophobicity or charges for neutral organics or ionic species, respectively (Karickhoff et al., 1979; Karickhoff, 1981; Weber et al., 1983; Voice and Weber, 1983; Chiou et al., 1983; Reuber et al., 1987). Partition coefficients based on hydrophobicity of neutral organics have successfully been correlated with octanol/water partition coefficients of the chemical and organic carbon content of the sediment (Karickhoff et al., 1981; Reuber et al., 1987; Karickhoff, 1985). The partition coefficient between air and water droplets is expressed as Henry's law constant and that between air and solid or soil particles can be estimated from data on the partition coefficient between water and suspended solids and on Henry's law constant. As discussed above this transport may be important for a limited number of chemicals only.

Usually in modelling, heterogeneous transport within environmental compartments is expected to be very fast and, therefore, equilibrium is assumed to exist (Harris et al., 1984; Mackay et al., 1986). The validity of this assumption is disputed by Karickhoff (1985) in so far as transport between water and suspended solids or sediment is concerned. In laboratory experiments, Karickhoff found that the efficiency of extraction with hexane of sediment-sorbed polynuclear aromatic hydrocarbons (and other chemicals) decreased with increasing duration of the incubation period from better than 90 per cent after approximately three minutes, to 20–40 per cent after several days of incubation. Similar experience led DiToro and Horzempa (1982) to conclude that part of the sorbed chemicals is highly resistant to desorption, e.g. for hexachlorobiphenyl, about 50 per cent. Karickhoff (1985) disagrees with this explanation and suggests that this phenomenon is caused by a diffusion-controlled process in the sorbed state where the sorbed chemical is slowly incorporated into either porous aggregates or sorbent components. Despite this, for modelling the fate of chemicals in the natural environment, the concept of equilibrium is probably acceptable for most chemicals since, in contrast to the laboratory situation, transport within the interstitial water or air in the environment is usually very slow, allowing ample time to attain equilibrium. However, in a situation such as rapid infiltration in soil after rainfall, equilibrium with sorbed chemicals in the aggregates is probably not achieved.

Transformation

Important transformation processes that can occur in the environment (see Figure 2.1) are biodegradation, phototransformation and hydrolysis. Biodegradation involves the transformation of chemicals by the action of bacteria and is very important in terrestrial and aquatic environments. Aerobic biodegradation requires the presence of oxygen while anaerobic biodegradation does not. Phototransformation involves a chemical reaction that takes place under the influence of sunlight. It occurs mainly in the atmosphere and in aquatic environments. It may be described as direct phototransformation, in which case the chemical reaction is induced directly by sunlight, or indirect phototransformation, in which case the chemical reaction is induced by photochemically formed reactive species, such as ozone. Hydrolysis is a chemical reaction that occurs mainly in water and soil. Essentially, these processes can be considered to be second-order rate processes involving the chemical and an environmental reagent. If the reagent can be considered constant under natural conditions the reaction becomes pseudo first-order. For modelling the fate of chemicals in the environment, daily, seasonally or even annually, averaged values for the environmental reagent are generally assumed and used to calculate the pseudo first-order rate constant.

AEROBIC BIODEGRADATION

Aerobic biodegradation can occur in water, wet soil and sediment. Second-order kinetic equations for biodegradation rates have been derived from the Monod equation for low substrate concentrations (Paris et al., 1981). These workers showed that this approach for estimating second-order rate constants was valid for three pesticides, the butoxyethyl ester of 2,4-dichlorophenoxyacetic acid, malathion and chloropropham in a variety of natural waters. However, the method should be used with caution. The compounds selected by Paris and co-workers were reported to have been degraded by hydrolytic enzymes present in most bacteria in the environment. Thus, the first-order rate constants may have been normalized with respect to concentrations of hydrolytic enzymes, rather than bacterial numbers. It follows that this approach is unlikely to be successful for compounds that are degraded by specific bacterial populations or where a period of adaptation to the substrate is required before biodegradation can begin.

Biodegradation in aquatic environments may be associated with bacteria attached to surfaces, rather than those suspended in the water. Bacteria attached to aquatic vegetation and those associated with the sediment are much more numerous than those associated with the water column (Costerton and Greesey, 1979). For example, bacterial concentrations in a sample of pond water were enumerated using agar plates and were estimated to be $10^7 \, l^{-1}$, whereas in the sediment they were $2 \times 10^{10} \, l^{-1}$ (Crossland and Bennett, 1984; Crossland et al., 1986). Under these conditions biodegradation of methyl parathion was primarily associated with the sediment rather than with the water column, and the overall rate of loss of this compound from the pond water was determined by its rate of transport to the sediment.

When chemicals are discharged into a flowing system biodegradation may occur in two phases (Apoteker and Thevenot, 1983). Bacteria, not previously exposed to a compound, may need to adapt to it before biodegradation can start.

There have been some indications that the rate of biodegradation of degradable compounds in the natural environment changes dramatically when their concentrations are below a certain threshold value (Paris *et al.*, 1975; Rittmann and McCarthy, 1980a, 1980b). For aerobic biodegradation this threshold value has generally been estimated to be below 1 mg l^{-1}. For instance, Rittmann and McCarthy (1980b) found the value for acetate to be 0.66 mg l^{-1} although the biotreatment system used might not be valid for environmental situations.

ANAEROBIC BIODEGRADATION

Anaerobic biodegradation mainly occurs in sediment but it can also occur in soil. The kinetics of this process in the environment are probably similar to the kinetics of aerobic biodegradation but are generally slower (Lawrence and McCarthy, 1970). Very fast degradation rates were observed in anaerobic sediment samples for methyl parathion (Wolfe *et al.*, 1986). However, it was suggested that this occurred through biotic as well as abiotic pathways.

The concept of the threshold value for anaerobic biodegradation has also been investigated by Rittmann and McCarthy (1980b). They found this value to be much higher for anaerobic than for aerobic degradation (for acetate 356 mg l^{-1} and 0.66 mg l^{-1}, respectively).

DIRECT PHOTOTRANSFORMATION

Direct phototransformation occurs mainly in the air and in the water compartments. Direct phototransformation of chemicals associated with solid surfaces, such as particles in the air and soil and plant surfaces, has also been reported.

The rate of this process depends on the rate of solar light absorption, i.e. the number of photons absorbed by the compound per unit of time, and on the quantum yield for the reaction, i.e. the fraction of the compound that is converted upon absorption of a photon (Zepp and Cline, 1977; Zepp, 1978; ECETOC, 1983; ECETOC, 1984). The rate of solar light absorption can be calculated from data for the light absorption spectrum of the compound, the solar spectrum and environmental parameters which determine the incident radiation, such as season, latitude, light absorption spectrum of the medium through which sunlight has to pass, and cloud cover.

A computer program has been developed (Zepp and Cline, 1977) and revised (Zepp, 1986) to calculate the rate of direct phototransformation for compounds dissolved in the water compartment. Predictions of the rate constant of direct phototransformation for pentachlorophenol and 3,4–dichloroaniline in outdoor pond systems proved to be within a factor of two of observed values (Wolff and Crossland, 1985; Crossland and Wolff, 1985).

A similar approach could be adopted for calculating rate constants of direct phototransformation of gaseous compounds in air. Proposals for measuring quantum yields for reactions in air have recently been formulated (ECETOC, 1983; OECD, 1987). However, it is generally accepted that for most chemicals indirect phototransformation in air is much faster than direct phototransformation (ECETOC, 1983; OECD, 1987).

Direct phototransformation of chemicals sorbed on airborne particles has been studied by several investigators (e.g. Gäb et al., 1977). The importance of this process, however, has been disputed, firstly because for most compounds present in the atmosphere only a very small fraction of the total is associated with solid particles (ECETOC, 1983) and, secondly, because the relevance of the kind of particles used during testing, namely silica gel or sand, to the large variety of particles that exist in the atmosphere is not clear (ECETOC, 1983; Parlar, 1986). Therefore, results obtained in such an experimental set-up should be interpreted with caution.

INDIRECT PHOTOTRANSFORMATION

Indirect phototransformation, occurring mainly in the air and water compartments, is the general term for the reaction with photochemically formed reactive species. The occurrence of these reactive species in surface waters, mainly at very low concentrations, has been demonstrated by various investigations. They are: singlet oxygen (Joussot-Dubien and Kadiri, 1970; Zepp et al., 1977; Wolff et al., 1981; Momzikoff et al., 1983), hydrogen peroxide (van Baalen and Marler, 1966; Kok, 1980; Cooper and Zika, 1983), ozone (Garland et al., 1980), superoxide (Cooper and Zika, 1983), OH radicals (Mill et al., 1980; Zepp et al., 1978a), RO_2 radicals (Mill et al., 1980), and hydrated electrons (Breugem et al., 1986; Zepp et al., 1987b).

Ideally, the concentration and reactivity of all of these reactive species should be known in order to model the fate of any given xenobiotic. However, the concentrations of individual reactive species are generally too low to have any significant effect on xenobiotic concentrations. Therefore, a test method for measuring the overall contribution of reactive species, called the sensitization factor, has been developed (Miller et al., 1980). This method involves determination of the concentration of sensitizers, measured as light absorption of natural water at wavelengths around 310–320 nm. A relatively large number of water samples may have to be processed to determine spatial and temporal variations of concentrations of sensitizers in natural waters.

Although in air the situation is of similar complexity, a critical review of the literature (ECETOC, 1983; Atkinson, 1985; OECD, 1987), showed that the contribution of the reaction with OH radicals overwhelms that of reactions with other reactive species. The reaction with ozone seems to be next in importance, but only for unsaturated compounds. Thus, for calculating the rate of indirect phototransformation of compounds in air, their reactivity towards OH radicals, and in some cases also towards ozone, should be measured or calculated (OECD, 1987).

HYDROLYSIS
Hydrolysis mainly occurs in the water and soil compartments, although for some compounds hydrolysis in air may also take place. Neutral, specific acid-catalyzed and specific base-catalyzed hydrolysis can be distinguished (Mabey and Mill, 1978; Smith *et al.*, 1978; Lyman *et al.*, 1982; OECD, 1981; Neely, 1985). The important parameters for the latter two types are the second-order rate constant and the concentration of H^+ ions or OH^- ions, usually expressed as the pH. For neutral hydrolysis the general kinetic description is further simplified. In this case only a pseudo first-order rate constant is required.

Environmental distribution and fate analysis

Many computer programs, and relatively simple rules and rating systems, are available which can be used in predicting distribution and fate of chemicals in the environment. These programs, rules and rating systems have been developed for a multitude of purposes which can be categorized as (1) global distribution and fate analysis, (2) area-related distribution and fate analysis and (3) source-related distribution and fate analysis. A small selection of computer programs, rules and rating systems is discussed below.

For many of the existing chemicals, it is not possible to make a proper distribution and fate analysis because physico-chemical data on these compounds are lacking or not easily available. Developments in this area and methods for predicting or estimating these data also are discussed below.

Global distribution and fate analysis

The main objective in global exposure assessment is to establish the potential environmental impact of chemicals on a global scale without identifying local sources or effects (Branson, 1980; Cairns, 1980; Neely, 1980; Schmidt-Bleek *et al.*, 1982; Klein *et al.*, 1984; Daniels *et al.*, 1985; Freitag *et al.*, 1985; Halfon and Reggioni, 1986; Kenaga, 1986). On the basis of such assessments chemicals might be selected for further testing and important data that have to be determined may be identified. In addition, on the basis of such analysis, chemicals can be classified. In these computer programs or models, which are usually multi-compartmental models, the environment is highly simplified (Figure 2.3); in most cases equilibrium is assumed to exist between compartments and components and annually averaged values for environmental properties are used, if any.

Area-related distribution and fate analysis

Even if chemicals do not cause unacceptable environmental problems on a global scale, local concentrations might exceed acceptable levels. In area-related exposure assessments the primary objective is to estimate cumulative environmental concentrations within a certain area, usually as a result of a multitude of sources. Such areas might be the ground-level atmosphere above certain regions, surface waters such as rivers, lakes, or seas and also groundwater.

In most cases the computer programs used represent single compartment grid models where grid series can vary from a few hundred metres to several tens of kilometres. Usually, the results of the distribution and fate analysis are given as iso-concentration lines. Regional average values for environmental properties are used as input. In Figure 2.4 a hypothetical graphical output for such a model for atmospheric contaminants is given.

The primary tool for area-related exposure assessment is monitoring of the chemical concentrations in the region. Predictive modelling is used to investigate the relation between sources, chemical properties and environ-

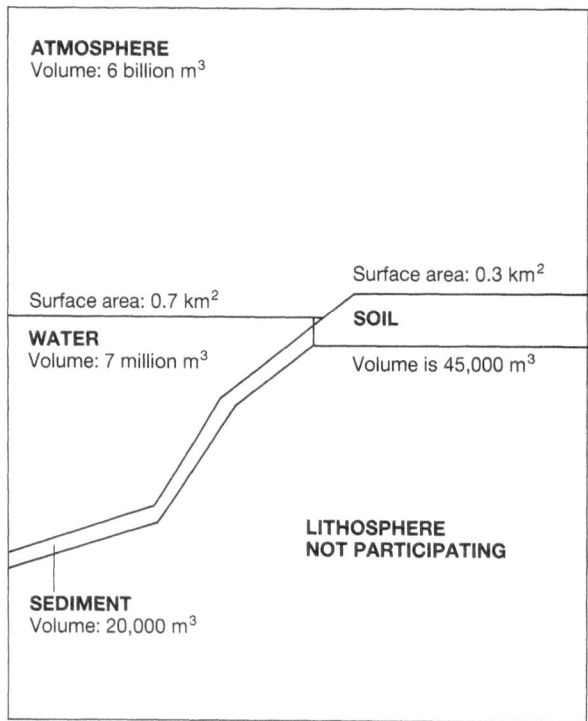

Figure 2.3   Global distribution and fate model of a highly simplified 'unit world'

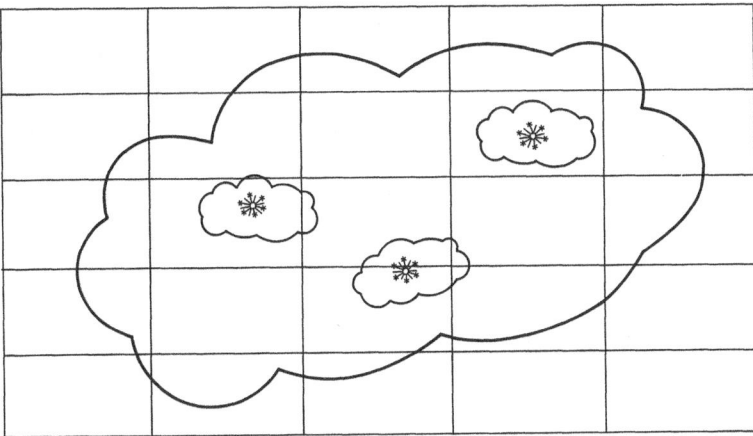

Figure 2.4   Hypothetical graphical output of an area-related grid model for atmospheric emissions

mental parameters on the one hand and observed concentrations on the other hand. When such relationships have been established and the results of predictive modelling appear to be satisfactory, predictions can be used to reduce the monitoring programme and they may also be used as a management tool, e.g. to assess the effect of additional new sources, or the effect of reduction of releases.

The numbers of this type of model and their complexity and application are increasing rapidly and, at present, they are used by local, national and international authorities in many industrialized countries. The first generation of these models was developed to investigate the effects of nutrients, mainly phosphate, on aquatic systems. The sources for these nutrients were the disposal of wastes containing phosphate-based detergents and runoff from agricultural land. Subsequently, aquatic fate models were developed to assess the exposure of the aquatic ecosystem to other types of chemicals. An example is EXAMS (Burns *et al.*, 1980; Burns and Cline, 1985) which has been used, for example, to assess the exposure in aquatic systems to pesticide runoff from agricultural land (Wood, 1986). Other models are used to calculate short and long term transport and transformation in the atmosphere and subsequent deposition on land or water of atmospheric contaminants such as SO_2 and SO_3 (Johnson, 1983; Renner *et al.*, 1985). Others may be used to investigate the potential surface and groundwater contamination in typical agricultural areas through the use of pesticides (Balek, 1983; Schuessler, 1986) or the impact of various effluent streams in estuaries (Harris *et al.*, 1984).

Investigations are now being started to evaluate the usefulness of expert computer systems in combining the often complex information from

Figure 2.5   Schematic representation of a local modelling exercise: plume calculations for atmospheric, aquatic and groundwater pollution

environmental quality models with expert knowledge on fate and effects of chemicals to assist authorities in impact assessments.

Source-related distribution and fate analysis

These models relate to one source or sometimes, within the bubble concept (Clean Air Act, 1955), to the combined sources from one industrial site. The sources are usually gaseous emissions, aqueous effluents or leachates from water disposal sites. Usually, the models are short-range and take only transport into account, the rate of transformation processes often expected to be too slow to influence the short-range fate.

In most cases the computer programs are single compartment models. Local average values for the environmental properties are used, often also taking into account their variability. In Figure 2.5, a schematic representation is given for a local modelling exercise.

Originally for industrial hygiene, but increasingly for environmental protection purposes, international, national and local authorities are establishing quality criteria for air, surface water and groundwater pollution. Usually, these criteria relate to chemical concentrations that are considered to be acceptable and, at present, much effort is directed to extending the number of these criteria. In Figure 2.6, a graphical representation is given of a model that is developed to estimate the exposure of humans to chemicals in contaminated soil. For many chemicals the criteria are based on toxicity data, but for chemicals that are of low toxicity, criteria may be based on physico-chemical properties and measures such as BOD and COD.

The primary tool for source-related exposure assessment is monitoring, which is often mandatory. Predictive modelling can be used to evaluate the influence of (1) varying environmental conditions, (2) changes in the emissions or effluents, and (3) changes in source geometry. When the reliability of the results of the predictive models has been demonstrated this may lead to (4) reduction or cessation of the monitoring programme.

GASEOUS EMISSIONS

The dispersion of gaseous emissions will initially be influenced by turbulence generated by jet flow or buoyancy. Gradually, this turbulence will decrease and dispersion will be dominated by environmental forces (List, 1982). The important parameters which influence jet flow and buoyancy are the effluent flow speed and flow direction, and the temperature and density of the effluent. The environmental forces are mainly due to the wind and the temperature-induced stability of the atmosphere which is usually expressed as the weather stability class (Hinrichsen, 1986; Gryning et al., 1987). Turbulence in the atmosphere near the earth's surface can be strongly influenced by surface roughness, heterogeneity of the surface, e.g. hills and buildings, and transitions, e.g. from land to water (Davis et al., 1986).

Programmes have been started in many countries to control air pollution. These programmes involve the listing of potentially hazardous air pollutants, the collection of background data and the development of standards, e.g. under the US NESHAP (National Emission Standard for Hazardous Air Pollutants), and the reduction in the use of certain chemicals (e.g. chlorofluorocarbons). These standards do not apply to the emissions themselves but to the resulting ambient concentrations, especially at

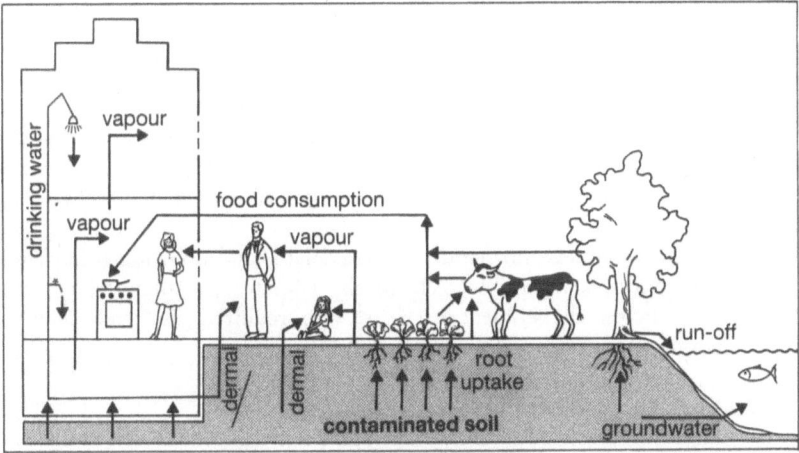

Figure 2.6 Graphical representation of a model to estimate the exposure of humans to chemical contaminants in soil

ground level. Predictive dispersion modelling is used to evaluate these ground-level concentrations. These dispersion models, usually gaussian plume models (Hinrichsen, 1986; Gryning et al., 1987), are simple and relatively easy to use, and have been constructed to predict ambient concentrations conservatively, i.e. worst-case assessment, such as the EPA model PTMAX. A disadvantage of these simple models is that they may not be applicable to complicated source geometries.

A further development is represented by the models such as the Dutch National Model and Extended Model in which weather statistics are included. They are used to predict the frequency of occurrence of certain concentrations at ground level. Thus, rather than a worst-case assessment these models indicate a frequency of the worst-case situation.

AQUEOUS EFFLUENTS

As previously described for gaseous emissions the dispersion of aqueous effluents will initially be influenced by jet flow or buoyancy and, subsequently, by environmental forces. In water the same parameters influence jet flow and buoyancy as in the atmosphere (Muellenhoff et al., 1985; Quetin and de Rouville, 1986). The environmental forces are influenced mainly by the speed of water currents while bottom characteristics are important only in shallow, fast-flowing waters.

Lists of potentially hazardous water pollutants have been compiled such as the list of 129 priority pollutants (US EPA) and the grey and black lists (EEC). Environmental Quality Standards (EQSs) are being formulated by international, national and local authorities. The EQS values do not necessarily apply at the point of discharge to the receiving water. A certain amount of dilution may be expected to occur initially. This will occur quickly and within a short range and, therefore, an instantaneous dilution of a factor of ten has traditionally been assumed. Recently more sophisticated estimation methods have been developed to calculate the initial dilution. In the US and elsewhere the concept of the mixing zone has been adopted in which higher concentrations than the EQS are permitted but beyond the mixing zone concentrations must be less than the EQS (EPA, 1982). Methods including computer programs for calculating the initial dilution within the mixing zone have been described (Muellenhoff et al., 1985).

For many chemicals no standards are available and even less is known about the effect of mixtures of chemicals. Therefore, in many countries toxicity data on the effluent itself are also required, for end-of-pipe regulation, permitting, and monitoring.

LEACHING FROM WASTE DISPOSAL SITES

Leaching from waste disposal sites through percolating rainwater has been recognized as a major threat to groundwater quality. It can also affect surface water quality (e.g. Niagara River and Lake Ontario, Canada). No methods or models are available for measuring or calculating chemical concentrations in percolating water except for specific homogeneous wastes (Sloot *et al.*, 1984). Therefore, extensive monitoring of groundwater quality at disposal sites is often mandatory and, increasingly, measures have to be taken to prevent percolated rainwater from reaching groundwater.

There are no universally accepted standards for groundwater quality at present. They may vary from no standards at all to drinking water quality standards (US, parts of Canada, the Netherlands).

Complementary to these considerations is the development of ground-water models, a variety of which are available (Prickett and Lonnquist, 1971; Balek, 1983; Pizzi and Sartori, 1985; Kemblowski, 1986; Leistra, 1986). There is general agreement that movement of a groundwater pollutant is mainly determined by partitioning of the compound between soil solids and water, molecular diffusion, groundwater flow and soil characteristics. Since movement is generally very slow, transformation, especially biodegradation, should also be taken into account in these model calculations.

Physico-chemical data required for environmental distribution and fate analysis

For many of the existing chemicals, it is not possible to make a proper distribution and fate analysis or exposure assessment because data are lacking or not easily available. Moreover, it is not possible, for both economic and technical reasons, to generate the necessary data for all these chemicals. Therefore, ways of priority-setting among existing chemicals are being discussed (Anon., 1985). Proposals include use of improved com-puterized data search facilities, property estimation methods and classifi-cation based on quantities produced and use patterns. It is envisaged that for certain selected chemicals missing data would have to be obtained.

For new synthetic organic chemicals legislation is in force or in prepar-ation (e.g. Toxic Substances Control Act, 1976; European Committees Council, 1979) that requires notification of those new chemicals which are not covered already by other, more demanding legislation such as for pesticides and food additives. The notification data are required to enable governmental authorities to determine environmental distribution and fate and to perform exposure assessments.

Minimum sets of physico-chemical properties that are needed for distri-bution and fate analysis of chemicals have been discussed by various authors (e.g. European Committees Council, 1979, Hushon *et al.*, 1983). These sets may vary depending on the degree of sophistication that is pursued.

The absolute minimum set of physico-chemical properties that is needed for an initial distribution and fate analysis is discussed below. However, even these data are not available for many chemicals and, more seriously, they are not always required for notification of new chemicals. Therefore, much research has been devoted to developing methods of estimating their value from data on other physico-chemical properties or from structural data. These estimation methods are also discussed.

Minimum data set for initial analysis

Potential exposure is primarily determined by production volume, use pattern, distribution, bioaccumulation, and persistence of the chemical. Production volume and use pattern are usually taken into consideration when deciding which data are required for a particular chemical. In the EEC Directive (European Committees Council, 1979), for example, only base set data are required when the production volume is above one tonne per year and depending on these data or, automatically, when the production volume is greater (more than 100 tonnes per year) more data are asked for (Schmidt-Bleek *et al.*, 1982).

The octanol/water partition coefficient can be used to estimate the bioaccumulation factor (Neely *et al.*, 1974; Neely, 1979; Veith *et al.*, 1979; Southworth *et al.*, 1980; Mackay, 1982; Schnoor, 1982; Oliver and Niimi, 1983; Spacie *et al.*, 1983; Chiou, 1985; Hawker and Connell, 1985; Hermans *et al.*, 1985; McCarthy *et al.*, 1985). It has been argued that, when the octanol/water partition coefficient is greater than 3 (Hushon *et al.*, 1983) or 4 (Veith, 1983), the compound may have a potential for bioaccumulation and, therefore, is a candidate for more extensive testing.

To assess a chemical's persistence in the environment, knowledge is needed, firstly, of its distribution between environmental compartments and, secondly, of its rate of transformation in relevant compartment(s). For such evaluations global environmental models have been developed (Mackay, 1979; Mackay and Paterson, 1981; Mackay and Paterson, 1982; Frische *et al.*, 1983; Mackay *et al.*, 1985).

Using these models the steady-state distribution of the chemical between the environmental compartments can be estimated (Mackay and Paterson, 1982). Then, on the basis of the results of these calculations a decision can be made on the transformation processes that should be studied, the dominant processes being biodegradation for the water and soil compartments and reaction with OH radicals for the air compartment. Subsequently, the results of these studies can be used as input for higher level calculations (Mackay and Paterson, 1982) to arrive at a measure of the global persistence of the compound. When, on the basis of this minimum set of data a chemical appears to be persistent, more data will be required to perform a more detailed assessment.

Estimation methods for the minimum data set

The absolute minimum data set of physico-chemical properties consists of (1) the octanol/water partition coefficient, (2) the vapour pressure and (3) the solubility in water, from which data it is possible to calculate the potential global distribution of the chemical. Further (4) the biodegradability in water and soil or (5) the rate of reaction with OH radicals in the atmosphere are required to estimate the global persistence. When it appears to be necessary, the set can be extended further.

An excellent handbook of estimation methods has been written by Lyman *et al.* (1982). However, AS developments are rapid many new or improved methods have become available since this handbook was issued. Using this handbook as its basis, a computer program was developed to estimate physico-chemical properties. This program (CHEMEST), updated regularly, is commercially available. Estimation methods for the absolute minimum set of data are now discussed in more detail.

OCTANOL/WATER PARTITION COEFFICIENT

The octanol/water partition coefficient can be regarded as a multiple purpose parameter. It has been correlated, often successfully, with almost every environmentally relevant physico-chemical property and thus, conversely, data on these properties may be used to estimate the value of the octanol/water partition coefficient itself. For this purpose, however, methods based on structural data are to be preferred (Lyman *et al.*, 1982).

In the early 1970s two fragmental addition methods were developed almost in parallel (Leo *et al.*, 1971; Rekker, 1977; Rekker and de Kort, 1979). A computerized version of a combination of these two methods is now commercially available, and includes a regularly updated database on experimentally determined values.

Isomers are known to have slightly different octanol/water partition coefficients. In general, these differences cannot be evaluated using the above methods. Very recently, however, a new method based on structural data has been developed that has the advantage of discriminating between all isomers (Camilleri, 1987). Additionally, this latter method gives slightly better results than the methods of Leo and Rekker and includes an estimation of the error. Research is being conducted to refine and extend this method.

SOLUBILITY IN WATER

An overview of available methods to estimate the solubility of a chemical in water has been presented by Lyman *et al.* (1982). Further refinements of the correlation between solubility and the octanol/water partition coefficient (Miller *et al.*, 1985) and between solubility and molar volume (Lande and Banerjee, 1981) have been proposed. No preferred method can be indicated. Selection should be based on the defined applicability of the methods and on available data. Preferably, several methods should be applied.

VAPOUR PRESSURE

Lyman et al. (1982) recommend two methods for organic chemicals that give reasonably accurate results at environmental temperatures, the Antoine equation for liquids and gases and the modified Watson correlation for liquids and solids. Both methods require the boiling point as input and, for compounds that are solid at environmental temperatures, the melting point as well. Correlations have been critically discussed by Mackay et al. (1982) who conclude that greater accuracy can be obtained by correlating classes of compounds separately. They developed equations that apply to hydrocarbons and halogenated hydrocarbons. An extensive database for measured vapour pressures is commercially available.

RATE OF BIODEGRADATION

Biodegradation is generally considered to be the most important process of transformation, usually leading to mineralization of organics in water, soil and sediment. Biodegradation has been studied in many different systems but, in most cases, at very high chemical and microbial concentrations, relative to the environment. Usually, mixed microbial populations have been used which have probably differed significantly between studies. Test data obtained after five to thirty days have been expressed in various ways, such as the amount of oxygen consumption, the amount of carbon dioxide evolution or the amount of residual parent compound at the end of the test period, but seldom as rate constants of microbial transformation. Therefore, Howard et al. (1987) concluded that not enough data are available at present to develop quantitative structure activity relationships (QSARs) and they proposed instead a system of collecting and evaluating biodegradation data.

A review of correlations, developed to date, has been presented by Boethling (1985). He suggested that correlations between biodegradation and structural parameters would be superior to correlations based on chemical properties, such as the rate of alkaline hydrolysis (Wolfe et al., 1980), because of associated experimental errors. He presented correlations between molecular connectivity indices and various biodegradation data sets. Dearden and Nicholson (1985) suggested that for biodegradation, being an enzymatic process, electronic and steric factors might predominate. They developed successful correlations for numerous congeneric series of compounds between biodegradation expressed as the 5 days BOD value and the difference of atomic charge across selected bonds, e.g. across the C-N bond for aliphatic and aromatic amines, and also between the 5 days BOD value and structural parameters, the so-called sterimol parameters. Freitag et al. (1985) explored correlations between the octanol/water partition coefficient and biodegradation in activated sludge expressed as a percentage of the compound that is mineralized. They found such a correlation to exist for some classes of compounds but not for others. This is in agreement with the findings of Vaishnav *et al.* (1987).

From the above it is clearly feasible to correlate biodegradation to structure-related data, such as connectivity indices, octanol/water partition coefficient and atomic charges, but only for congeneric series and, most likely, only if the same microbial transformation pathway applies. From mechanistic studies such as performed by van der Linden and Huybrechtse (1967) and van der Linden and van Ravenswaay Claasen (1971), it can be concluded that rates of biodegradation are primarily dependent on rates of enzymatic processes and types of bonds within the molecule and only slightly on hydrophobicity. Therefore, successful correlations based on hydrophobicity and topological data can be formulated and used but only for congeneric series which are transformed by the same enzymatic process. Much more research is needed to arrive at quantitative structure activity relationships that are independent of these restrictions.

RATE OF REACTION WITH OH RADICALS IN THE ATMOSPHERE

In the atmosphere reaction with OH radicals is considered to be the main transformation process (ECETOC, 1983; Atkinson, 1985; OECD, 1987). Various methods of predicting rates of this process in the atmosphere have been discussed at the OECD Ad Hoc Expert meeting (OECD, 1987). Of these methods those based on the assumption of fragmental additivity are preferred as they are highly accurate and least expensive. The assumption of fragmental additivity is based on the observation that the reactivity of each fragment in a molecule, such as $-CH_2$ or $-OH$, appears to be constant, dependent on its own structure and only slightly on the structure of adjacent fragments. By adding the contributions of the various fragments in a molecule, taking into account the influence of adjacent fragments, the rate of reaction of the total molecule is obtained. The method described by Atkinson (1985) is expected to have an accuracy of 30 per cent. Values are not yet available for all molecular fragments, such as those containing nitrogen and sulphur. However, by assuming these to be zero a minimum value for the whole of the molecule can be calculated.

Conclusions

There has been much progress, especially in the last decade, in developing methods for predicting and evaluating the fate of chemicals in the environment. One of the most significant developments was the mathematical model EXAMS (Burns and Cline, 1985) since this represented, for the first time, an attempt to incorporate all of the transport and transformation processes, together with relevant environmental parameters, in a single model that could be used for the environmental management of aquatic systems. It has been demonstrated that predictions of this and other mathematical models are reasonably reliable provided that model inputs can be measured, or estimated, with sufficient accuracy.

Validation of such models has involved comparison of model predictions with chemical distributions in real-world or experimental systems. For this purpose reference chemicals have been chosen whose physico-chemical properties are relatively well established. However, for many organic chemicals, their physico-chemical properties are insufficiently known to be reliable inputs for distribution and fate models. Therefore, further work is needed to characterize these physico-chemical properties, either by direct measurement or by methods of estimation based on chemical structure or other available data.

References

American Petroleum Institute (1982), *Review of Groundwater Models*, (Washington, DC: American Petroleum Institute).

Anon. (1985), *Proceedings of the Workshop on Environmental Modelling for Priority Setting among Existing Chemicals*, 11–13 November 1985, GSF-Bericht 40/85 (Munich: Gesellschaft für Strahlen und Umweltforschung).

Apoteker, A. and Thevenot, D. R. (1983), 'Experimental simulation of biodegradation in rivers. Oxygen, organic matter and biomass concentration changes', *Water Res.* 17, pp. 1267–73.

Atkinson, R. (1985), 'Kinetics and mechanisms of the gas-phase reactions of the hydroxyl radical with organic compounds under atmospheric conditions', *Chem. Rev.* 85, pp. 69–201.

Baalen, C. van and Marler, J. E. (1966), 'Occurrence of hydrogen peroxide in sea water', *Nature* 211, pp. 951–5.

Balek, J. (1983), 'Topic 3: State of the art of mathematical modelling of agricultural impact on groundwater pollution', *Environ. Geology* 5, pp. 27–32.

Beek, W. J. and Mutzall, K. M. (1975), *Transport Phenomena* (London: John Wiley and Sons).

Boethling, R. S. (1985), 'Application of molecular topology to quantitative structure-biodegradability relationships', *Environ. Toxicol. Chem.* 5, pp. 797–806.

Branson, D. R. (1980), 'Prioritization of chemicals according to the degree of hazard in the aquatic environment', *Environ. Health Perspectives* 34, pp. 133–8.

Breugem, P., van Noort, P., Velberg, S., Wondergem, E. and Zijlstra, J. (1986), 'Steady state concentrations of phototransient hydrated electron in natural waters', *Chemosphere* 15, pp. 717–24.

Burns, L. A. and Cline, D. M. (1985), *Exposure Analysis Modelling System*. Athens, GA 30613, Environmental Research Laboratory, Office of Research and Development, US Environmental Protection Agency. Report EPA/600/3–85/038.

Burns, L. A., Cline, D. M. and Lassiter, R. R. (1980), *EXAMS: An Exposure Analysis Modelling System*, Preliminary Draft Document (February 1980), available through R. R. Lassiter, United States Environmental Protection Agency, Environmental Research Laboratory, College Station Road, Athens, GA 30613.

Cairns, J. Jr (1980), 'Estimating hazard', *Bioscience* 30, pp. 101–7.

Camilleri, P. (1987), Shell Research Limited, Sittingbourne Research Centre, Sittingbourne, Kent ME9 8AG, UK. Personal communication.

Chiang, H-C. and Sill, B. L. (1985), 'Entrainment models and their application to jets in turbulent cross flows', *Atm. Environ.* 19, pp. 1425–38.

Chiou, C. T. (1985), 'Partition coefficients of organic compounds in lipid–water systems and correlations with fish bioconcentration factors', *Environ. Sci. Technol.* 19, pp. 57–62.

Chiou, C. T., Porter, P. E. and Schmedding, D. W. (1983), 'Partition equilibria of nonionic compounds between soil organic matter and water', *Environ. Sci. Technol.* 17, pp. 227–31.

Clean Air Act 1955 (US), last modified in 1983.

Colenbrander, G. W. (1980), *A Mathematical Model for the Transient Behaviour of Dense Vapour Clouds.* 3rd Intl. Symp. on Loss Prevention and Safety Promotion in the Process Industries, Sept. 1980, Basle.

Cooper, W. J. and Zika, R. G. (1983), 'Photochemical formation of hydrogen peroxide in surface and groundwaters exposed to sunlight', *Science* 220, pp. 711–12.

Costerton, J. W. and Greesy, G. G. (1979), 'Which populations of aquatic bacteria should we enumerate?', in J. W. Costerton and R. R. Colwell (eds), *Native Aquatic Bacteria: Enumeration, Activity and Ecology,* ASTM STP 695 (Philadelphia: American Society for Testing and Materials).

Crossland, N. O. and Bennett, D. (1984), 'Fate and biological effects of methyl parathion in outdoor ponds and laboratory aquaria', *Ecotoxicol. Environ. Saf.* 8, pp. 471–81.

Crossland, N. O., Bennett, D. and Wolff, C. J. M. (1987), 'Rate of 2,5,4'-trichlorobiphenyl in outdoor ponds and its uptake via the food chain compared with direct uptake via the gills in grass carp and rainbow trout', *Ecotoxicol. Environ. Saf.* 13, pp. 225–38.

Crossland, N. O., Bennett, D., Wolff, C. J. M. and Swannell, R. P. J. (1986) 'Evaluation of models used to assess the fate of chemicals in aquatic systems', *Pestic. Sci.* 17, pp. 297–304.

Crossland, N.O., and Wolff, C. J. M. (1985), 'Fate and biological effects of pentachlorophenol in outdoor ponds', *Environ. Toxicol. Chem.* 4, pp. 73–86.

Daniels, S. L., Hoerger, F. D. and Moolenaar, R. J. (1985), 'Environmental exposure assessment. Experience under the Toxic Substances Control Act', *Environ. Toxicol. Chem.* 4, pp. 107–17.

Davis, P. A., Reimer, A., Sakiyama, S. K. and Stawson, P. R. (1986), 'Short-range atmospheric dispersion over a heterogeneous surface-I. Lateral dispersion', *Atmos. Environ.* 20, pp. 41–50.

Dearden, J. C. and Nicholson, R. M. (1985), *The Use of Structure–Activity Relationships to Predict Biodegradability.* Presentation on the Joint International Symposium: Physico-Chemical Properties and their Role in Environmental Hazard Assessment. University of Kent, Canterbury, 1–3 July 1985.

DiToro, D. M. and Horzempa, L. M. (1982), 'Reversible and resistant components of PCB absorption–desorption isotherms', *Environ. Sci. Technol.* 16, pp. 594–9.

ECETOC (European Chemical Industry Ecology and Toxicology Centre) (1983), *Experimental Assessment of the Phototransformation of Chemicals in the Atmosphere,* Technical Report No. 7 (Brussels: ECETOC).

ECETOC (European Chemical Industry Ecology and Toxicology Centre) (1984), *The Phototransformation of Chemicals in Water: Results of a Ring-test,* Technical Report No. 12 (Brussels: ECETOC).

EPA (Environmental Protection Agency) (1982), *Modifications of Secondary Treatment Requirements for Discharges into Marine Waters,* Federal Register, November 26, 1982, 47(228):pp. 53666–85 (Washington, DC: US Environmental Protection Agency).

European Committees Council (1979), Annex VII to the European Committees

Council Directive 79/831/EEC of 18 September 1979 amending for the 6th time the Council Directive 67/348/EEC of 27 June 1967 on the Approximation of the Laws, Regulations, and Administrative Provisions Relating to the Classification, Packaging and Labelling of Dangerous Substances.

Fischer, H. B., List, E. J., Koh, R. C. Y., Imberger, J. and Brooks, N. H. (1979) in *Mixing in Inland and Coastal Waters*, p. 11 (New York: Academic Press).

Freitag, D., Balhorn, L., Geyer, H. and Korte, F. (1985), 'Environmental hazard profile of organic chemicals. An experimental method for the assessment of the behaviour of organic chemicals in the ecosphere by means of simple laboratory tests with 14C labelled chemicals', *Chemosphere* 14, pp. 1589–1616.

Frische, R., Klöppfer, W., Rippen, G. and Günther, K. (1983), 'The environmental model segment approach for estimating potential environmental concentrations. I. The model', *Ecotoxicol. Environ. Saf.* 8, pp. 352–62.

Gäb, S., Schmitzer, J., Thamm, H. W., Parlar, H. and Korte, F. (1977), 'Photomineralization rate of organic compounds adsorbed on particulate matter', *Nature* 270, pp. 332–3.

Garland, J. A., Elzerman, A. W. and Penkett, S. A. (1980), 'The mechanism for dry deposition of ozone to seawater', *J. Geophys. Res.* 85, pp. 7488–90.

Geankoplis, C. J. (1972), *Mass Transport Phenomena*, (New York: Holt, Rinehart and Winston).

Genuchten, M. Th. van and Alves, W. J. (1982), *Analytical Solution of the One-Dimensional Convective-Dispersive Solute Transport Equation*. Technical Bulletin 1661, United States Department of Agriculture, Agriculture Research Services.

Gryning, S. E., Holtslag, A. A. M., Irwin, J. S. and Sivertsen, B. (1987), 'Applied dispersion modelling based on meteorological scaling parameters', *Atmos. Environ.* 21, pp. 78–89.

Halfon, E. and Reggioni, M. G. (1986), 'On ranking chemicals for environmental hazard', *Environ. Sci. Technol.* 20, pp. 1173–9.

Harris, J. R. W., Bale, A. J., Bayne, B. J., Mantoura, R. F. C., Morris, A. W., Nelson, L. A., Radford, P. J., Uncles, R. J., Weston, S. A. and Widdows, J. (1984), 'A preliminary model of the disposal and biological effect of toxins in the Tamar estuary, England', *Ecological Modelling* 22, pp. 253–84.

Hartley, G. S. and Graham-Bryce, I. J. (1980), *Physical Principles of Pesticide Behaviour*, Vol. I, pp. 325–31 (London: Academic Press).

Hawker, D. W. and Connell, D. W. (1985), 'Relationship between partition coefficient, uptake rate constant, clearance rate constant and time to equilibrium for bioaccumulation', *Chemosphere* 14, pp. 1205–19.

Hermans, J., Busser, F., Leeuwanch, R. and Musch, A. (1985), 'Quantitative correlation studies between the acute lethal toxicity of 15 organic halides to the guppy (*Poecilla reticulata*) and chemical reactivity towards 4-nitrobenzylpyridine', *Toxicol. Environ. Chem.* 9, pp. 219–36.

Hinrichsen, K. (1986), 'Comparison of four analytical dispersion models for near-surface releases above a grass surface', *Atmos. Environ.* 20, pp. 29–40.

Howard, P. H., Hueber, A. E. and Boethling, R. S. (1987), 'Biodegradation data evaluation for structure/biodegradability relations', *Environ. Technol. Chem.* 6, pp. 1–10.

Hushon, J. M., Klein, A. W., Strachan, W. J. M. and Smidt-Bleek, F. (1983), 'Use of OECD premarket data in environmental exposure analysis for new chemicals', *Chemosphere* 12, pp. 887–910.

Johnson, W. B. (1983), 'Interregional exchange of air pollution: model types and applications', *J. Air Pollut. Cont. Ass.* 33, pp. 563–74.

Joussot-Dubien, J. and Kadiri, A. (1970), 'Photosensitized oxidation of ammonia by singlet oxygen in aqueous solution and in water', *Nature* 227, pp. 700–2.

Karickhoff, S. W. (1981), 'Semi-empirical estimation of sorption of hydrophobic pollutants on natural sediments and soils', *Chemosphere* 10, pp. 833–46.

Karickhoff, S. W. (1985), 'Sorption phenomena', in W. B. Neely and G. E. Blau (eds) *Environmental Exposure from Chemicals*, Vol. I. (Boca Raton, Fl: CRC Press).

Karickhoff, S. W., Brown, D. S. and Scott, T. A. (1979), 'Sorption of hydrophobic pollutants on natural sediments', *Water Res.* 13, pp. 241–8.

Kemblowski, M. (1986), 'A boundary element-random walk model of mass transport in groundwater', *J. Hydrology* 85, pp. 305–18.

Kenaga, E. E. (1986), 'Assessing chemical hazards', *Environ. Sci. Technol.* 20, pp. 660–2.

Klein, W., Geyer, H., Freitag, D. and Rohleder, H. (1984), 'Sensitivity of schemes for ecotoxicological hazard ranking of chemicals', *Chemosphere* 13, pp. 203–11.

Kok, G. L. (1980), 'Measurements of hydrogen peroxide in rainwater', *Atm. Envt.* 14, pp. 653–5.

Lande, S. S. and Banerjee, S. (1981), 'Predicting aqueous solubility of organic nonelectrolytes from molar volume', *Chemosphere* 10, pp. 751–9.

Lawrence, A. W. and McCarthy, P. L. (1970), 'Unified basis for biological treatment design and operation', *J. Sanitary Eng. Div.*, Proc. Am. Civil Eng. SA3, 757.

Leistra, M. (1986), 'Modelling the behaviour of organic chemicals in soil and groundwater, *Pestic. Sci.* 17, pp. 256–64.

Leo, A., Hansch, C. and Elkins, D. (1971),'Partition coefficients and their uses', *Chemical Reviews* 71, pp. 525–615.

Lewis, W. K. and Whitman, W. G., (1924), 'Principles of gas absorption', *Ind. Eng. Chem.* 16, pp. 1215–20.

Linden, A. C. van der and Huybrechtse, R. (1967), 'Induction of alkane oxidizing and alpha-olefinepoxydizing enzymes by a non-hydrocarbon in a pseudomonas', *J. Microbiol. Serology* 33, pp. 381–5.

Linden, A. C. van der and van Ravenswaay Claasen, J. C. (1971), 'Hydrophobic enzymes in hydrocarbon degradation'. *Lypido* 6, pp. 437–43.

Liss, P. S. and Slater, P.G. (1974), 'Flux of gases across the air–sea interface', *Nature* 247, pp. 181–4.

List, E. J. (1982), 'Turbulent jets and plumes', *Annual Review of Fluid Mechanics* 14, pp. 189–212.

Lyman, W. J., Reehl, W. F. and Rosenblatt, D. H. (1982), *Handbook of Chemical Property Estimation Methods. Environmental behaviour of organic compounds* (New York, McGraw-Hill).

Mabey, W. and Mill, T. (1978), 'Critical review of hydrolysis of organic compounds in water under environmental conditions', *J. Phys. Chem. Ref. Data* 7, pp. 383–414.

McCarthy, L.S., Hodson, P. V., Craig, G. R. and Kaiser, K. L. E. (1985), 'The use of quantitative structure-activity relationships to predict the acute and chronic toxicities of organic chemicals', *Environ. Toxicol. Chem.* 4, pp. 595–606.

Mackay, D. (1979), 'Finding fugacity feasible', *Environ. Sci. Technol.* 13, pp. 1218–23.

Mackay, D. (1982), 'Correlation of bioconcentration factors', *Environ. Sci. Technol.* 16, pp. 274–8.

Mackay, D., Bobra, A., Chan, D. W. and Shiu, W. Y. (1982), 'Vapor pressure correlations for low-volatility environmental chemicals', *Environ. Sci. Technol.* 16, pp. 645–9.

Mackay, D. and Paterson, S. (1981), 'Calculating fugacity', *Environ. Sci. Technol.* 15, pp. 1006–14.

Mackay, D. and Paterson, S. (1982), 'Fugacity revisited', *Environ. Sci. Technol.* 16, pp. 654A–60A.

Mackay, D., Paterson, S., Cheung, B. and Brock Neely, W. (1985), 'Evaluating the environmental behavior of chemicals with a level III fugacity model', *Chemosphere* 14, pp. 335–74.

Mackay, D., Paterson, S. and Schroeder, W. H. (1986), 'Model describing the rates of transfer process of organic chemicals between atmosphere and water', *Environ. Sci. Technol.* 20, pp. 810–16.

Mackay, D. and Yeun, A. T. K. (1983), 'Mass transfer coefficient correlations for volatilization of organic solutes from water', *Environ. Sci. Technol.* 17, pp. 211–17.

Mill, T., Henry, D. G. and Richardson, H. (1980), 'Free-radical oxidants in natural waters', *Science* 207, p. 886.

Miller, G. C., Zisook, R. and Zepp, R.G. (1980), 'Photolysis of 3,4-dichloroaniline in natural waters',. *J. Agric. Food Chem.* 28, p. 1053.

Miller, M. M., Wasik, S. P., Huang, G. L., Shiu, W. Y. and Mackay, D. (1985), 'Relationships between octanol-water partition coefficients and aqueous solubility', *Environ. Sci. Technol.* 19, pp. 522–8.

Momzikoff, A., Santus, R. and Giraud, M. (1983), 'A study of the photosensitizing properties of sea water', *Mar. Chem.* 11, pp. 1–7.

Muellenhoff, W. P., Soldate Jr., A. M., Baumgartner, D. J., Schuldt, M. D., Davis, L. R. and Frick, W. E. (1985), *Initial Mixing Characteristics of Municipal Ocean Discharges.* Vol, I – *Procedures and Applications*, NTIS, PB86–137478.

Neely, W. B. (1979), 'Estimating the rate constants for the uptake and clearance of chemicals by fish', *Environ. Sci. Technol.* 13, pp. 1506–10.

Neely, W. B. (1980), *Chemicals in the Environment* (New York: Marcel Dekker).

Neely, W. B. (1985) 'Hydrolysis', in W. B. Neely and G. E. Blau. (eds), *Environmental Exposure from Chemicals*, Vol. I, pp. 157–74, Ch. 7 (Boca Raton, Fla: CRC Press).

Neely, W. B., Branson, D. R. and Blau, G. E. (1974), 'Partition coefficients to measure bioconcentration potential of organic chemicals in fish', *Environ. Sci. Technol.* 8, pp. 1113–15.

OECD (1981), OECD *Guidelines for Testing of Chemicals* (OECD, 2. rue Andre-Pascal, 75775 Paris-Cedex 16, France).

OECD (1987), *Draft OECD Test Guideline on Photochemical Oxidative Degradation in the Atmosphere.* Revised by the Ad Hoc Expert meeting, Berlin, Reichstag, January 1987 (Berlin: Umweltbundesamt, Bundesrepublik Deutschland).

Oliver, B. G. and Niimi, A. J. (1983), 'Bioconcentration of chlorobenzenes from water by rainbow trout: Correlation with partition coefficients and environmental residues', *Environ. Sci. Technol.* 17, pp. 287–91.

Paris, D. F., Lewis, D. L. Barnett, J. T. and Baughmann, G. L. (1975), *Microbial Degradation and Accumulation of Pesticides in Aquatic Systems.* Report no. EPA–660/3–75–007 (Athens, GA 60613: US Environmental Protection Agency).

Paris, D. G., Steen, W. C., Baughmann, G. L. and Barnett Jr., J. T. (1981), 'Second-order model to predict microbial degradation of organic compounds in natural waters', *Applied and Environmental Biology* 41, pp. 603–9.

Parlar, H. (1986), personal communication.

Pizzi, G. and Sartori, L. (1985). 'Interconnected groundwater system simulation (IGROS) – Description of the system and a case history application', *J. Hydrology* 75, pp. 255–85.

Prickett, T. A. and Lonnquist, C. G. (1971), 'Selected techniques for groundwater resource evaluation', *Illinois State Water Survey Bulletin 55.*

Puttock, J. S. (1986), *Gravity-Dominated Dispersion of Dense-Gas Clouds.* IMA Conference on Stratified Flow and Dense Gas Dispersion. Chester, April 1986.

Quetin, B. and Rouville, M. de. (1986), 'Submarine sewer outfalls – A design manual', *Mar. Poll. Bull.* 17, pp. 133–8.

Rekker, R. F. (1977), *The Hydrophobic Fragmental Constant* (Amsterdam, Elsevier).

Rekker, R. F. and de Kort, H. M. (1979), 'The hydrophobic fragmental constant; an extension to a 1000 data point set', *Eur. J. Med. Chem.-Chimica Therapeutica* 14, pp. 479–88.

Renner, E., Ratzlaff, U. and Rolle, W. (1985), 'A lagrangian multi-level model of transport, transformation and deposition of atmospheric sulfur dioxide and sulfate', *Atmospheric Environment* 19, pp. 1351–9.

Reuber, B., Mackay, D., Paterson, S. and Stokes, P. (1987), 'A discussion of chemical equilibria and transport at the sediment-water interface', *Environ. Toxicol. Chem.* 6(10), pp. 731–9.

Rittmann, B. E. and McCarthy, P. L. (1980a) 'A model of steady-state biofilm kinetics', *Biotechnol. Bioeng.* 22, pp. 2343–7.

Rittmann, B. E. and McCarthy, P. L. (1980b) 'Evaluation of steady-state biofilm kinetics', *Biotechnol. Bioeng.* 22, pp. 2359–66.

Schmidt-Bleek, F., Haberland, W., Klein, A. W. and S. Caroli, (1982), 'Steps towards environmental hazard assessment of new chemicals', *Chemosphere* 11, pp. 383–415.

Schnoor, J. L. (1982), *Field Validation of Water Quality Criteria for Hydrophobic Pollutant nr. 302.* Aquatic Toxicology and Hazard Assessment: Fifth Conference. ASTM STP 766, eds J. G. Pearson, R. Foster and W. E. Bishop. American Society for Testing and Materials.

Schuessler, L. (1986), 'A new tool for improving water quality', *Farm. Chemicals International* 1, pp. 52–4.

Sloot, H. A. van der, Piepers, O. and Kok, A. (1984), *A Standard Leaching Test for Combustion Residues*, Report No. 31 (Petten, the Netherlands: Bureau Energy Onderzoek Projecten (BEOP)).

Smith, J. H., Mabey, W. R., Bohonos, N., Holt, B. R., Lee, S. S., Chou, T. W., Bombergen, D. C. and Mill, T. (1978), *Environmental Pathways of Selected Chemicals in Fresh Water Systems, Part II: Laboratory Studies*, EPA–600/7–78–074 (Springfield, Virginia 22161: National Technical Information Service).

Southworth, G. R., Keffer, C. C. and Beauchamp, J. J. (1980), 'Potential and realized bioconcentration. A comparison of observed and predicted bioconcentration of azaarenes in the fathead minnow (*Pimephales promelas*)', *Environ. Sci. Technol.* 14, pp. 1529–34.

Spacie, A., Landrum, P. F. and Leversee, G. J. (1983), 'Uptake, depuration and biotransformation of anthracene and benzo(a)pyrene in bluegill sunfish', *Ecotoxicol. Environ. Saf.* 7, pp. 330–41.

Thibodeaux, L. J. and Scott, H. D. (1985), 'Air/soil exchange coefficients', in: W. B. Neely and G. E. Blau (eds) *Environmental Exposure from Chemicals*, Vol. I. (Boca Raton, Fla: CRC Press).

Toxic Substances Control Act 1976 (US), pp. 94–469.

Vaishnav, D. D., Boethling, R. S. and Babeu, L. (1987), 'Quantitative structure-biodegradability relationships for alcohols, ketones and alycyclic compounds', *Chemosphere* 16, pp. 695–703.

Veith, G. D. (1983), United States Environmental Protection Agency, Duluth Research Laboratory, personal communication.

Veith, G. D., Defoe, D. L. and Bergstedt, B. V. (1979), 'Measuring and estimating the bioconcentration factor of chemicals in fish', *J. Fish. Res. Board Canada.* 36, pp. 1040–8.

Voice, Th. C. and Weber Jr., W. J. (1983), 'Sorption of hydrophobic compounds by sediments, soils and suspended solids, I: Theory and background', *Water Res.* 17, pp. 1433–41.

Weber, W. J. R., Voice, Th. C., Pribazari, M., Hunt, G. E. and Ulanoff, D. M., (1983), 'Sorption of hydrophobic compounds by sediments, soils and suspended solids, II: Sorbent evaluation studies', *Water Res.* 17, pp. 1443–52.

Wolfe, N. L., Kitchens, B. E., Macalady, D. L. and Grundl, T. J. (1986), 'Physical and chemical factors that influence the anaerobic degradation of methyl parathion in sediment systems', *Environ. Toxicol. Chem.* 5, pp. 1019–26.

Wolfe, N. L., Paris, D. F., Steen, W. L. and Baughman, G. L. (1980), 'Correlation of microbial degradation rates with chemical structure', *Environ. Sci. Technol.* 14, pp. 1143–4.

Wolff, C. J. M. and Crossland, N. O. (1985), 'Fate and effects of 3,4-dichloroaniline in the laboratory and in outdoor ponds, I: Fate', *Environ. Toxicol. Chem.* 4, pp. 481–7.

Wolff, C. J. M., Halmans, M. T. H. and van der Heijde, H. B. (1981), 'The formation of singlet oxygen in surface waters', *Chemosphere* 10, pp. 59–62.

Wolff, C. J. M. and van der Heijde, H. B. (1982), 'A model to assess the rate of evaporation of chemical compounds from surface waters', *Chemosphere* 11, pp. 103–17.

Wood, W. P. (1986), Environmental Protection Agency, Office of Toxic Substances, 401 M Street, SW, Washington DC. 20460, personal communication.

Zepp, R. G. (1978), 'Quantum yields for reaction of pollutants in dilute aqueous solution', *Environ. Sci. Technol.* 12, p. 327.

Zepp, R. G. (1986), personal communication.

Zepp, R. G., Braun, A. M., Hoigné, J. and Leenheer, J. A. (1987b), 'Photoproduction of hydrated electrons from natural organic solutes in aquatic environments', *Environ. Sci. Technol.* 21, pp. 485–90.

Zepp, R. G. and Cline, D. M. (1977), 'Rates of direct photolysis in the aquatic environment', *Environ. Sci. Technol.* 11, p. 359.

Zepp, R. G., Hoigné, J. and Bader, H. (1987a), 'Nitrate-induced photooxidation of trace organic chemicals in water', *Environ. Sci. Technol.* 21, pp. 443–50.

Zepp, R. G., Wolfe, N. L., Baughmann, G. L. and Hollis, R. C. (1977), 'Singlet oxygen in natural waters', *Nature* 267, pp. 421–3.

3 Basic toxicological considerations

G. M. RAND

Introduction

This chapter summarizes information which is fundamental to the study of toxicology, a critical part of the information base necessary for the effective management of chemicals. This involves an understanding of chemical/ biological interactions and biological mechanisms. Since the amount of information on these subjects is voluminous and constantly changing, this chapter directs its attention to general terminology and principles applicable to a myriad of chemical substances to gain an appreciation of chemical –biological phenomena. The chapter also describes current testing procedures used to assess the toxicity of chemical substances to aquatic organisms, wildlife and mammalian species (also see Chapter 4 on ecotoxicology).

Toxicology is concerned with the study of the adverse effects of naturally occurring and man-made (anthropogenic) substances on living systems. The nature and mechanisms of toxic lesions and spectrum of biological changes, coupled with the diversity of substances in the environment, make toxicology a broad and challenging multidisciplinary science. Therefore, in order to understand the toxicological activities of any substance, it is important to have a background in biology, chemistry, pathology, physiology, biochemistry, pharmacology, public health and statistics. A toxicologist is typically specialized in one or all of the following three areas: toxicity testing (laboratory and/or field); understanding the mechanisms of toxic action; and determining whether a given substance is marketable for some special use (e.g. drug, pesticide) based on readily available toxicity data.

Specialists in toxicology may be active in one or more divisions of toxicology. *Environmental* toxicology is concerned with the effects of environmental substances on humans exposed via the atmosphere or soil during occupational (i.e. workplace) or recreational activities. *Ecotoxicology* is a new subdivision of environmental toxicology involving study of the adverse effects of substances on populations and communities of non-human species. The latter science is rapidly evolving; many testing techniques have not yet been adequately developed and validated as they have in *mammalian toxicology*. In mammalian toxicology, testing is routinely conducted with rats, mice, guinea pigs, rabbits and dogs depending on the endpoint(s) being studied. For example, rabbits are used in dermal toxicity studies while rats and mice are test models used to define the general oral toxicity of substances. Mammalian toxicology studies are designed to yield data that can be used to evaluate the risk to humans as a result of exposure to chemical substances. Ecological toxicity assessments are more difficult to conduct than mammalian toxicity assessments because the ultimate concern is the protection of a myriad of species in the environment which may be affected not only by foreign substances (e.g. pesticides, inorganic pollutants) but a host of other elements (e.g. water quality, type of soil, temperature) as a result of the complexities and interactions of aquatic and terrestrial communities; in short, it deals with whole natural ecosystems. *Forensic toxicology* is the branch that is concerned with the medical and legal aspects of the adverse effects of substances on humans and animals. It involves the acquisition of information concerned with diagnosis, treatment and condition of exposure to a substance in order to elucidate cause–effect relationships between exposure to a substance and harmful effects. *Clinical toxicology* is the medical science that deals with biological anomalies caused by intentional or accidental poisoning by toxic substances and drugs. Attention is therefore directed at treatment of intoxication and the development of new techniques to diagnose and treat poisoning (Klaassen *et al.*, 1986).

Toxicity, hazard and risk

Toxicity is the relative potential of a substance (toxicant) or combination of substances for producing injury or harm to living organisms. Toxicants may be classified according to their target organ (liver, kidney), their use (pesticide, solvent), their source (synthetic, biological toxin), their physical state (gas, liquid), or their chemistry (aliphatic hydrocarbon, chlorinated hydrocarbon, etc.).

Toxicants enter the environment either deliberately or accidentally from the following: *nonpoint sources* such as agricultural runoff from land, contaminated groundwater and bottom sediments, urban runoff, dredged sediment disposal, and atmospheric fallout, and *point sources* such as discharges (effluents) from manufacturing plants, hazardous disposal sites, and municipal waste water treatment plants.

A highly toxic substance may adversely affect an organism exposed to a very small quantity or concentration of it while a substance with low toxicity may not produce any effect unless the organism is exposed to large quantities or concentrations. Toxicity cannot be adequately defined' unless reference is made to: the amount of material to which an organism is exposed, the manner in which the organism is exposed (e.g. mammalian species – inhalation, ingestion, injection; aquatic species – water, ingestion), extent of exposure (e.g. single or multiple exposures), the type and severity of injury, and the time needed to produce the injury. An important principle in toxicology is that no substance is completely safe and conversely no substance is entirely harmful. The major factor that determines whether a substance is potentially harmful or safe is the relationship between the concentration or dose (quantity) of the substances to which an organism is exposed and the duration of exposure. A measure of the severity of the response resulting from the exposure is the concentration (or dose)–response relationship. For any substance, contact with a biological membrane or system may not produce an adverse effect if the concentration of the substance is below some minimal effective (threshold) level. This implies that all substances are capable of producing an adverse effect if a high enough concentration of the substance comes into contact with a biological membrane or system (Snyder, 1984).

Toxicity tests (discussed below) are typically conducted with experimental organisms (e.g. rats, mice, fish, invertebrates, plants) under standardized, reproducible conditions and are used to compare the toxicity of different substances. The information generated from toxicity tests may be used to define the hazard, safety and risk of substances to humans, fish, birds and other organisms in the environment. The latter terms are used widely in toxicology and are discussed below.

Hazard is used to indicate the likelihood that a substance will cause an adverse effect (injury) under conditions in which it is produced or used. Hazard is considered the product of exposure and toxicity (Macek, 1986). *Risk* is a statistical concept defined as the expected frequency or probability of harmful effects resulting from a specified exposure or combination of exposures to known or potential concentrations of a substance. A substance is considered safe if the risks associated with its exposure are judged to be acceptable. Estimates of risk may be expressed in absolute or relative terms. Absolute risk is the excess risk due to exposure. Relative risk is the ratio of the risk in the exposed population to the risk in the unexposed population. *Safety* is thus the reciprocal of risk or the practical certainty that adverse effects or injury will not result from exposure to a substance when used in the quantity and the manner proposed for its use. The prescribed safety of chemicals may be viewed as a socially acceptable risk and is a matter of judgement. Such judgements are based on an evaluation of benefits and risks. High risks may be socially acceptable when there are no reasonable alternatives in life-threatening situations (Morgan, 1984).

The factors to be considered in assessing whether a risk is acceptable or unacceptable are: (1) benefits of the substance and specific uses, (2) possible alternative substances, (3) extent of use and potential human and environmental exposure, (4) potential exposure of endangered or threatened species and (5) socioeconomic considerations if the substance is eliminated from the market (see Chapter 12).

Factors which affect toxicity

Adverse effects are produced only if a substance or its metabolites come into contact at appropriate biological receptor sites at a high enough concentration and for a sufficient length of time. The toxic manifestation is dependent upon the exposure situation, the biological system and the physical and chemical properties of the substance. Therefore to define the hazard or toxicity of any substance, it is important to understand not only the effect and the dose (or concentration) needed to produce the effect but also the characteristics of the substance, the exposure and the target species. The factors related to exposure that affect toxicity are the route of administration or manner in which contact with the substance is made, and the length and frequency of exposure.

Exposure route

Substances may gain access to the body either intentionally or by accident by a variety of routes. In mammalian species substances may enter the body through the skin, lungs, mouth or gastrointestinal tract. It is usually necessary to obtain toxicity test data by different routes of administration. Toxicity tests are therefore often conducted on mammalian species dosed by oral, inhalation, dermal (percutaneous), ocular and parenteral routes. The common parenteral routes consist of injection of a substance into the skin (intradermal), beneath the skin (subcutaneous), in the muscle (intramuscular), into the blood of the veins (intravenous), into the spinal fluid (intrathecal) or into the abdominal fluid (intraperitoneal). Inhalation and topical exposure through the skin are the most common types of accidental exposure that occur in the industrial environment while oral ingestion is the most common route for accidental or intentional poisoning. As a general rule, for the usual routes by which man or other mammalian organisms may be exposed, absorption of substances will be most rapid when administered by inhalation, less for intraperitoneal, subcutaneous, intramuscular and intradermal and then followed by oral (gavage) and dermal (topical) application. This order may be modified depending on physico-chemical properties of the substance, the vehicle and the microenvironment of the absorbing surface. The rate of absorption is a major factor determining the rate at which signs of intoxication will occur. When the rate of detoxification, excretion or recovery from injury exceed the rate at which a chemical is absorbed, it is possible that toxic signs observed by one route of administration will not be detectable by another route for which absorption is slower. Comparison of toxicity data by different routes of exposure for a substance provides information about the absorption of the substance. For example, if the lethal quantity of a substance is similar for oral, topical and intravenous administration the assumption is that the substance is absorbed rapidly. On the other hand, when the lethal quantity via topical administration is significantly higher than the oral lethal quantity, it is assumed that outer layers of the skin provide an effective barrier to the transfer of chemicals (Lu, 1985).

Aquatic organisms may be exposed to substances when they are present in the water, sediment, or food. Substances which are water-soluble are more readily available to aquatic organisms from the medium than water-insoluble substances which are absorbed or bound. Substances in food items may be ingested and absorbed through the gastrointestinal tract. Substances which are absorbed may dissociate into the water from their bound state over time and enter the organisms through the body surface and gills.

In general, the route of exposure may affect kinetic processes such as absorption, distribution and biotransformation and may also determine the toxicity of a substance.

Length and frequency of exposure

Adverse or toxic effects may be produced in the laboratory or in the natural environment when organisms are exposed to substances for different time periods. There are four different exposure categories: acute, subacute, subchronic and chronic. In acute exposure, organisms come into contact with a substance delivered either in a single administration or in multiple administrations within a short period of time, generally hours to a few days. In aquatic toxicity testing acute exposure generally refers to exposure of ninety-six hours or less in water. In mammalian toxicity testing acute exposure refers to exposures of less than twenty-four hours and exposure routes may be through oral intubation (gavage), dermal application and intravenous, intraperitoneal or subcutaneous injection. Acute exposures to substances that are rapidly absorbed generally produce immediate effects, but they may also produce delayed effects similar to those resulting from longer term exposure. Repeated continuous exposure can be divided into three categories: subacute, subchronic and chronic. Subacute exposure involves daily or frequent exposure of mammalian organisms to a substance over a period of up to one month. Subchronic exposure in mammalian toxicology refers to exposure of one to three months and chronic for more than three months. During chronic exposures, organisms are exposed to low concentrations, or a dose, of a substance delivered or administered either continuously or periodically. Chronic exposure to substances may induce rapid, immediate effects similar to acute effects, in addition to effects that develop slowly. Acute exposure, in general, involves a short period of time compared to the life span of an organism, whereas a chronic exposure may involve the entire generation time. However, in aquatic toxicology acute exposure is a larger portion of the life cycle of many aquatic invertebrates (e.g. *Daphnia*) compared to the vertebrates (i.e. fish). Exposures of aquatic organisms that are intermediate in duration (a month to several months) and are less than a complete generation time and include exposure during sensitive early stages of development are subchronic exposures. Studies on subchronic exposure in aquatic toxicology are called early-life stage, critical life stage or embryo–larval tests.

The frequency of exposure may also affect toxicity. For example, one acute exposure to a single concentration of a substance may have an immediate adverse effect on an organism, while two successive exposures cumulatively equal to the single acute exposure may have little or no effect. This may occur as a result of detoxification and excretion of the substance between exposures or acclimation of the organism to the substance. However, if there is minimal metabolism the substance may not be easily transformed and excreted and may remain in the organism, eventually producing a chronic effect. Toxicity which results from chronic exposure occurs if the substance bioaccumulates, if the effects are irreversible or if there is little or no recovery between exposures.

Factors related to the organism

Species differ in their susceptibility to different chemicals. Rates and patterns of metabolism and excretion can affect susceptibility. In the natural environment behavior may play a role by enabling organisms to avoid exposures to toxic substances. Differences in reaction to substances among rats, mice, fish and other organisms of different strains also result from genetic factors. Evidence for genetic selection in the natural environment has been observed in various organisms after exposure to pesticides.

Dietary factors may also influence toxicity, by producing changes in body composition, physiological and biochemical functions and the nutritional state of the organism.

Immature organisms in general appear to be more susceptible to chemical substances than are adult organisms. This may be due to differences in the degree of development of detoxification mechanisms between young and adult organisms. Differences in rates of excretion of toxic chemicals may also be involved in age-dependent toxicity effects. Organism size and surface area to volume ratios also play a role (Lu, 1985, Klaassen *et al.*, 1986).

In order to evaluate the potential toxicity of a given substance, tests are traditionally conducted with healthy organisms. However, test organisms that are in poor health or are stressed such as by previous exposures or concurrent exposure to other substances may be more susceptible to a toxic substance. Tolerance is the state or ability of an organism to elicit a decreased response to a toxic dose of a substance because of prior exposure to the same substance or a structurally similar substance. Tolerance is a descriptive term and is the result of two mechanisms: (1) failure in translocation of the substance, that is, enhanced excretion or metabolic alteration lowers the concentration of the substance at the site of toxic action and (2) reduced response of the affected biological system (target site, target organ or target tissue) to the substance.

True tolerance is only the decreased reaction between the substance and the biological effector site(s). Tolerance should not be used to describe the decreased response of a living system to the first dose of a substance. This may be described as resistance and is probably the result of possession of an enzyme system which inactivates the substance. Similarly, immunity may be developed to certain substances such as polysaccharide toxins, as a result of the development of a specific antibody system which reacts with the specific toxin on the next exposure. This reaction results in an inactivation of the toxin and the animal develops a resistance to the toxin and not a tolerance to it. True tolerance is then a result of an altered effector site, for a substance. For those substances possessing similar structure and biological activity, cross-tolerance may occur where the living system develops a tolerance for one substance and also develops a tolerance for another similar substance.

In general, tolerance has significance in toxicology because it represents another mechanism by which certain organisms are protected against the harmful effects of substances. Tolerance or cross-tolerance between similar chemicals is an immunity to the adverse effects of substances which may be responsible for variation in reactions between members of a species.

Environmental factors

Environmental factors may influence the toxicity of a substance. For aquatic organisms these factors are associated with the bioavailability of the substance in the water, such as dissolved oxygen, pH, temperature and dissolved solids (Klaassen *et al.*, 1986).

Factors related to the chemical

The toxicity of any substance is strongly influenced by its composition. Contaminants or chemical impurities in a sample may be more toxic than the substance itself. If the contaminants are unknown or their biological activity unsuspected, toxicity tests may lead to erroneous results concerning the primary substance. In contrast, toxicity tests on highly purified samples may not detect the toxic action of contaminants present in samples used commercially. Furthermore, for many substances used in manufacturing or accidentally released into the environment, specifications of purity may not be standardized. The purity may vary from batch to batch so that the results may not be reproducible. The identity and purity of the substances are extremely important in toxicity testing. Toxicity tests conducted with purified samples of a substance may not accurately predict the hazard associated with exposure to the substance when it is released into the environment. The other factors which affect a substance's toxicity are its physical and chemical properties which include solubility, partition coefficient, vapor pressure and pH. These factors will affect bioavailability, persistence, transformation, distribution and fate in the environment (see Chapter 2 for additional information).

Substances which are soluble in water do not require organic solvents or carriers for preparation of stock solutions for toxicity tests. Furthermore, water-soluble substances tend to be more widely and homogenously distributed than insoluble substances in aquatic and terrestrial environments. The organic solvent/water partition coefficient, expressed as $\log_p$, and pK are properties important in the determination of absorption and distribution of a substance in organisms and in the development of appropriate extraction and assay procedures for the substance. The extent of ionization of an organic substance will affect its passage through lipoidal membranes. The unionized lipid-soluble form of an organic substance will readily pass through a biological membrane until equilibrium is reached. Lipid and water-insoluble substances gain access to the interior of a cell via a special carrier molecule which is present in the membrane. Organic solutes of low molecular weight (e.g. organic acids) readily pass through most cell membranes by simple diffusion whereas ions (e.g. sodium, potassium) are selectively partitioned by an active transport mechanism.

The pH characteristics of the substance and its photochemical properties are important for determining the stability of a substance and how it should be stored. Furthermore, the toxicity of certain substances may change with an increase or decrease in pH. The pH and temperature of a solution are especially critical in aquatic toxicology since in the field the pH and temperature of water may vary geographically and seasonally so that exposure and hazard may differ drastically. Toxicity tests (discussed later) in the laboratory should be conducted at different pHs and temperatures. Many substances also undergo photochemical reactions that lead to either more or less toxic products. Special precautions should be exercised to avoid reactions during preparation and storage of test solutions or diets and during the analysis of tissues, water (for aquatic tests) and metabolic reaction mixtures.

Vapor pressure of a substance is important for both laboratory and environmental considerations. A substance which has a high vapor pressure and is volatile must be handled carefully in the laboratory when mixed in a liquid medium or in the diet. Frequent chemical analysis of the liquid and diet, as well as frequent preparation of solutions and diets and/or restricted exposure periods to limit the time for vaporization loss are necessary to provide accurate estimates of intake in feeding studies for substances with low boiling points. In the aquatic environment, substances with high vapor pressure and low water solubility will tend to volatilize into the atmosphere; the fraction remaining in the water may absorb to abiotic and biotic components of the aquatic system. If in fact, these substances are persistent, that is, not significantly degraded, they can accumulate in the environment to toxic levels (Sprague, 1985).

The molecular structure of a substance determines its biological activity and consequent toxicity. The nature of the toxic action of a substance will depend to a large extent upon the functional groups in the molecule. Understanding the reactions these functional groups undergo at the active site of critical endogenous molecules provides an indication of the nature and site of toxic action that may be expected. The study of the type of substance that will react with a specific target is referred to as the study of structure-activity relationships (SAR). These studies are used to define the limits of variation in structure that are consistent with the production of a specific biological effect. If sufficient SAR studies are conducted, a hypothesis may be developed concerning the most likely target(s) or receptor(s) of the foreign substance. These studies are also important for the development of a hypothesis concerning the structure of the target.

All substances are potentially capable of producing adverse effects on living systems. Mechanisms by which these effects are produced vary from a generalized destruction of numerous cells and tissues to specific action on single enzyme systems. Substances like strong acids or bases that produce generalized destruction of all living cells are nonselective (nonspecific) in their mode of activity. In contrast to nonselective induced destruction of cells, many substances in toxicology are selective in their action in that they affect one type of cell or tissue without harming others that are in close contact. If the affected and nonaffected cells are in different organisms, the substance is species-specific in its selective activity. If the cells are in the same organism, the substance may or may not be species-specific, but it will be selectively active or toxic on certain types of cells or tissues within a species. The selective toxicity is produced at concentrations far below those needed to produce generalized destruction of all cells. Selective toxicity or activity is a result of the biological diversity and variability in response of cells and tissues to foreign substances. This diversity may prevent the extrapolation of the effects of one substance in one species from results of studies in other species. Two mechanisms are responsible for the selective action of a substance. The first type is one that involves factors concerned with alteration of the effective concentration of the substance at effector sites. This is accomplished by selective absorption, translocation, biotransformation and excretion. The second mechanism involved in selective toxicological activity of substances involves the presence or absence of specific target or receptor sites in the exposed system which are susceptible to the substance. The reaction between the substance and the target may or may not affect the viability of the living system. This will depend on whether the effect is vital to the function of the system (Klaassen et al., 1986).

Types of effects

The spectrum of biological effects following exposure to a substance is broad. The multitude of effects may be categorized by systems, by chemical groups, by types of adverse responses and by the action of the substance (direct *v.* indirect). These classifications are designed to consider the adverse effects of drug therapy. Drugs are a special class of substances which are intentionally administered to animals or humans in specific doses with the objective of eliciting some desirable effects. Drugs may at the same time produce side-effects which are undesirable and which consequently limit their usefulness. Although the side-effects of drugs may be dose-dependent and avoidable, a decrease in their dose may not result in a therapeutic effect.

Some non-drug chemicals are also intentionally administered in specific doses to organisms but are not intended for wide distribution to all species. Examples of such substances are nutrients, vitamins, food additives (preservatives), and pesticides which are limited for use on certain species. With drugs used for therapeutic purposes, the dose is adjusted to eliminate or minimize adverse effects but with pesticides, for example, the dose is only adjusted upwards to eliminate certain unwanted species.

Although certain chemicals are not intended for exposure to living systems, exposure to these substances are unavoidable. Therefore any dose (or concentration) of these substances which is capable of producing an adverse effect would be considered an overdose. For any substance there is a dose or concentration that, if presented in sufficiently low concentrations, will produce no effects in the exposed population. This concentration is considered an acceptable non-toxic concentration. Those members of the population that respond atypically to the acceptable dose or concentration of a substance are classified as being hypersensitive. This type of response is normal and is to be expected in certain members of a uniform population of organisms with every substance. In the hypersensitive population of organisms all doses, even the safe ones, are excessive.

An effect can typically be measured on a graded scale of severity and its magnitude is directly related to dose. Several effects do not exhibit a classical gradation and are only expressed as 'occurring' or 'not occurring'. These effects are usually called quantal effects. Examples of quantal effects are death or occurrence of a malignant tumor.

A distinction should be made between toxic effects based on the exposure period. Acute effects are those that occur rapidly after a single administration or short-term exposure to a substance but acute effects may also appear after repeated or prolonged exposure. Generally, acute effects are relatively severe. The most commonly measured acute effect is lethality or mortality. A substance is classified as acutely toxic if it produces mortality in 50 per cent or more of an exposed population of test organisms in a short period of time, such as ninety-six hours (in aquatic species) or up to twenty-one days (in mammalian species). Chronic or longer term toxic effects may also occur as a result of a single exposure but most often they are a consequence of repeated or prolonged exposures. Chronic effects may arise from accumulation of a substance or its metabolites, or from a summation of acute effects. The latency period (or time-to-occurrence) for the expression of effects may be long if the concentration or dose is low. Chronic effects may be lethal or sub-lethal. A common lethal chronic effect is failure of exposed organisms to produce viable young. The most frequent sub-lethal effects are behavioral changes (e.g. attraction/avoidance responses), physiological changes (e.g. growth, development), biochemical changes (e.g. enzyme activity, ionic levels), and histological changes (e.g. pathologies of the liver or hepatopancreas).

Some sub-lethal effects may indirectly result in death. For example, certain behavioral effects may influence an organism's ability to find food, a mate, or avoid a predator. Ultimately some of these effects may lead to death. Certain sub-lethal effects may have little or no effect on the organism because they are reversible or cease with time. Most sub-lethal effects go unnoticed in laboratory studies which measure the effects of acute exposure to a substance. The sub-lethal effects of a substance are more accurately defined during chronic exposures, especially when the exposures cover the developmental period or most of a generation period.

Biological effects may be elicited during or immediately after exposures to a substance, or they may be delayed for some time. Carcinogenic effects of substances typically occur after a latency period (time-to-occurrence) of twenty or more years in humans and six or more months in aquatic species. The manifestation of toxic effects in organisms is a function of the condition of the biological species, the severity and length of exposure and the properties of the substance. Substances which are easily metabolized (or biotransformed) will have a short half-life in the organism because they will be excreted rapidly. Substances with a short half-life should not produce delayed effects. In the environment, substances may have a short half-life because they are easily transformed by physical, chemical and biological processes. However, some metabolites of organic compounds are stored and can be toxic if rapidly mobilized, and others are simply more toxic after metabolism, e.g. some higher molecular weight PAHs.

The specific effects of substances may be either reversible or irreversible in nature. The degree of regeneration and type of tissue following pathological damage will determine the reversibility of the effect. Because of the ability of certain tissues to regenerate (e.g. liver, lung, skin) many injuries are reversible. On the other hand injuries to the central nervous system are irreversible. In the environment effects may be reversible if the organism can escape the toxic exposure situation and find a toxicant-free environment. The damage to an organism may, however, be severe and eventually lead to death. The reversibility or irreversibility of effects may be studied in the laboratory with a recovery study by removing organisms from the exposure situation. In the case of aquatic organisms the test species are transferred from a medium containing the toxicant to a medium that is free of the toxicant. With mammalian organisms in the laboratory, administration of the toxicant is terminated.

The types of effects in organisms can also be differentiated according to the general site of action of the toxicant. If at the first site of contact between the organism and the toxicant a response occurs, these are called local effects. An example of a local effect is a skin or gill reaction caused by a corrosive or caustic material. For a substance to have adverse effects at a site after its initial entry, it must access the organism through the circulatory system. Systemic effects are those that require absorption and distribution of the substance to a site distant from the original contact or entry site. Most substances produce systemic effects; highly reactive substances typically do not. Some substances produce both local and systemic effects. The central nervous system is the target organ most often involved in systemic toxicity while muscle and bone are the least often involved targets for systemic effects (Lu, 1985).

In assessing the multitude of effects and substances that may produce these effects it is important to consider that in the natural environment organisms are not exposed to a single substance but to a number of different substances which form mixtures. Exposure to mixtures may result in toxicological interaction. Interactions may occur in a number of ways. For example, interaction between substances may occur as a result of changes in absorption, distribution, biotransformation or excretion of one or more of the interacting substances. Furthermore, exposure to two or more substances may result in a biological response quantitatively or qualitatively different from that expected from the action of each of the substances alone. The exposures to two or more substances may be sequential or simultaneous, and the responses may be a simple summation of the individual responses or may be smaller or greater in magnitude.

Many processes may be involved before an organism shows a response to a substance. These processes may differ depending on the organism and the environment. In the aquatic environment, a substance may interact with other constituents in the water; for example, pH will affect the dissociation of acids and alkalis, and humic acids will form complexes with certain heavy metals. These processes should be clearly understood since they may have a profound effect on the response of the organism and lead to misleading conclusions about the type of joint action of the substances present. It is therefore important that the biological availability of the toxic constituents in a mixture be known before conclusions are reached on their joint action. Secondly, physiological processes within an organism such as absorption, transport and distribution by the circulatory system, biotransformation, accumulation and excretion, may all affect the concentration of substances and their metabolites present in the body and thus the quantity available at the site or sites of action. These processes may also be influenced by other water quality characteristics in the aquatic environment of the organism. Thirdly, when an organism is exposed to two or more potentially toxic substances, interaction between different physiological processes within the organism may occur, including those affecting the chemicals' absorption, binding to plasma proteins, distribution and transport, release from tissues, metabolism and elimination, all of which may contribute to the response of the whole organism (e.g. death, delayed or slower growth).

The concurrent presence of two or more foreign substances will often yield a toxic response which is said to be simply additive; that is, the combined effects of two or more substances equal that expected when considering the substances individually $(6+5=11)$. The effect may also be greater than additive and is therefore said to be synergistic, that is, the combined effect of two or more substances is greater than the sum of each individual substance by itself $(6+6=15)$. If the effect is less than additive, it is said to be antagonistic. In the latter case several substances are administered and either one substance affects the actions of the other substance or each affects the actions of the other substance $(2+4=5, \ 2+0=1, \ 2+(-2)=0)$ (Rand and Petrocelli, 1985).

Concentration (dose)–response relationships

In a homogenous population of cells, tissues, organs or organisms, differences still may exist between individuals. These differences become obvious when exposed to a toxic agent or a life-threatening stress. If the agent is capable of producing death or some sub-lethal adverse effect and these could be quantified, then an experimental study would show that not all organisms respond to the same dose or concentration of the agent in a quantitatively identical way. The response to exposure to the same dose would range from very intense in some organisms to minimal or no response in other organisms. At the same dose some organisms may die and many would survive with apparently no deleterious effects. These differences are due to biological variation which is generally small for organisms of the same species of similar age and health and generally greater between species or different life history stages. What is generally considered as an all-or-none response actually applies only to a single member of the test group and is actually a graded response when the entire test group is considered. In essence, each member has a unique 'tolerance' and requires a certain dose before responding. In principle there exists a low dose to which none will respond (no effect) and a high dose to which all will respond.

In toxicity testing, the purpose is to estimate as precisely as possible the range of concentrations (doses) of a substance that produces some readily observable, quantifiable response in groups of the same test species under controlled conditions. The results of exposure to a substance are plotted on a graph that relates the proportion of the organisms in test groups respond-

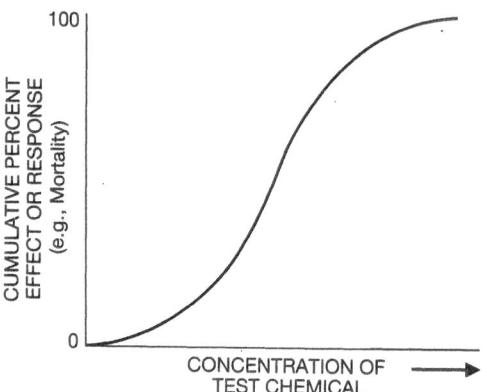

Figure 3.1 Typical form of the concentration–response curve

ing with a quantal effect to the concentration (or dose) of the test substance. This correlation is the concentration (or dose)–response relationship. Concentration (dose)–response curves (Figure 3.1) demonstrate the relation between dose and the magnitude of a graded effect, either in an individual or in a population. Most individuals in a population respond closely to a central dose (concentration), and a few will respond only at very low or at very high levels. This leads to a frequency distribution for the individuals responding as a function of concentration (or dose). A frequency distribution describes the random occurrence of individuals with different susceptibilities. The figure shows a normal frequency distribution; the curve is symmetrical around a central point. The cumulative frequency distribution provides the sigmoid S-shaped curve. The dose–response relationship has a cumulative frequency distribution because individuals which respond at a low dose also respond at higher doses. The frequency of responders at a specific high dose includes all those at that and all lower doses. In general these curves have upper and lower asymptotes, since at all concentrations below some minimum (threshold) value, no measurable adverse response will be elicited, while at all concentrations above some maximum value most or all of the test group will be adversely affected. For each effect there will usually be a different dose–response curve, either in shape or in location on the concentration axis.

In mammalian toxicity testing the exposure takes place by direct introduction of a measured amount (dose) of test substance either onto (dermal application) or into (feeding or injection) the test organisms. The dose is therefore known and the response or effect can be correlated with it. In aquatic toxicity testing, the test organisms are typically exposed to test substances in water in which the plant or animal lives, thus producing a test

concentration. Generally there is no measure of the amount of the substance in the organism, although the quantity of the substance in the water
is measured. The response in aquatic toxicology is thus correlated with the
exposure concentration, hence the term concentration–response relationship.

The dose (or concentration)–response relationship is one of the most
fundamental concepts in toxicology. This relationship is based on two
assumptions: (1) a particular response (or effect) is the result of exposure to
a specific chemical substance, and (2) the severity or degree of the response
is a function of the concentration of the substance to which the organism is
exposed.

In assessing the dose–response relationship, it is also important to have a
precise means of expressing the toxicity and a quantitative method of
measuring it. There are a number of endpoints or criteria for effects that
could be used to compare exposed organisms (treated) to unexposed
(control) organisms. The best endpoints are those associated with molecular damage as a result of exposure. However, this is typically difficult to
achieve since the molecular events may not be determined even for well-
known substances. As an alternative, criteria for toxicological effect should
be selected that are relevant, observable, measurable, biologically
significant and reproducible. Many sub-lethal effect measurements may be
used such as growth, number of offspring produced, enzyme activity,
condition of the blood (clinical hematology), pulmonary responses, or
changes in behavior.

Although there are many endpoints which have toxicological relevance,
the initial test measurement for a new substance utilizes lethality (or
mortality) as an index. Measurement of lethality is important, unequivocal
and quantifiable especially to estimate the relative potency of a new
substance in comparison to other toxic substances. Lethality studies enable a
comparison of substances with different modes of action and they also
indicate whether additional toxicity studies should be conducted. Mortality
and survival are typical effect criteria in short-term (acute) exposure tests.
Information from lethality tests are quantal; that is, the organisms live or
die (all-or-none response). Since detailed observations including gross
pathology are typically gathered during lethality studies using different
routes of administration, one can also obtain insight into the potential
target organ(s) and the possible mode of action (Snyder, 1984).

Sub-lethal effect criteria used in long-term (chronic) exposure studies are
quantitative and graded in nature. They are measured not in terms of
incidence but rather in some unit of response so that treated and untreated
(control) test organisms can be compared to determine whether the differences between them are statistically significant. In a quantal test it is
sufficient to merely determine in every organism whether or not the
selected response (effect) has occurred, while in a quantitative test the extent
of the response for each organism must be measured. A series of graded

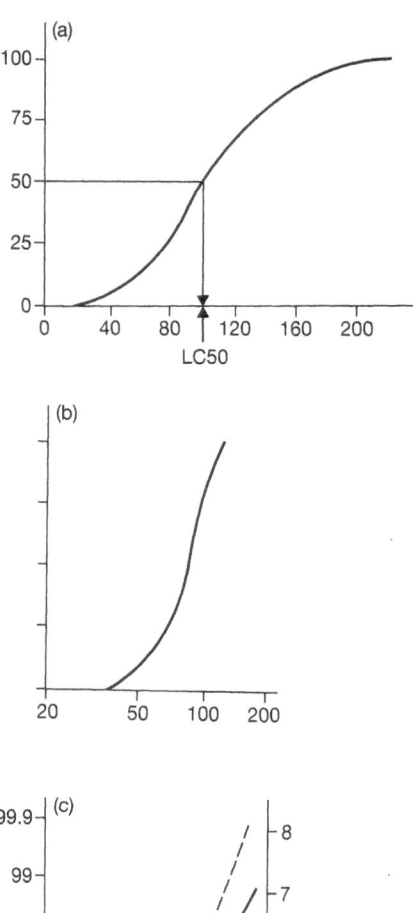

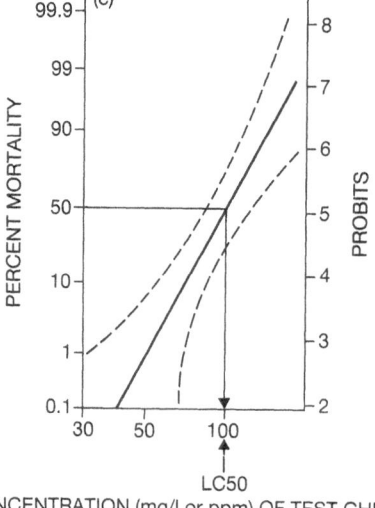

Figure 3.2 Mortality in a fish population exposed to a range of concentrations of a chemical in water

measurements will be generated for each concentration since the responses of many organisms are assessed.

Regardless of the effect or response measured, the relationship between degree of response and dose (or concentration) of substance almost always assumes a classic sigmoid or S-shaped form. In Figure 3.2 the Y-axis represents percent mortality and the X-axis represents concentration (or dose) of the chemical substance (toxicant); both increase with increasing distance from the origin. These graphs represent the results of acute toxicology studies in which groups of the same test species were exposed to different concentrations of a chemical substance for a defined time period (e.g. ninety-six hours). In Figure 3.2(a), the mean percent mortality for each test group has been plotted on an arithmetic scale against the concentration producing that mortality. Each point on the curve represents the mean response of a group of test organisms to a specific concentration of a chemical substance. Each mean response has a biological variation above and below this value as a result of different responses of individual organisms. The least variability in the curve is at the 50 per cent level of response. The concentration (or dose) at which 50 per cent of the organisms respond after some duration of exposure (e.g. twenty-four hours, ninety-six hours) is used as an index of toxicity.

For a new chemical substance acute toxicity tests are conducted to estimate the relative toxicity of the material in comparison to other known toxic agents. This estimate is measured as the LC50 or LD50 which is the median lethal concentration (or dose) of a substance expected to produce mortality in 50 per cent of the test population within a specified time period. When effects (or responses) other than mortality are measured, such as a behavioral deficit, the expression EC50 or ED50 is used. This is simply the effective concentration or dose of a substance expected to elicit a sub-lethal effect (i.e. behavioral or physiological) in 50 per cent of a test population of organisms in a given time period.

The LC50 (or LD50) can be interpolated from the curve in Figure 3.2(a) by drawing a line across from the 50 per cent mortality point (Y-axis) to the curve and then drawing a vertical line from the point of intersection with the curve to the X-axis. The vertical line intersects the X-axis at the LC50 or LD50 value. These are estimates of the quantity of a substance that may produce a specific effect (mortality) in 50 per cent of a population of the same species held under similar conditions.

In a normally distributed sigmoid curve, mortality approaches 0 per cent as the concentration is decreased and 100 per cent as the concentration is increased; theoretically it never passes through 0 and 100 per cent. The middle portion of the curve, in the region between 16 per cent and 84 per cent, is linear. These values represent limits of 1 standard deviation (SD) of the mean in a normally distributed population of organisms.

In a normally distributed population, the mean ±1 SD include 68.3 per cent of all members of the population. Thus 16 per cent of the population will be responders at concentrations (or doses) equal to or less than the mean −1 SD, and 84 per cent will be responders at doses equal to or less than the mean +1 SD. It may also be calculated that approximately 95.5 per cent of the population will respond within a dose range given by the ±2SD and approximately 99.7 per cent will respond between the mean ±3SD (Galbor *et al.*, 1985).

Figure 3.2(b) uses the same data as (a) but the dose information is transformed to a logarithmic scale. The curve now approaches a straight line. Figure 3.2(c) represents the same data with percent mortality on the commonly used probit scale plotted against logarithm of concentration. The probit transformation is easily derived using log-probability paper to express the dose–response relationships as a linear function for log-normal distributions of an effect. The concentration–response data are normally distributed.

Confidence limits are illustrated as dotted lines on both sides of the solid line in Figure 3.2(c). These limits indicate the range or scatter of observed values within which the concentration (dose)–response line would be expected to fall 95 per cent of the time, if nineteen out of twenty samples

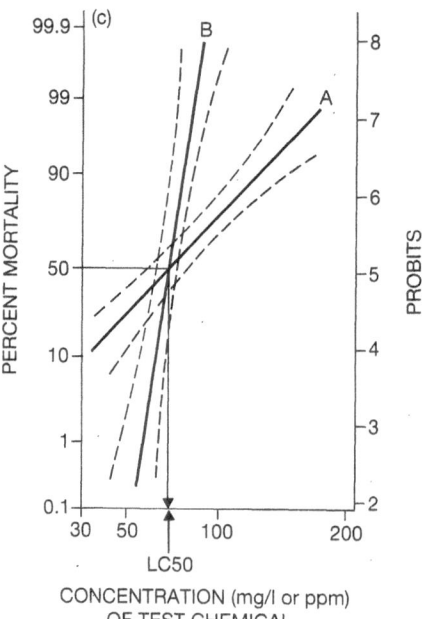

Figure 3.3 Hypothetical concentration–response curves for two chemicals, A and B, demonstrating differences in slope

were taken at random from the same test population under similar conditions.

Figure 3.3 shows two concentration response curves on a log-probability graph for two chemical substances. Both LC50s are the same but the test systems respond in a different manner. Hence, the slopes of the curves should be calculated to enable one to differentiate potential differences in toxicological mechanism. For example, the flat slope of line A indicates that mortality increases by small increments as the concentration of the substance increases. An effect is thus observed slowly with this substance. On the other hand, the steep slope of line B indicates that large increases in mortality are associated with small increases in concentration. The slope may thus be an index of the range of sensitivity or 'margin of safety' to a chemical substance. The 'margin of safety' is the magnitude of the range of doses involved in moving from a no-effect dose to a lethal dose.

Flat slopes, as for chemical A, indicate slow absorption, rapid excretion or detoxification, or delayed toxicity. Steep slopes, as for chemical B, indicate rapid absorption and rapid onset of effects. There is a considerable range of concentrations (or doses) involved between a no effect and a 100 per cent nonreversible effect for substance A. For substance B, the range is less, in fact it will be difficult to administer a dose so small that the effect would be less than lethal. Therefore, substance A has a greater 'margin of safety' than substance B.

Toxicity testing

Toxicological tests are conducted to define the conditions that must be present for a biological system to respond to a chemical substance and to describe the nature of the effects produced. The conditions that must be present and the consequent biological effects may be easily attainable with small quantities of a highly toxic substance or they may be totally unattainable as is the case with large quantities of an innocuous substance. The biological effects of a substance may be of such minor significance that the organism can still carry on its normal function unless acted upon by an additional stress.

Biological testing methodology in toxicology relies on the need for as much data as possible on the potential effects of a chemical substance to assess safety to humans although minimal data sets are often identified. Testing methodology relies on species that have extrapolative values for humans such as mice, rats, guinea pigs, rabbits and monkeys. Several methods involve the use of non-mammalian species, including bacterial and cell cultures. The effects of chemical substances are not studied directly on humans until after a substance is used unless in limited clinical trials as is the case with new drugs. Substances such as pesticides may have widespread use and indirectly impact not only humans but other species in the environment.

Therefore, biological methods in toxicology also involve the use of non-target species such as aquatic organisms and wildlife (e.g. birds) since they may be accidentally exposed to toxic substances (Bourdeau *et al.*, 1990).

The principles of toxicity testing are based on the assumption that an effect or reaction is only produced if a living biological system has contact or is exposed to a given substance at a high enough concentration for a sufficient length of time. The effect of the reaction may be on the function and/or structure of the biological system. An effect on function may not be accompanied by a change in structure. An effect may also be reversible if exposure is terminated or reduced, or it may ultimately lead to death. The methods in toxicology are used to detect changes in the structure and function of living systems, how substances are absorbed, metabolized and excreted by individual organisms, the conditions under which effects are produced and the mechanisms of toxic action (Klaassen *et al.*, 1986).

Various principles of toxicity testing have become apparent through the development of methodology. These are

- in order for a substance to produce an effect, it must come into contact with the living system. Exposure is an essential part of toxicity.
- there will be some concentration of each substance at which and below which there will be no effect on a living system, and there will be some higher concentration of each substance at which and above which an effect will be present. Between the no-effect concentration and the significant-effect concentration will be a range of exposure concentrations of each substance that will produce both lethal and sub-lethal effects on some systems.
- similar species will be affected similarly; cells with similar metabolic systems and functions in different species will be affected similarly by a chemical substance, providing equivalent exposures occur.
- changes in the molecular structure of a substance may influence the activity of that substance (Rand and Petrocelli, 1985).
- the toxicity of a chemical substance may be affected by type of animal diet or by abiotic factors such as changes in temperature, pH, dissolved oxygen (as may be the case with aquatic organisms); all non–chemical factors may be toxic alone.

The first principle is that for a substance to have an effect at a site other than its initial application, it must gain access to the internal milieu of the organism. The substance must be translocated across membranes and carried to various parts of the body in the bloodstream. Substances which are not absorbed through the skin, gastrointestinal tract or gills are typically innocuous if applied to the skin, gills or ingested except for whatever direct action they would exert on the site of application. If however, the substance was injected into the bloodstream or inhaled into the lungs and then translocated within the body, it could potentially affect an unlimited number of sites. The methods used for a toxicological evaluation for man should thus involve administering test substances to experimental animals by different routes of exposure. The most common route by which the substance would come in contact with a living system under conditions of practical or accidental use of the substance should be included in the testing program. For aquatic organisms, the most typical route for test exposure is obviously water. In aquatic toxicity testing, test substances may also be mixed with food or sediment. For birds, test substances are typically administered directly or mixed in the diet.

The substances of toxicological interest are those most readily available to the organism. A substance does not have to be water-soluble to be translocated; it can move across a membrane with a carrier, be immersed in a lipoidal medium or be biotransformed to derivatives which have different aqueous characteristics.

As was previously discussed two types of data may be expected following exposure to substances: (1) all-or-none data in which a response (or effect) is either present or absent (i.e. yes or no) and (2) a graded response type of data in which the response can be measured in terms of incidence or intensity. The second type of data can be quantified in terms of percent of control (or background) or incidence of occurrence in a group of test organisms.

Toxicity tests are generally conducted on groups of experimental animals or microbial preparations. The amount of substance which is under consideration is typically administered in different concentrations (or doses) to the different groups. Toxicology test data are therefore obtained by exposing different groups of organisms to progressively increasing concentrations of a substance, in which each group contains a specific number of animals. When data are generated from these studies in the laboratory, the data consist of responses from each group of organisms from each treatment, including all controls that are used.

Depending on the organisms being exposed and the type of study, quantities of the test substance are administered over different durations of time; the quantity is often referred to as 'dose' or 'concentration' of the substance. If the substance is solid it may be administered to a rat, mouse, or dog in terms of weight (e.g. milligrams of substance per kilogram weight of animal). If the solid is dissolved in water or any solvent the quantity may be defined in terms of a given volume or concentration of a substance in a solution (e.g. 1 milligram of substance per liter of water or 1 mg/l = 1 ppm or 1 part per million). If the test substance is a liquid it may be expressed in terms of weight or volume of the substance. If the substance is a gas it may be expressed in terms of volume or concentration of the gas in a mixture of gases. Quantity may be defined per unit weight or per unit of body surface area of test animal. The duration over which the substance is administered is also sometimes part of the terminology. Examples of dose are grams (g) or milligrams (mg), of substance per kilogram (kg) of body weight (BW) of animal or per square meter of body surface area, milliliters of solution per kilogram of body weight or per square meter of body surface area, or parts per million (ppm) or millimoles per liter of a gas in air. These examples are typically used in mammalian toxicology. As previously discussed, 'concentration' is used in aquatic toxicology to denote quantity of substance in water. In mammalian toxicology, concentration is used to refer to quantities of substance in food (dietary studies) and in the air (inhalation studies). Concentration is also used to express quantity of substance in soil, sediment or tissue (e.g. milligram of substance per kilogram of sediment or 1 ppm = 1 mg/kg). Weight-specific concentrations are generally expressed on the basis of tissue wet weight and sediment wet or dry weight.

The second principle is that toxicity tests are designed not only to demonstrate the presence or absence of toxic effects but also to estimate the no-effect level or limits of safety associated with the use of a substance. Each toxicological test must define what is meant by an 'effect', the effects to be studied, the confidence that can be placed on the result of the 'test procedure', and the time over which the effects will be studied.

Most substances under certain conditions of use will produce toxic effects in a living system; most toxicity tests can be designed to show these effects. It is customary in a good toxicity test that toxicity be elicited in at least the group of organisms exposed to the highest dose (or concentration) of substance. In any experiment, especially one that demonstrates a positive toxic effect, the data must include background information on an untreated or negative control (untreated and/or treated with a solvent/carrier) if the toxicity is to be considered a result of the substance. Groups of control animals should be the same size as treated groups and treated similarly except for exposure to the substance under investigation. The test substance should also be analyzed and characterized as to purity and contaminating substances. The test substance should be from the same batch throughout experimentation and it should be properly stored. The test substance must be analyzed before, during and after testing to confirm analytical stability and purity. The numbers of animals used in all groups including controls should be high enough to yield statistically significant results. If this is the case the results of treated groups can be compared to the results from controls. The number of animals needed to establish a cause-effect relationship is directly dependent on the incidence of the toxic effect in the controls compared to the test animals.

A good toxicity test on a substance will also show at one low dose (or concentration) level that there are no toxic effects on the test organisms. The highest tested concentration where there is no effect is the NOEL or 'no-observable-effect-level'.

If the data of a well-designed test show negative results, this test should include one group of organisms which demonstrate positive effects for the same type of toxicity under investigation. In this test it is necessary to add a positive control. A positive control is a group which consists of organisms similar to the test groups and treated in an identical manner except that this group is not exposed to the test substance but to a substance known to produce a similar type of toxicity, i.e. a reference toxicant. The positive control is used to determine the health and sensitivity of the organisms, to compare the relative toxicities of substances by using the control as an internal standard, to perform inter-laboratory calibrations, and to evaluate the reproducibility of test data with time.

It is generally not possible to conduct a toxicity testing program to establish the safety limits of chemical substances with absolute confidence. Extrapolation of dose (or concentration) response curves outside the range of test results is prone to error, and extrapolation of mammalian toxicology data to man in either a qualitative or quantitative manner is subject to inaccuracies because of the limited knowledge of responses of different mammalian species to the same chemical substance. In ecotoxicology the objective is to define the potential toxicity of a substance or mixture to different aquatic and wildlife species in the environment. Therefore extrapolation is difficult in ecotoxicology because testing in the laboratory can only include a few of the many existing species. Neither can laboratory testing take into consideration all of the environmental interactions occurring naturally which may affect the exposure to and the toxicity of the substance. A comparison of mammalian toxicology and ecotoxicology is summarized in Table 3.1. Please refer to Chapter 4 for further discussion of ecotoxicology.

The third testing principle specifies that cells with similar functions and biochemical pathways will be affected similarly by a substance whether the cells are in the same or different animal species. Receptor sites for the substance may also be identical in different species. This principle becomes important when the toxicity data generated on one substance from one species of animal will be extrapolated to another species.

Cells with different functions and different biochemical pathways may or may not react similarly if the mode of action of the substance is on a biochemical mechanism. If in fact cells with different functional characteristics are affected similarly by the same quantity of the same substance, then it will be assumed that the mode of action was common in the cells. If different quantities of a substance are required to produce effects on different cell types, it will be assumed that the mode of action is different between the cells. This is in fact the basis for selective toxicity: two cell types do not respond similarly to similar quantities of the same substance.

The mode of action of the substance on a cell may be on a mechanism or function which is vital to the cell. If the cell can not substitute or compensate for the affected mechanism or function, it may die. Because of the selective action of many substances on physiological, behavioral or biochemical mechanisms common to cells in different species, data from toxicity tests serve some value in predicting effects on man and the environment.

A final toxicity testing principle is that changes in the molecular struc-

Table 3.1 Comparison of mammalian toxicology and ecotoxicology differ in many respects

Mammalian toxicology	Ecotoxicology (aquatic/avian toxicology)
Objective: to protect humans	Objective: to protect populations of many diverse species, and natural ecosystems
Must almost always rely on animal models since experimentation with humans is not feasible	Can experiment directly on species of concern, or appropriate surrogates
Species of interest (man) is known; thus, degree of extrapolation is certain	Not able to identify and test all species of concern; thus, degree of extrapolation is uncertain
Test organisms are warm-blooded (body temperature is relatively uniform and nearly independent of environmental temperature); thus, toxicity is predictable	Test organisms (aquatic) live in a variable environment and most are cold-blooded (body temperature); thus, toxicity may not be predictable
The dose of a test chemical can usually be measured directly and accurately, and administered by a number of routes	The 'dose' is known only in terms of the chemical's concentration in water (for aquatic organisms) and the length of exposure to it; the actual 'absorbed dose' is sometimes determined experimentally using bioconcentration and metabolism studies
Extensive 'basic' research has been conducted; emphasis has been on understanding mechanisms of toxic action	Little 'basic' research has been conducted; emphasis has been on measuring toxic effects and generating relative toxicity data, with an eye toward regulatory needs
Test methods are well developed, their usefulness and limits well understood	Most test methods are either relatively new or generally not formalized (standardized); their usefulness in many cases is being determined

ture of a substance may affect its biological action on a living system. This is a basic extension of the fundamental concept that all chemical effects on a component of a living system can be related to some physical–chemical or biological–chemical interaction. In order for a reaction to occur, the biological molecule must have the appropriate receptor sites. As previously discussed, structure–activity relationships are used to study the type of chemical structure that will react with a particular biological receptor. These studies are important in the design of toxicity tests for compounds which are similar; each compound should be tested for all possible effects and extrapolation should not be relied upon. If the purity of a substance changes during a testing program, all prior testing conducted with different purities may have to be repeated for the new formulation. Impurities are typically materials or intermediates in the manufacture of the substance and may alone be responsible for the toxicity to experimental animals.

Kinds of toxicity tests

The extent of testing for a particular chemical substance is dependent on the intended use of the substance. Substances to be used directly for human intake either as drugs or food additives require extensive toxicological testing with animals, including human clinical trials. If a substance is to be used as a pesticide either for agricultural or household application, extensive toxicological testing would be required although human exposure only occurs by accident.

Toxicological testing of a substance in the laboratory usually follows a tiered approach, progressing from simple acute tests to more complex and sophisticated chronic tests. Although the tests are designed to detect and assess different types of toxicity, the general test design is similar: careful control of conditions such as temperature, photoperiod, diet, and dissolved oxygen and pH (the latter two in the case of aquatic organisms). Organisms are held in cages or chambers (aquatic organisms in glass tanks) and exposed to various concentrations or doses of the test substance. In mammalian toxicity studies rats, mice, guinea pigs, rabbits or dogs are exposed to the test substance. The chemical and physical properties of the substance determine the most likely route of exposure. Solids and liquids are administered to mammalian species by oral, dermal and dietary routes. Volatile liquids are also administered in the atmosphere to animals. Injection of the test substance is used to bypass membranes or barriers and also to obtain a high dose of the substance rapidly; it is the least common route used. The most common routes of exposure to humans are through the body surfaces and through oral and respiratory routes.

In general, aquatic organisms are exposed in the laboratory to the test substance when it is mixed in water or in sediment. Solids and liquids are easy to administer; volatile substances require special test systems. In aquatic tests the controls and treated organisms can be exposed to the dilution water (controls) or solutions containing the test substance (treated) by four techniques. In a static test the organisms are exposed to the test substance in still water for the duration of the test. In a recirculation test, exposure is similar to a static test except that the test solutions and control water are pumped through an apparatus, such as a filter, to maintain water quality but not reduce the concentration of the test substance; water is returned to each test chamber. A renewal test is similar to a static test because it is conducted in still water, but the test solutions and control water are renewed periodically by transferring the test organisms to chambers with freshly prepared material or by removing and replacing the material in the original containers. In a flow-through test, test solutions and control water flow into and out of the chambers on a once-through basis in which the organisms are maintained either intermittently or continuously. Metering pumps or diluters control the flow of dilution water and test substance in order that the proper proportions of each will be mixed. Static and flow-through techniques are the most widely used for acute tests. However, for chronic tests flow-through techniques are preferred because there is a greater degree of assurance that the concentrations of the substance and conditions to which the organisms are exposed during the test period have remained constant. Cause–effect relationships can thus be more easily established (Rand and Petrocelli, 1985).

In mammalian toxicity testing, the species are standardized but in ecotoxicology only a few species are standardized and other species could vary in the testing program. Several criteria should be considered in selecting organisms for ecological toxicity testing (see also Chapter 4): species that represent a broad range of sensitivities should be used; widely available and abundant species should be considered; species should be studied which are indigenous or representative of the ecosystem that may receive the impact; species that are recreationally, commercially or eco-logically important should be included; species should be amenable to routine maintenance in the laboratory; and adequate background data on a species should be available to assist with data interpretation. Species selected for testing may differ depending on the ecosystem of interest so that site-specific considerations are important. For example, when assessing the impact of a substance on cold water streams, a trout would be the selected test model, while for warm water streams a sunfish would be chosen. There is no standard test species that can be used for all ecosystems. The kinds and numbers of species will depend on the complexity of the ecosystem under study. Some organisms typically used in aquatic toxicity testing include the water flea, fathead minnow, bluegill sunfish, rainbow trout, mysid shrimp and sheepshead minnow, the first four for freshwater toxicity testing and the last two for saltwater testing.

Avian toxicity tests are a relatively new area (see also Chapter 4). The extent of bird toxicity testing programs, like all toxicity programs, depends on the intended use of the substance. Programs may include studies on acute and chronic toxicity and field programs in the natural environment. The field studies are primarily conducted for pesticides in order to assess the impact of a typical application or applications of the substance at realistic use rates on a specific crop or crops. In the laboratory, studies are conducted with mallard duck and bobwhite quail, and more recently songbirds have been incorporated into laboratory programs. The substance in acute studies is administered either orally (gavage) or in the diet. For chronic reproduc-tion studies the substance is mixed in the diet and mallard duck and/or bobwhite quail are test organisms.

There are many laws and regulations governing the introduction, use and transportation of new substances like pesticides, food additives, drugs and chemicals. Although there is uniformity in the objectives of these laws, there are differences in their approaches to assessing safety. These differ-ences are reflected in the number and types of tests and in the procedures used to generate toxicity data for risk-benefit analysis. Guidelines for acute through chronic toxicity testing have been generated by the World Health Organization and the Organization for Economic Cooperation and Devel-opment (OECD), and in the United States, by the US EPA for registration of pesticides under the Federal Insecticide Fungicide and Rodenticide Act

(FIFRA) and for testing new chemicals when required under the Toxic Substances Control Act (TSCA) (Foster, 1985; OECD, 1986).

A general outline of the types of mammalian and ecotoxicity testing are illustrated in Tables 3.2 and 3.3. The basic premise behind animal toxicity tests is that knowledge of the effects produced by the substances in the laboratory and their threshold concentrations can be applied when properly qualified, to humans and other species in the environment. This is obviously easier to support with species other than man since those being tested in the laboratory are similar to or the same as species being protected in the environment. If one considers dose per unit body surface, adverse effects in humans occur typically in the range of those in laboratory species. However, on a body weight basis humans are more sensitive than laboratory species by at least a factor of ten. Safety factors are thus applied to toxicity tests to derive relatively safe dosages for humans. Safety factors are also used in ecotoxicology. The second premise is that exposure of laboratory species to substances in high doses (concentration) is a valid means of detecting potential hazards of low doses to man and other species. This is based on the dose–response relationship in that the number of individuals eliciting an effect will increase as the dose (or exposure) increases. In the design of experimental model systems, it is required that the number of animals used in experiments with toxic substances will always be small compared with the size of human or ecological populations which may be at risk. To compensate for the small groups of animals, relatively large doses (or concentrations) are used so that effects will occur frequently enough to be detected. Detecting low incidences of effects at low level exposures in experimental animals would require hundreds of thousands of animals. It is for this reason that small groups of animals are exposed to large doses and then biological models are used in extrapolating the results to estimate risk at low doses. The latter premise is especially important in cancer studies with laboratory animals.

Data usage

Toxicity tests provide a database that has a variety of applications including:

- industrial decisions on product development, manufacture and sale.
- registration of products to satisfy regulatory requirements.
- permitting requirements for the discharge of municipal and industrial wastes.
- hazard and risk evaluations for humans and the environment.
- prosecution and defense in environmental litigation.

Developing a new chemical or pesticide necessitates determining its impact on human health and the environment. Industrial decisions to

Table 3.2 Types of mammalian toxicity tests

Type of test	Route of administration	Species	Number of administrations	Observation period	Observations, endpoints
Acute[a]	Oral	Rat, mouse, dog	1	14–21 days	LD50, bodyweight
	Dermal	Rabbit, rat, guinea pig	1	14–21 days	LD50, bodyweight
	Inhalation	Rat, mouse	1 exposure of 4 hours	14–21 days	LC50, bodyweight
Skin[a] and eye irritation	Directly to eye or skin	Rabbit	1	3–7 days	Severity of irritation
Sensitization[a]	Intradermal injection/ topical application	Guinea pig	Several over 2 weeks followed by a challenge exposure after additional 2 weeks	24, 48 hours after each application	Erythema and edema and other lesions are scored
Subacute[a]	Oral, dermal, dietary, inhalation	Rat, mouse, dog	14 over 2 weeks	14 days	Bodyweight, clinical chemistry, set dose levels for longer term studies
Subchronic[b]	Oral, inhalation, dietary, dermal	Rat, mouse, dog	Daily	90 days in rodents, 1 year in dogs	Bodyweight, mortality, urinalysis, blood chemistry, clinical chemistry, gross- and histopathology, organ weight, no observable effect level
Chronic[c]	Dietary	Rat, mouse	Daily up to 2 years	Up to 2 years	Same as in subchronic study including tumor incidence
Reproduction/ teratology Phase I (general fertility and reproductive performance)	Oral	Rat, mouse, rabbit	60 days for males and 14 days for females prior to mating and continued through lactation	Prior to mating through weaning	Pre and post implantation embryo lethality; weanlings autopsied for gross and visceral abnormalities
Phase II (teratology study)	Oral	Rat, mouse, rabbit	Females only during organogenesis (rat and mouse, days 6–15; rabbits, days 6–18 of gestation)	Up until time of delivery (rat and mouse, day 21; rabbit, day 31)	Number of resorptions, dead and resorbed fetuses, fetus weights, skeletal and visceral fetal abnormalities
Phase III (perinatal/ postnatal study)	Oral	Rat, mouse, rabbit	Female from 15th day of gestation through lactation and weaning	Up until weaning	Birth weight, survival, growth of offspring

Table 3.2 *continued*

Type of test	Route of administration	Species	Number of administrations	Observation period	Observations, endpoints
Mutagenicity (genotoxicity)	Variable	Cell cultures, bacteria, yeast, rodents, *Drosophila*	Variable	Variable	Gene mutation, chromosomal alterations, primary DNA damage

[a] Observations include signs of intoxication, behavioral changes and other pharmacotoxic signs such as diarrhea, lethargy, morbidity, salivation, tremors, convulsions, etc. Gross necropsy and histological examinations are conducted in some cases.

[b] Subchronic studies are used to characterize adverse effects resulting from repeated exposure over a portion of the life span of the animal (extending to approximately 10 per cent of the life span). These studies also may yield information on cumulative effects, latency period for development of toxicity, the dose-response relationship and the doses to be used in the chronic study.

[c] Chronic studies are conducted similarly to subchronic studies but the exposure is longer. These studies are used to determine the carcinogenic potential of a substance.

proceed with the development of a new chemical will depend on its potential biological effects. Many new chemicals do not survive the initial stages of development because they are persistent in the environment, bioaccumulate, or are toxic to either mammalian species and/or ecological species.

Toxicity test data are also used to evaluate the potential hazard resulting from the discharge of municipal and industrial wastes (effluents) into the environment. Permits to discharge may in fact require 'end-of-pipe' bio-monitoring of industrial discharges. Biomonitoring required in permits may entail the conduct of periodic aquatic toxicity tests, mutagenicity testing and even rodent toxicity tests to ascertain the potential effects of the wastes on the environment and human health.

Laboratory toxicity tests provide data to calculate the lowest observed effect level or concentration (LOEL or LOEC) and the no observed effect level or concentration (NOEL or NOEC). The NOEL or NOEC is the highest level or concentration tested at which no significant deleterious effects are observed. The objective is first to determine the safe level for laboratory species and then determine the safe level for humans and the environment. The important extrapolation is from dosage studies in the laboratory to low dosage in humans and the environment (Maki and Bishop, 1985).

The classical method used for setting 'safe levels' for chemicals to which humans and ecological species may be exposed is to reduce the NOEL or NOEC with a safety or application factor that takes into account variation due to intra- and inter-species differences. For example, for human extrapolation a safety factor of ten is often used when chronic human exposure studies are available. The safety factor of ten takes into account the wide range of variability within the human population. A safety factor of 100 is

Table 3.3 Types of ecological toxicity tests

Type of test	Exposure system	Species (fresh or salt water)	Length of exposure	Observation period	Observations, endpoints
Aquatic[a]					
Acute	Static or flow-through	Fish, invertebrates, Algae	24–96 hours	24–96 hours	Fish (mortality), invertebrates (mortality or immobility), algae (growth), LC50 and EC50
Chronic	Flow-through	Invertebrate (Daphnia)	21–28 days	21–28 days	Growth, mortality/survival, number young produced, MATC[b]
	Flow-through	Fish (fathead minnow, rainbow trout)	6 months–2 years depending on species	6 months–2 years depending on species	Growth, reproduction, spawning success, survival of larvae and fry, behavior, MATC[b]
Early-life stage	Flow-through	Fish	1–2 months depending on species	1–2 months	Incubation time, hatching time for embryos, time from hatch to feeding, per cent normal/abnormal larvae, total per cent of deformed fish, per cent of deformed fish, per cent survival of fish, length and weight of fish, MATC[b]
Biocon-centration	Static or flow-through	Fish or invertebrate	Variable exposure period followed by exposure in untreated medium (depuration period)	Variable	Mortality, con-centrations of substance in tissue and water, BCF[c]
Avian[a]					
Acute	Oral (gavage)	Mallard duck, bobwhite quail	1	14–21 days	Mortality, signs of intoxication, LD50
	Dietary	Mallard duck, bobwhite quail	Variable, 5 day treatment followed by 3-day untreated diet	14–21 days	Mortality, signs of intoxication, LC50
Chronic	Dietary	Mallard duck, bobwhite quail	Approximately 6 months	Approxi-mately 6 months	Reproductive parameters, NOEC

used for chemicals when only chronic animal studies are available and a safety factor of 1000 is used when there are no chronic exposure data.

It is evident that the NOEL or NOEC is a subthreshold level and the safety factor procedure is not appropriate for substances that elicit nonthreshold effects. For example, one molecule of a carcinogen may in theory result in tumorigenicity. Alternative procedures have thus arisen to estimate risks, particularly in humans. Almost every aspect of life exposes both people and the environment to risks. The toxicologist needs to determine the risks and benefits associated with the use of a substance in order to establish an acceptable level of risk. Mathematical models are used to estimate the effects of low, chronic exposure levels when test data are not available. These models presently provide a rough estimate of the magnitude of cancer risk to the human population. In order to conduct an environmental risk assessment, the ecological toxicity test data must be evaluated in light of the environmental fate data (hydrolysis, photolysis, bioavailability, absorption/desorption)(see Chapter 2). Environmental risk assessments are difficult to conduct because of the complexities of aquatic and terrestrial environments and the difficulty in extrapolating from one ecosystem to another. This is further coupled with the fact that in ecotoxicology (see Chapter 4) many test methodologies are still being developed and adequately validated, and in some cases they have not yet met the needs of a growing and demanding society (Goldberg and Frazier, 1989).

Summary

Toxicology functions within the regulatory framework as a discipline whose main role is to predict effects in humans and the environment and to derive safe, low level, ambient exposure concentrations for chemical substances. When the level of uncertainty is high scientific decisions are usually conservative. Often clear-cut assessments of substances cannot be extrapolated from the readily available scientific data and policy decisions are made with incomplete knowledge. Ultimately most decisions are combinations of both science and policy. The point at which science moves into policy is not evident, but it does take place and new toxicological understanding plays an important and continuous role.

In mammalian toxicology a substantial database exists for constructing a

---

Notes to Table 3.3

[a] The observations and endpoints listed are examples only and are not exhaustive.

[b] The MATC (maximum acceptable toxicant concentration) is the estimated threshold concentration of a substance within a range defined by the highest concentration tested at which no significant deleterious effect is observed (NOEC) and the lowest concentration tested at which some significant deleterious effect is observed (LOEC).

[c] The BCF (bioconcentration factor) is the ratio of the concentration of substance accumulated in tissues of organisms to the concentration in water (under equilibrium).

paradigm to compare the likelihood of responses from rats/mice to humans. This is related to a better understanding of physiological/biochemical mechanisms and pharmacological responses in mammalian species versus fish and wildlife. As a result the correlation between responses from chemical exposure in fish and birds in the laboratory to those that might realistically occur in the field is not clearly understood.

Ultimately, it will be important to know how toxicology is used in the regulatory process. This will require a better understanding of basic cellular mechanisms involved in responses to noxious chemicals.

References

Bourdeau, P., Somers, E., Richardson, G. M., and Hickman, J. R. (eds) (1990), *Short-Term Toxicity Tests for Non-genotoxic Effects*, SCOPE, no. 40, (New York: John Wiley and Sons).

Foster, R. B. (1985), 'Environmental legislation', in Rand and Petrocelli, *op. cit.*, pp. 587–600.

Galbor, R. D., Lavin, P. T., Dehta, C. R. and Schoenfeld, D. A. (1985), 'Statistical analysis', in Rand and Petrocelli op. cit., pp. 110–23.

Goldberg, A. M. and Frazier, J. M. (1989), 'Alternatives to animals in toxicity testing', *Scientific American* 261 (2), pp. 24–30.

Klaassen, C. D., Amdur, M. O. and Doull, J., (eds) (1986), *Casarett and Doull's Toxicology – The Basic Science of Poisons* (3rd edn) (New York: Macmillan Publishing Company).

Klaassen, C. D. and Doull, J., (1986), 'Evaluation of safety = toxicologic evaluation', in *Casarett and Doull's Toxicology – The Basic Science of Poisons* (3rd edn) (New York: Macmillan Publishing Company).

Lu, F. C. (1985), *Basic Toxicology: Fundamentals, Target Organs and Risk Assessment* (Washington DC: Hemisphere Publishing Corporation).

Macek, K. J. (1986), 'Perspectives on the application of hazard evaluation to effluents', in H. L. Bergman, R. A. Kimerle and A. W. Maki (eds) *Environmental Hazard Assessment of Effluents*, (New York: Pergamon Press).

Maki, A. W. and Bishop, W. E., (1985), 'Chemical safety evaluation', in Rand and Petrocelli, *op. cit.*, pp. 619–35.

Morgan, M. G. (1984), 'Uncertainty and quantitative assessment in risk management', in *Assessment and Management of Chemical Risks*. ACS Symposium 239, pp. 113–30. (Washington DC: American Chemical Society).

OECD (Organization for Economic Cooperation and Development) (1986) *Existing Chemicals – Systematic Investigations* (Paris: OECD).

Rand, G. M. and Petrocelli, S. R., (eds) (1985), *Fundamentals of Aquatic Toxicology – Methods and Applications* (New York: Hemisphere Publishing Corporation).

Snyder, R. (1984), 'Basic concepts of the dose–response relationship', in J. V. Rodericks and R. G. Tardiff, (eds), *Assessment and Management of Chemical Risks*. ACS Symposium, Series 239, pp. 37–56 (Washington DC: American Chemical Society).

Sprague, J. B. (1985), 'Factors that modify toxicity', in Rand and Petrocelli, *op. cit.*, pp. 124–63.

4 *Ecotoxicological considerations*

P. J. SHEEHAN

Introduction

Ecotoxicology is an extension of the science of toxicology as well as a subdiscipline of the science of ecology. Whereas toxicology is the study of the effects of poisons on individuals with the emphasis on human health, ecotoxicology is the study of the effects of toxic chemicals on wild populations and ecological systems. In toxicology, the focus of studies is on the mechanisms of action and effects of chemicals on biochemical processes, cells, tissues, organs, and the ultimate measure of impact is the survival or death of the exposed individual. In ecotoxicology, the effects of chemicals may be expressed as direct toxicity to exposed individuals or as the degradation of the environment to the detriment of the individual. The significance of effects in ecotoxicology are ultimately assessed in terms of impacts on structure and function at the population, community, and ecosystem levels. The present emphasis within the regulatory communities of most countries has been on adverse effects on individual species. This chapter presents a broader view of ecological impacts associated with chemical exposures.

Ecology is the scientific study of the interactions that determine the abundance and distribution of organisms (Krebs, 1972). A large number of chemical, physical, and biological factors impinge upon the success of organisms. Toxic chemical exposures are just one of the many factors that shape the structure and function of ecosystems. Not all factors structuring ecological interactions are well understood. Therefore, the ecotoxicological significance of an actual or projected chemical release is most easily assessed if the magnitude of the impact is substantial, that is, if the influence of the toxic chemical overwhelms other major chemical, physical and biological factors controlling population success. If the magnitude of impact is not substantial, ecotoxicological effects may go unnoticed.

It is clear that in some cases there will be insufficient information on the chemical or mixture of chemicals and on ecological interactions to predict the type and magnitude of impacts on exposed ecosystems. The most sensitive populations and the dependence of other species on these sensitive taxa may be unknown. Furthermore, the types and magnitudes of exposures may be unknown. In those cases where ecological changes from chemical exposures can be measured, their significance may be more clearly assessed in terms of impacts on commercial species, species which play critical roles in the community such as 'keystone' predators and pollinators, and in terms of effects on system functions like primary productivity, decomposition of complex wastes, and the cycling of essential nutrients.

There are numerous examples where chemical impacts on a single population may have little effect on the functioning of the ecosystem. If the abundance of preferred prey is reduced by toxic stress, often the population of predators can switch their choice of food without showing adverse effects. Pollutant-sensitive species may be replaced by opportunistic competitors without affecting ecosystem biomass or productivity. These examples point out that it is not always possible to characterize the response of the ecosystem to chemical perturbation, solely from the knowledge of effects on component parts. An ecotoxicological assessment must integrate single-species data with analysis of community structure and dynamics and ecosystem function.

A substantial amount of information has been accumulated which demonstrates that the impacts from toxic chemical exposures can be felt at the community and ecosystem levels and that the testing of individual species and the monitoring of individual populations are often inadequate to assess larger scale impacts.

The objective of this chapter is to evaluate the utility of the ecotoxicological perspective in the management of toxic or potentially toxic chemicals. As a basis for this assessment, some concepts of ecotoxicology are defined. The critical issue of distinguishing ecotoxic effects is discussed. Field monitoring procedures are examined as a basis for quantifying the magnitude of ecological effects and as a mechanism for assessing the ecological significance of detected impacts. Toxicity testing is evaluated as a tool for predicting ecotoxic impacts and, in conclusion, ecological risk assessment is examined.

Some principles of ecotoxicology

Concentration–response

A cornerstone of toxicology is the dose–response relationship. Exposure (dose)–response relationships are also keys to assessing the ecological impacts of toxic chemicals. A generalized model of ecological response to dose or environmental concentration is presented in Figure 4.1. This general model describes a curve crossing three zones of response: unmeasurable, measurable at the population level and measurable at the community or ecosystem level, as the concentration of the toxic substance increases. Figure 4.1 illustrates two important concepts:

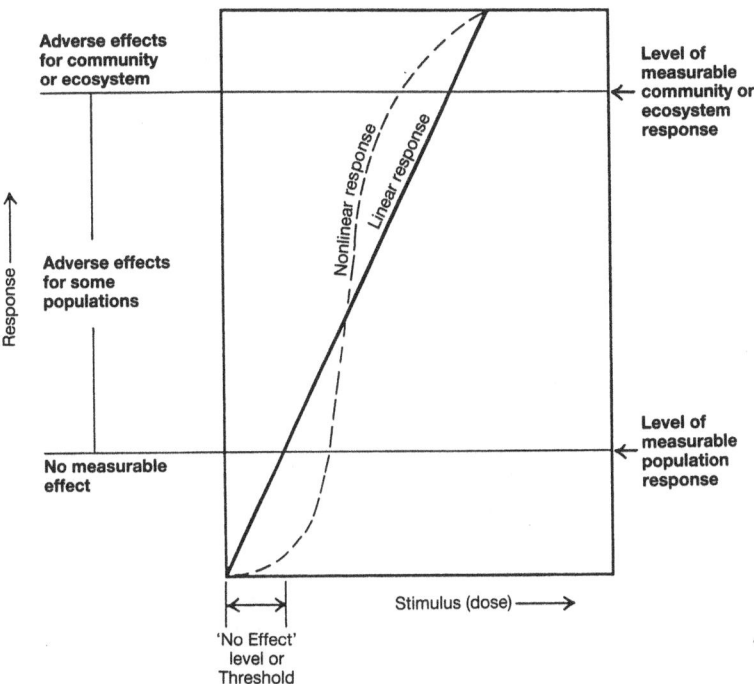

Figure 4.1 A generalized model of ecological response to dose or concentration

1 Organisms in an ecosystem exhibit differing susceptibilities to toxic
   chemicals. The severity of response will be directly related to the level
   of chemical exposure. At a low level of exposure to the specific
   chemical, a few individuals of a few populations may respond to either
   the direct toxic insult or an adverse change in their environment. At a
   high level of exposure to the same chemical, a large portion of the
   population and more populations may respond adversely to the insult
   and changes may be measured at the community and ecosystem levels.
2 A 'no effect' level or threshold dose must be exceeded before a
   measurable population response will occur. Defining the threshold for a
   particular response, however, may be difficult. There is no equivalent
   in ecotoxicology of the no-threshold response model for carcinogenesis
   in humans. A threshold dose is assumed to exist even if it is beyond our
   ability to estimate it.

Response variables

A number of measures of effect may be used to quantify changes in
populations, communities and the ecosystem exposed to chemicals. The

selection of appropriate response variables will be situation specific. In most cases, more than one variable will be required to clearly describe ecosystem changes. A summary of response variables is presented in Table 4.1 and additional information pertaining to the use of the variables is provided in the text.

POPULATIONS

A population is a group of individuals of a single species. Ecological responses to chemical exposures may be expressed at the population level in the following terms (Sheehan 1984a): reduced abundance; altered distribution; changed age structure; altered gene pool; altered behavioral patterns.

Chemical injury to individuals resulting in premature death and reduced reproductive success and recruitment are ultimately reflected in lower population abundance and altered distribution of the exposed populations. The numbers of individuals or biomass of a population are the most commonly measured indices of population well-being. Monitoring studies of the effects of a chemical concentration gradient on animals and plants have provided clear examples of concentration-related effects on exposed populations (e.g. Winner et al., 1975, 1980, and Sheehan and Winner, 1984). Controlled ecosystem studies using in situ enclosures have experimentally demonstrated the inverse relationship between the abundance of sensitive populations and toxicant concentrations. Stephenson and colleagues (1986) concluded from a series of limnocorral enclosure studies that the response to methoxychlor of various groups of plankton generally appeared to be exposure dependent, with Cladocera and Chlorophyta exhibiting the greatest sensitivity. In many cases where moderate concentrations of a toxic chemical reduces the abundance and distribution of sensitive populations, the abundance and distribution of tolerant populations increase due to their release from competition. For example, rotiferan populations have been shown to increase with reductions of cladocerans and copepods in enclosures treated with pesticides (Kaushik et al.,1985; Stephenson et al.,1986). This provides a good example of why it is usually difficult to assess the actual, in situ impact of toxic compounds solely on the basis of laboratory bioassays using isolated populations of single species.

For those populations that have identifiable year or size classes, the impact of chemical stress can be assessed through studying changes in population age structure. The influence of natural variability in growth rates, fecundity and recruitment must be considered in any analysis of chemical effects on age (size) structure. For example, Notini (1978) reported that the changes in size distribution of the common mussel (Mytilus edulis) for the years 1971–6 reflected the recovery of the population from the severe stress of an oil spill. During the first two years following the spill, only a few larger mature organisms and a few recruits were found. By 1976, the population structure was composed of a complete size range of mussels with the greatest portion in the mid-range, reflecting normal recruitment and age distribution.

The effects of chemical stress on a population gene pool has received limited attention in monitoring investigations, although a large body of literature exists which indicates that long-term sub-lethal chemical stress selects for some measure of tolerance in exposed populations. Luoma (1977) suggested that under chemical stress, the probability of an opportunistic species developing resistant populations is greater than that of more specialized species. Therefore, more simplified communities may result from chronic chemical stress through the elimination of species not having resistant genotypes.

Chemical stress altering the behavioral patterns of a population may have substantial repercussions on the ultimate success of the population. The significance of altered behavioral patterns is particularly obvious where chemical pollutants have altered the migration of anadromous fish and thereby reduced spawning and recruitment success. Sprague *et al.* (1965) reported that copper pollution in the northwest Miramichi River prevented Atlantic salmon (*Salmo salar*) from reaching their spawning grounds. Lorz and McPherson (1977) reported that copper also reduced the success of downstream migration of juvenile Coho salmon (*Oncorhynchus kisutch*).

COMMUNITIES

Communities are assemblages of populations structured by biotic interactions and the constraints of their physical and chemical environment. The structure of a community is defined by the abundance and biomass of all their populations and their spatial, taxonomic and trophic organization. The integrated response of the component populations to the presence of toxic pollutants will be reflected in alterations of the structural characteristics and dynamics of the stressed community (Sheehan, 1984b).

Structural changes may be visualized as an information network reflecting environmental conditions but not demonstrating the external mechanisms or internal interactions which brought about the reorganization (Cairns *et al.*, 1972). Changes in community structure are, therefore, after the fact evidence of the integrated responses of populations to the chemical insult.

Structural characteristics and the numerical indices dependent on them provide various types of information which differ in ecological value. For instance, it is more informative to know the taxonomic or trophic composition of a community than merely its biomass or the abundance of organisms. Also, structural indices do not necessarily follow similar patterns of change under conditions of induced stress. Hellawell (1977) described several possible alterations in a community which would be reflected only in biomass, in biomass and relative dominance, or in biomass dominance and composition. Because of the differences in (1) the value of

Table 4.1 Summary of population, community and ecosystem response variables and their application to assessment of chemical effects

Response variable	Measurement	Comments
Impacts on populations		
Reduced abundance	Numbers of individual or biomass of population versus chemical exposure	Compare to abundance of reference population in uncontaminated area or control population in enclosure study
Altered distribution	Presence/absence or commonness/rareness of a population versus chemical exposure	Analysis must account for life history and other ecological and environmental factors structuring population distribution
Changed age structure	Frequency distribution of age or size class versus chemical exposure	Can be used to assess recruitment success or recovery
Altered gene pool	Electrophoretic analysis of genotypic frequencies versus chemical exposure	Opportunistic species more likely than specialized species to develop resistant populations
Impacts on community and ecosystem structure and dynamics		
Population extinction	Absence of a population known to have existed prior to chemical exposure	Loss of 'key' commercial or ecological populations is most easy to interpret
Changed community composition	Species list, indicator species or indicator assemblages versus chemical exposure	Best to know what taxa are absent as well as what taxa present; trophic organization may provide insights into effect on feeding relationships
Dominance switches	Relation of abundance versus chemical exposure	Abundance under chemical stress often depends on the opportunistic life histories of species
Changed diversity	Margalet, Simpson or Shannon diversity indices versus chemical exposure	Applicable only to gross levels of chemical pollution due to conflicting data from richness and evenness components and insensitivity to moderate levels of pollution; species richness may be a more consistent index of chemical stress
Changed similarity	Coefficient of similarity, quotient of similarity, or percentage similarity versus chemical exposure	More consistent and sensitive than diversity indices in assessing chemical effects on community composition
Reduced abundance/biomass	Total number or biomass of individuals in community versus chemical exposure	Do not provide much information on the ecological character of the system but least expensive variables to measure
Altered spatial structure	Vertical and horizontal patterns versus chemical exposure	Only of value in assessing pollution effects on non-mobile communities such as forests which exhibit a distinct spatial structure
Stability fluctuations — inertia	50% change in species composition or richness versus chemical exposure	Is a measure of the system's resistance to pollution
— elasticity	Recovery time to reach 85% similarity to the original composition	Is a measure of the system's ability to recover from chemical stress

Table 4.1 continued.

Response variable	Measurement	Comments
— amplitude		Is a measure of the maximum amount of damage from which a system can recover in a specific time
— hysteresis	Sperman's rank correlation coefficient comparing disappearance and reappearance of species	A measure of the degree to which an ecosystem's pattern of recovery is not the reversal of the pattern of species loss
Impacts on ecosystem functions		
Reduced organic decomposition	Decomposition rate of plant litter or reference organic substrates versus chemical exposure	Effects may not be obvious until some time after exposure
Reduced nutrient conservation	Net loss of essential elements in mass balance studies versus chemical exposure; nutrient spiralling length in stream studies	Of most obvious importance in terrestrial ecosystems
Reduced primary productivity	^{14}C assimilation rate or other methods of measuring plant growth versus chemical exposure	Long-term reductions in primary productivity are the most obvious index of a functionally stressed ecosystem
Reduced ecosystem production	Net gain in ecosystem production versus chemical exposure, e.g. net oxygen evolution in an aquatic system; Odum's index of power may be appropriate	Integrates chemical effects on primary productivity and energetic costs
Altered food web and functional regulation	Changes in predator–prey or consumer–consumed interactions versus chemical exposure	The loss of a predator indirectly through the elimination of prey and the dramatic increase in a population after the removal of a predator or competitor are examples of chemical alteration of functional regulation

information provided by these indices, (2) the ease with which they are measured or calculated and (3) the sensitivity of their response to stress, certain criteria are essential in order to evaluate the usefulness of structural characteristics in monitoring pollutant effects.

The ideal community index would have the following properties (see Sheehan, 1984b):

- sensitivity to the stressful effects of pollution;
- general applicability to various types of ecosystems;
- capability to provide a continuous assessment from unpolluted to polluted conditions;
- independence of sample size and ease of measurement or calculation;

- ecological meaningfulness;
- ability to distinguish the cyclical and natural variability of the system.

Sensitivity refers to the level of chemical stimulus to produce a measurable community response (see Figure 4.1). Applicability refers to the utility of the index if applied to different population assemblages in different types of ecosystems. Ecological meaningfulness refers to our ability to conclude that change is adverse to the community.

The final criterion has to do with the problem of separating chemically induced responses from natural changes. This critical problem will be addressed in 'Distinguishing Ecotoxic Effects' (see pages 91–96).

Chemical impacts at the community level may be expressed in terms of the following structural characteristics (see Sheehan 1984b; Kelly and Harwell 1989): reduced abundance or biomass; changed community composition; dominance switches; changed diversity or similarity patterns.

Abundance and biomass are the most simple indices of community response. Because of their simplicity these indices are cost effective to measure, but do not provide much information on the ecological character of the community. Abundance and biomass are useful as quantitative screens for community-level impacts. These indices are more informative if reported along with other data.

Reductions in density and complete extinction of sensitive populations due to the stress of toxic chemical exposures are primary factors altering community structure. Although it is very difficult to evaluate the loss of any one species, it is easier to appreciate the disappearances of those that fulfill unique roles in the community such as pollinators and key predators. Adverse effects of pesticide spraying for spruce budworm control have been demonstrated on forest pollination in New Brunswick (NRCC, 1981a). Direct toxic impacts on pollinating species led to changes in the reproductive performance of various insect-pollinated plants.

Decreases in the number of species (species richness) in a community are a useful measure of the severity of chemical stress and a consistent measure of community recovery with the abatement of chemical pollution. Species richness is quantifiable for all known organisms but measured values are dependent upon the number and size of samples taken as well as life history factors (Rosenberg, 1977).

Changes in community composition and dominance patterns, as influenced by chemical stress, reflect the responses of a number of individual populations interacting within their specific tolerance requirements. Community composition is generally assessed as a partial list of species or indicator species present in monitoring studies or multispecies toxicity tests. The satisfactory use of the indicator species or assemblage concept requires information on the species absent, as well as those present in the stressed community (Hellawell, 1977).

Dominance is an integral part of defining community response. A species is considered to be dominant if it is present in the greatest abundance, or has the largest individuals, or has the greatest impact on community dynamics. Although dominance implies a position of advantage in community inter-actions, it does not necessarily correlate with tolerance to chemical stress. Abundance (dominance) under pollutant stress often depends upon having an opportunistic life history. A switch in dominance away from key commercial or ecological populations would signal an adverse community impact.

The sensitivity of shifts in composition and dominance to changes in the level of chemical exposure appear to differ with both the community type and toxic chemical. Therefore no generalization seems to be universally applicable concerning the sensitivity of taxonomic composition to toxic concentrations of a variety of pollutants. For specific communities and pollutants, some consistent trends have been reported. For example, the sensitivity of marine crustaceans to oil pollution is well documented (NAS, 1989) as is the tolerance of small flagellates in the phytoplankton communi-ties to triazine herbicides (deNoyelles et al., 1982; Sheehan et al., 1986).

Decreased taxonomic diversity has been used as an indicator of gross environmental deterioration. The concept of diversity is based on the number of species present and the distribution of individuals among these species (evenness). A number of numerical indices have been proposed. Many of these are summarized in Sheehan (1984b). Diversity indices have been widely used in chemical pollution studies. However, diversity does not appear to be clearly related to moderate levels of chemical stress when 'normal' controlling factors are more robust than toxic chemicals in structuring the community. Richness and evenness components can be contradictory and diversity indices may be inconsistent with other bio-logical indices (Ford, 1989). For all these reasons, the use of numerical diversity indices in pollution evaluation should be limited to the demon-stration of gross changes.

There is evidence to indicate that similarity indices are more sensitive and therefore more indicative of structural differences than are diversity measures at low levels of chemical exposure (Sheehan, 1984b). Similarity measures are therefore suggested as a single index to characterize structural response through time with pollutant input and abatement.

ECOSYSTEMS

The ecosystem is the fundamental unit of ecology and includes both the biotic community and its abiotic environment. The analysis of the eco-system as a unit in ecotoxicology is based on the premise that the system as a whole possesses characteristics which not only reflect the integrated response of component populations but, in addition, provides a more comprehensive picture of ecosystem status.

Ecological response to chemical insult may be expressed at the eco-system level in the following terms (see Sheehan, 1984c; Kelly and

Harwell, 1989): altered stability; reduced organic decomposition; modification of nutrient cycles; reduced primary production or ecosystem production; altered trophic dynamics; altered successional patterns.

Ecosystems are not stable naturally. They do not have a single stable equilibrium structure, nor does recovery of a perturbed ecosystem always follow a predetermined pattern. However, ecosystem characteristics do persist within certain bounds, despite small perturbations. Certain measures of ecosystem dynamics can be used to describe system response to chemical stress. These are described in Table 4.2 and are discussed in greater depth in Sheehan (1984b), Westman (1985) and TRI (1987).

Table 4.2   Measures of ecosystem dynamics and examples of their application

Characteristic	Definition	Example: Ecosystem subjected to oil spill
Inertia (ecological buffering capacity)	Resistance to change	Amount of oil that must accumulate over a given area in time period to cause a given level of ecosystem damage (such as local extinction of species x and y)
Elasticity	Rapidity of restoration of a stable state following disturbance	Time required to recover initial structure or function following ecosystem damage (e.g. restoration of populations x and y)
Amplitude	Zone from which the system will return to its original state	Maximum amount of oil that can accumulate in an area such that damage sustained can be fully repaired (e.g. restoration of populations x and y)
Resilience	Zone from which the system can return to a stable configuration, perhaps different from the original	Maximum pollutant level after which populations will stabilize at some level
Hysteresis	Degree to which path of restoration is an exact reversal of path of degradation	Degree to which pattern of secondary succession is not an exact reversal of the pattern of retrogression experienced following impact (e.g. were the last species to disappear the first ones to return?)
Malleability	Degree to which stable state established after disturbance differs from the original steady state	Degree to which new climax ecosystem resembles the initial climax state (e.g. how closely do the species composition and equitability of new climax state resemble the old?)

Source: After Holling (1973) and Westman (1978).

To assess the resistance to change or buffering capacity of an ecosystem, the degree of chemical contamination may be related to a measure of damage such as a 50 per cent change in species richness or percentage similarity in taxonomic composition or a detectable reduction in functional capability (e.g. 25 per cent reduction in primary productivity).

The elasticity of an ecosystem is a measure of its ability to recover within acceptable limits after being chemically stressed. Although restoration time is not a simple concept, it may be estimated as the time taken to reach 85 per cent similarity to the original taxonomic composition (Westman 1978) or to some desired level of functional similarity (see response model present in 'Community and Ecosystem Toxicity Testing', pp. 103–10). Recovery from disturbance appears faster in aquatic systems than in terrestrial systems. Oviatt et al. (1984) estimated a recovery of seven to ten years for Narragansett Bay based on clean sediment covering polluted sediment; however, Saldarriga (1987), for tropical forests damaged by slash and burn agriculture, estimated a minimum of many decades for recovery.

Amplitude is a measure of the maximum amount of perturbation from which an ecosystem can recover. This measure is a requisite in establishing protective standards for an ecosystem. However, the relationship between system response amplitude and chronic or successive inputs of toxic chemicals is not well documented and there are few data on the magnitude and duration of exposures that an ecosystem can withstand successfully in the long term. In one example, Baker (1973) concluded that approximately twelve successive crude oil spills over a fourteen-month period would exceed a salt marsh grass system's threshold of self repair.

Hysteresis is a measure of the degree to which an ecosystem's pattern of recovery is not the pattern of species loss. Hysteresis may be measured with techniques like Spearman's rank correlative coefficient to compare the disappearance and reappearance of species in relation to chemical release and abatement (Westman, 1978).

Assessing changes in the characteristics of ecosystem functions such as energy flow and nutrient cycling can provide quantitative data on ecosystem levels of response to chemical perturbations. The analysis of functional processes in chemical pollution studies has a relatively short history and is dependent on the development of ecosystem theory and the collection of sufficient baseline data to establish normal process rates. Changes in primary productivity, ecosystem production and the ratio of production to respiration have been used to assess the effects of toxic chemicals in aquatic and terrestrial ecosystems (e.g. Giddings and Eddlemon, 1978; Legge, 1980). Chemical effects on organic decomposition rates and nutrient conservation have been most often assessed in terrestrial ecosystems (e.g. Hutton, 1984).

Measuring the functional responses of ecosystems to chemical stress is essential to establishing key 'breakpoint' values. In this context, breakpoint defines a level of reduction in functional capability which is considered unacceptable. Being able to define breakpoints is a necessary step in establishing criteria to protect the ecosystem as a unit.

Time–response relationships

The time lag between the initiation of chemical exposure and the observation of impacts can differ substantially for population, community and ecosystem response variables. This time–response relationship is shown in Figure 4.2. Long-term impacts, as opposed to transient impacts, can only be quantitatively demonstrated for exposure and observation periods in the months to years range for most response variables.

Response lag time is related to the generation time of individual populations and to the time required for feedback from ecological interactions once chemically perturbed. Communities dominated by organisms like algae and bacteria with short generation times show relatively rapid response to chemical pollution. However, Stockner and Antia (1976) warn that exposures of twenty to forty days are necessary in plankton-dominated aquatic systems to demonstrate long-term habituation rather than a short-term shock response to toxic chemicals. The identification of inhibition of microbial function as being either reversible, but with an extended period

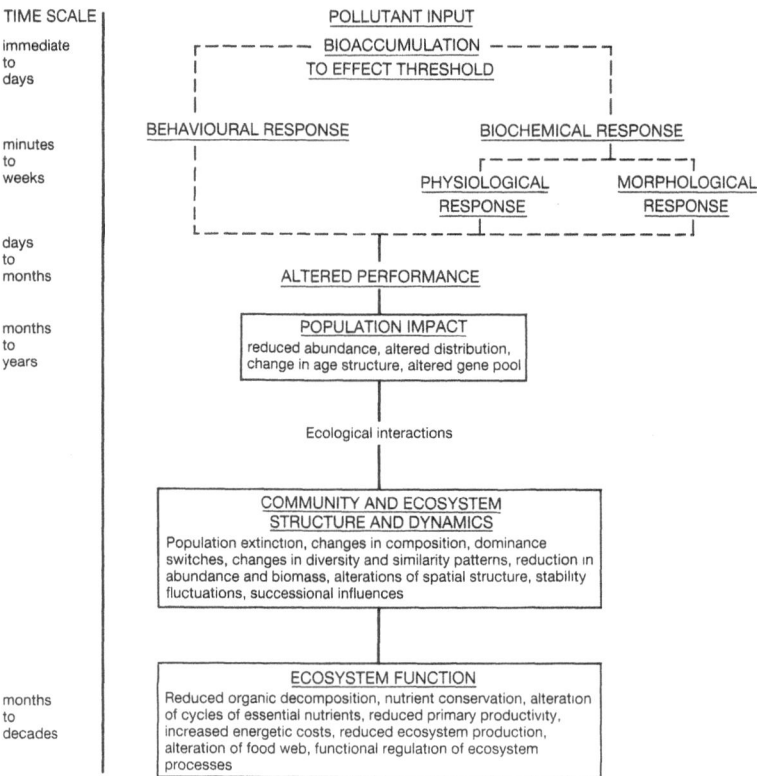

Figure 4.2   A conceptual chronology of induced ecological effects following exposure to toxic pollutants

of delay before recovery, or persistent, causing a significant deficiency in activity, is critical to the recognition of adverse impacts in both field studies and controlled toxicity tests.

Obviously longer periods are necessary to demonstrate impacts on larger species with longer generation times. The key point is that neither data from monitoring investigations nor controlled toxicity tests of short duration can effectively document ecologically relevant concentration–response relationships associated with low–level chronic exposures.

Direct versus indirect effects

Contrary to standard toxicological studies where all effects on individual organisms are directly related to the toxic chemical, ecotoxicological studies are concerned with both direct toxicity and indirect adverse ecological interactions. For example, studies of the success of duck populations in the Canadian prairies have suggested that low recruitment may be partially related to the effects of pesticides on food and cover rather than the direct toxic effects of these compounds on waterfowl (Sheehan et al., 1987). Nesting female ducks require dense protective cover to maintain good hatch success. Herbicide application in nesting areas prior to nesting can decrease the probability that ducks will nest successfully. Also, the aerial application of insecticides during the reproductive period can reduce invertebrate prey available for the ducklings during the critical post-hatch period when they are obligate invertebrate feeders. Hunter et al. (1984) showed that duckling growth rates (correlated with survival) were reduced significantly in controlled ecosystem experiments with carbaryl applied to ponds used by foraging mallard ducklings.

This example serves to point out that indirect effects should be considered in ecotoxicological assessments. In most past assessments, indirect ecological consequences of chemical exposures have not been considered. This oversight has contributed to our inability to explain many adverse ecological events believed to be associated with chemical exposures. Predicting responses beyond direct toxicity will require increased emphasis on microcosm, mesocosm and field scale studies.

Distinguishing ecotoxic effects

There are fundamental difficulties involved in the quantitative examination of ecosystems. These difficulties stem from the inherent biological variability of natural systems as related to the cyclic, stocastic or successional changes in populations. Our ability to describe natural biological phenomena are limited by the practical constraints of sampling. We are at times unable to apply appropriate designs to resolve natural population patterns in experimental and monitoring studies.

The problem is how to decide whether an observed change in some system response variable represents a deviation caused by the presence of a toxic chemical or is part of the natural fluctuations or cycles inherent in the ecosystem. The question of distinguishing ecotoxic effects is central to managing chemicals to protect the ecological system. Yet our understanding of the key ecological and statistical issues has in many cases been inadequate. Hurlbert (1984) described experimental design problems in ecological studies and presented data on the substantial misuse of inferential statistics. In many cases adequate baseline measurements are not available to allow us to define with confidence the types and sizes of inherent fluctuations. As significant as the problem of lack of baseline data is the improper design of laboratory and field experiments including problems of clearly defining testable hypotheses.

Experimental design considerations

Green (1984) stated the problem quite succinctly: 'What we need is good *a priori* design of environmental studies, validated by preliminary sampling, so that the results effectively displayed speak for themselves.' Green defines a logical flow to be followed in all environmental studies: purpose − question − hypothesis − model − sampling design − statistical analysis − test of hypothesis − interpretation and presentation of results. The logic in this progression, unfortunately, is forgotten at times.

A general statement of hypotheses to be applied may be as follows:

- H_0 (null hypothesis)   There is no biological response to the introduced chemicals, i.e. any biological response is caused by natural variation within the ecosystem.
- H_1 There are biological responses attributable at least partially to one or to the additive effects of more than one of the chemicals introduced into the ecosystem.
- H_2 There are biological responses attributable at least partially to interaction effects among the introduced chemicals or between them and their byproducts.

These hypotheses must be expressed in terms of a model which relates the biological response variables to predictor variables. The ultimate predictor variable in any model is 'error'. Statistical tests are done to quantify our uncertainty about the truth of alternative hypotheses. Estimate of the error term in any statistical model is essential to that process.

Principles of experimental design can be found in some statistical texts (Gilbert, 1987) and are well discussed in the context of pollution investigations by Green (1979), Hurlbert (1984) and Carney (1987). The following components of good experimental design need to be emphasized: control treatments, replication of treatments and randomization and/or interspersion of treatments.

A control treatment is a treatment against which one or more treatments may be compared. Controls may reduce confusion in terms of temporal change or procedural effects. Replicate controls are used to assess natural variability and perhaps to assess the influence of carriers or adjuvants added along with the chemicals to be tested. Appropriate controls for field experiments with toxic chemicals can be easily identified conceptually; however, the practical selection of control populations or ecosystems is difficult. There are problems in establishing the independence of populations or systems, and in the logistics and allocation of effort to monitor replicate controls to account for temporal variability and experimental effects. Perhaps the most difficult of design problems is associated with monitoring studies to determine the effects of point source chemical releases into large ecosystems. For example, finding adequate control populations for comparison with estuarine populations exposed to toxic effluents is a very complex problem.

The main purpose of replication or random repetition of treatments is to supply an estimate of variability (error) by which the significance of treatment and control comparisons can be judged. Without true replication (i.e. replication of treatments), valid statistical comparisons of chemically contaminated and uncontaminated ecosystems cannot be made. Hurlbert (1984) coined the term 'pseudo-replication' to define studies where the error term in tests related to impacts was from replication of analyses, counts or field samples within single treatments rather than from true treatment replicates. Pseudo-replication has unfortunately been commonplace in impact studies and is particularly evident in point source monitoring studies where there is *de facto* no treatment replication. Where there is a single unreplicated treatment and control, two approaches have been suggested to identify effects. Hurlbert (1984) suggests a straightforward presentation of quantitative and observational data to demonstrate gross changes with no use of inferential statistics. He cites the Hubbard Brook deforestation experiment as a good example of this type of optimal impact study. Green (1984), on the other hand, argues that error may be defined in terms of year-to-year variability in response variables. This term may be used to statistically compare exposed and control ecosystems. This latter approach is dependent on adequately supported studies to supply the long-term data.

The need for treatment replication, and the practical problems in applying replicated designs in large-scale field studies, point to the importance of manipulative studies and microcosm and mesocosm studies where replication is logistically feasible and can be included in experimental designs (Sheehan et al., 1986). Replication reduces the effects of random variation, thereby increasing the precision of the estimate of response. There have been a number of reports on replicability of microcosm test systems. The emphasis on replicability in some cases has been on the degree to which systems are similar (e.g. Harte et al., 1980) rather than on the more obvious question: given the observed variability among experimental units, how many should be assigned to each treatment to achieve the desired level of detection?

Natural variability and the level of detectable change

The variability in natural ecosystems is determined in baseline studies. The period of observation must be sufficient to encompass seasonal and cyclic changes. Obviously, in practical terms, no observation period will be sufficiently long to accurately describe successional changes.

Few long-term biological baseline data bases are available for ecosystems. The Continuous Plankton Recorder Programme of the North Sea and North Atlantic is a good example of an extensive data bank. Some 300 species of aquatic plants and animals in water along commercial shipping routes have been surveyed on a regular basis since 1948 (Colebrook, 1978). Most examples of monitoring programs to describe changes from chemical pollutants are based on much shorter observation periods. Legge and colleagues described an eight-year program of remote sensing, controlled laboratory fumigation experiments and detailed field studies to determine the effects of sulfur gas on foliar accumulation of sulfur and essential nutrients, soil changes, pine tree physiology and forest productivity (Legge, 1982).

The time trend of extinction of brown trout (*Salmo trutta*) populations associated with lake acidification in Norway was developed from surveys of fishermen and lake owners over a forty-year period (Muniz and Leivestad, 1980). This example demonstrates that long-term observations are also necessary to clearly demonstrate population changes for larger consumers in aquatic ecosystems. Obviously, the length of the observation period to define variability in the populations with rapid turnover rates will be substantially shorter than the multi-year studies described for some terrestrial and aquatic populations. For instance, trends in microbial populations may be adequately described in a year or less.

A second aspect of monitoring natural variability is the frequency of sampling. Annual sampling is adequate to define trends in woody tree production, and monthly sampling to define trends in aquatic insect production. Weekly sampling is sufficient to describe population trends in phytoplankton and zooplankton but may not provide sufficient resolution of trends in the bacterial community. A resolution time of hours may be necessary to define fluctuations in atmospheric contaminants.   These examples serve to point out the obvious need to design studies not only to account for the period of time over which observations must be made to define variability but also to consider the frequency with which observations need to be made to adequately measure and account for fluctuations.

A final consideration is the size of sample necessary to achieve a given level of reliability of data. The degree of effort put into quantification depends on the question being asked. If one wishes to be able to detect only gross changes between polluted and unpolluted areas, few samples are required. If one wishes to detect a small change between chemically exposed and unexposed systems, a substantial number of samples is necessary. At this point, it is useful to note that the increase in sampling effort is to improve the precision of the measurement (reduce the error) and not the accuracy of the measurement. We assume that the sample is representative and accurate, that is, close to its true value. This is not necessarily true as pointed out by Hynes and colleagues when they demonstrated that as much as 80 per cent of the stream macroinvertebrate fauna is not subject to collection when the typical surface-substrate samplers are used in habitats where deep burrowing forms are common (Hynes *et al.*, 1976). The aquatic macroinvertebrate community has been the focus of many chemical pollution studies. Some good information and insights on the magnitude of the problem of detecting chemical pollution effects come from evaluating benthic macroinvertebrate sampling data. Benthic macroinvertebrates, like many other communities, are clumped in their distribution. Clumping or aggregation is the basis for significant dependence of the sample variance on the sample mean. High variability in replicate samples results in low precision of estimates of means based on a small number of samples and consequently a low sensitivity of the study to detect perturbation effects.

Needham and Usinger (1956), and later Chutter and Noble (1966) and Resh (1979), in analyzing the same data set for a stream riffle community, concluded that several hundred samples would be required for a precise estimate of the mean of total numbers or wet weight (95 per cent CL ± 5 per cent). These analyses indicate that attempts to achieve this level of precision are in most cases impractical. Allan (1984) presented some further data describing the level of detectable differences in macroinvertebrate numbers based on his sample data and the size of the mean. A summary of his findings is presented in Table 4.3. Approximately ten to twenty replicate samples per treatment would be required just to detect a doubling or halving of invertebrate densities. Thirty to sixty sample replicates may be required to detect a 50 per cent difference in the mean. A greater number of samples is required in both cases if the observed mean is small.

It is worth noting that some precise information on aquatic macroinvertebrates can be gathered with few samplings. Needham and Usinger (1956) pointed out that as few as two to three samples are adequate to collect at least one representative of each of the common taxa of invertebrates with a 95 per cent probability.

Available data suggest that there is substantial spatial and temporal variability in biological communities in lake and marine ecosystems. Replicate sampling in these types of aquatic systems typically yields coefficients of variation for biomass and flux rate variables of the order 15–100 per cent (Sanders, 1985). For example, based on biweekly data on annual variations in a small marine bay, more than fifty samples would be required to

Table 4.3   The number of replicate macroinvertebrate samples required to distinguish means that differ by 25%, 50% or 100% at the 0.05 significance level

Mean values	Difference to be detected (%)	Number of replicates required[a]		
		Total fauna	*Baetis bicaudatus*	*Ephemerella coloradensis*
Lowest observed mean	100	11	39	624
	50	32	124	2,461
	25	106	451	9,591
Mean = 10	100	12	20	16
	50	36	60	48
	25	119	202	158
Mean = 50	100	11	18	14
	50	31	54	42
	25	104	178	140
Highest observed mean	100	11	18	14
	50	31	52	42
	25	102	173	140
[a] Variance $(\log[x+1])$		0.059	0.105	0.079
Highest mean		330.1	119.2	63.2
Lowest mean		43.7	1.5	0.12

Source: Modified from Allan (1984).

determine with 95 per cent probability a 100 per cent change in the mean of chlorophyll *a* or ATP levels as estimates of planktonic biomass (Naiman and Siebert, 1977). The results from field enclosure studies indicate that in general a two to threefold change in population levels is required to detect pollutant effects on zooplankton density with three replicate enclosures (Lawson and Grice 1977). This magnitude of change also appears to be required to identify statistically significant changes in field monitoring data (Wiebe and Holland, 1968; Kerfoot and Demott, 1980).

There is evidence to suggest that a small number of replicate samples is required to distinguish the effects of chemicals on functional variables describing the metabolic response of aquatic communities and ecosystems. Selected data from microcosm and mesocosm studies are summarized in Table 4.4. A 50 per cent change in microcosm production or respiration can normally be detected with as few as three experimental replicates. Six to eight replicates may be required to detect a 25 per cent change in mean production with 95 per cent probability. Algal primary productivity rate and bacterial uptake rate of organic substrates are more variable in experimental systems. Three to four replicates would be required to distinguish a 100 per cent change in the mean rate of microbial response.

There have been two recent reviews of replicability in aquatic multispecies test systems (Sanders, 1985; Giesy and Allred, 1985). Both reviews point to the need for sufficient replication of treated systems to achieve the desired level of detection in toxicity tests. These papers also summarize data on the replicability of a number of biotic and abiotic variables in microcosm and mesocosm systems that are not considered in this section.

Environmental monitoring and assessment

Environmental monitoring and assessment are essential to determine if chemicals released into the environment have caused ecological impacts and to determine if improvements in effluent quality have in fact produced ecological benefits.

Environmental monitoring programs normally have two components: (1) the monitoring of chemical and physical factors, including the concentrations of toxic chemicals and environmental conditions in space and time at contaminated and uncontaminated locations, and (2) the monitoring of biological response variables to identify ecological changes resulting from the cumulative impact of toxicants under natural conditions.

Three reasons have been cited for the need for biological evidence of contaminant effects (Cairns and van der Schalie, 1980):

1 Many chemical compounds produce adverse biological reactions at concentrations below present analytical capabilities.
2 Potential toxicants are rarely present in isolation from each other. Generally toxicants are present in effluents and natural systems as mixtures and the biological impact of a mixture cannot adequately be estimated from a series of chemical analyses alone as chemicals react in various ways with organisms and with each other.
3 Water quality (i.e. hardness, dissolved oxygen, pH, etc.) has a marked influence on the expression of toxicity. It is, therefore, a combination of toxicants, water quality and the organisms present that produces definitive data on harm. As a consequence, merely knowing the concentration of a chemical is not likely to produce useful management information.

Table 4.4 The number of replicate microcosm or mesocosm systems required to distinguish means that differ by 25%, 50% or 100% at the 0.05 significance level

Response variable	Difference to be detected (%)	Number of replicates required	References
Ecosystem production	50	3	Abbott, 1966; Giddings and Eddlemon, 1978; Leffler, 1981; Sheehan et al., 1986
	25	6–8	
Ecosystem respiration	50	3	Abbott, 1966; Giddings and Eddlemon, 1978; Leffler, 1981; Sheehan et al., 1986
	25	3–7	
Primary productivity	100	4	Kuiper, 1977; Sheehan et al., 1986
	50	13	
	25	51	
Acetate uptake	100	3	Sheehan et al., 1986
	50	11	
	25	43	

In addition to these, one additional factor has been noted:

4   Ecological interactions as well as chemical and physical reactions give structure to the biological community.

The essence of biological monitoring is that one cannot protect the quality of a natural system without obtaining information directly about the condition of the biota and the ecosystem.

Considerations for monitoring studies

Not all chemical, physical and biological variables can be monitored in a study. Typically, very few variables are measured and the duration of monitoring is short, often less than one year. Practical constraints, time, money, and manpower have often dictated the scope of monitoring programs. The ecosystems monitored are generally determined by size or location without consideration of external influences which might impact system response to chemical exposure.

An appropriate selection of monitoring variables should be based on the study hypothesis and the type, stage of development, and general properties of the ecosystem as well as on the known properties of the toxicant or effluent.

As stated previously, the appropriate question or hypothesis is the key element of any toxic chemical impact study. This fact cannot be over-emphasized. Where possible, a stated hypothesis on chemical impacts should be tested as part of the monitoring study design. However, in those cases where conditions are not applicable to an experimental design providing an acceptable 'error' term, a test of hypothesis cannot be performed. Many short-term, point source monitoring studies, where single contaminated and uncontaminated systems are compared, fall into this untestable category. In these cases, the question of differences in the two ecosystems cannot be appraised with inferential statistics. If differences or changes are substantial, reporting of sampling data should provide a convincing case for chemical impacts. If only moderate differences are found between ecosystems, detailed supporting information on the species affected and on the released chemicals will be necessary to make a definitive link between chemicals and impacts.

Ecotoxicological monitoring studies should be based on the appropriate ecological theory and characteristics. Ecosystem factors to be considered include the following: biological energy flow; material or nutrient cycling; ecological regulation and key ecological interactions; populations and communities likely to be exposed; presence of species of commercial interest; developmental stage of the ecosystem (seasonal and successional).

Energy and materials movement can be related to specific functional aggregations. For instance, algal and macrophytic primary producers provide a large portion of the energy base in lake ecosystems but may provide only a limited energy base in many stream ecosystems where nutrition for consumers comes primarily from allochthonous material (tree leaves and organic debris). Therefore, emphasis on monitoring the quantitative effects of chemicals on algal and macrophyte communities and their primary consumer communities to determine impact on energy flow is well supported for most lake but not stream ecosystems. Analysis of chemical effects on bacterial decomposers and macroinvertebrate consumers would be more applicable to interpreting changes to energy flow and materials movement in chemically exposed stream ecosystems.

The functional role of species in regulating ecological processes is not well understood. The consumer complex acts to control the producer and decomposer components of the ecosystem, and predator populations are limited by the reduction in energy as it passes through the food web and by the availability of essential materials. Thus, there are physical and chemical limits to the length of the food chain and any further elaboration is biologically regulated. This control is achieved through a variety of competitive and trophic interactions and through feedback from specific interdependent relationships. Where the role of functional regulators has been defined, such as microbes and macroinvertebrates in energy processing in a heterotrophic stream (Cummins, 1974), the effects of chemicals on regulation of the flow of both energy and materials can be assessed.

Monitoring studies have seldom examined the effects of toxic chemicals on key population interactions. Much of the work in this area has come from pest control studies. Consumer–consumer interactions are exemplified by species population explosions upon elimination of natural predators due to pesticide exposures (Pimentel, 1972). Chemically induced perturbations have been observed to be passed up the food chain where reduced primary productivity leads to reduced numbers of primary consumers and reduced growth of secondary consumers. These interactions are more likely to be observed in ecosystems where prey switching behavior is unlikely because the selection of prey is limited or specialized feeding relationships are strongly developed.

Limits to interpretive strength that can be identified for any community analysis are associated with the functional redundancy of each taxonomic group included in the analysis. Resh and Unzicker (1975) have argued that the identification of species is critical to the interpretation of community response data. Although species identity is critical to describing population interactions, community functions such as primary productivity may be assessed with little knowledge of specific taxonomic composition of the algal community (Sheehan et al., 1986).

Analysis of distributions of species which are commercially important to man are often included in monitoring studies. The reasons for this interest are obvious. However, monitoring data on commercial species by itself may not contribute substantially to interpreting changes in chemically exposed ecosystems.

The identification and monitoring of 'keystone' species is likely to provide useful data on community and system-level responses. A keystone species not only indicates community characteristics but also plays a role in controlling community structure or function out of proportion to its numbers or size. Linkage between species, population interactions, food web compartmentalization, and trophic complexity are essential elements for the selection of 'target' species when elements of community structure or function are integrated into monitoring studies.

Lastly, knowledge of the stage of successional development of the ecosystem and its condition within seasonal cycles can aid in the selection of populations to be monitored. Ecosystems in the earlier stages of succession are dominated by opportunistic species and are largely governed by density-independent population regulation. The successional state of a system has been described as an inverse function of stress (Regier and Cowell, 1972) and less developed ecosystems are thought to be more resistant to pollutant effects (Grassle and Grassle, 1974; Leppakoski and Lindstrom, 1978).

Change can only be detected and measured in the populations or communities surveyed. There is danger in concentrating monitoring efforts on restricted sections of the ecosystem. A measure of response in one community may not be directly applied to predict what is occurring in other communities. For example, adverse chemical effects on primary producers may or may not carry over to consumers. If only primary productivity is measured, effects on the consumer communities remain undocumented. Restrictive approaches which look only at an indicator species or pollution index are likely to provide incomplete data for interpreting ecosystem response to stress. An ecosystem monitoring approach which includes measures of producer, consumer and decomposer response provides a much higher probability of successful ecological interpretation and quantification of chemically induced ecological changes than does the often applied pollution index approach.

Monitoring approaches

Cairns and colleagues published a series on biological monitoring procedures including reviews on early warning systems, methods based on community structure and function, as well as future needs (Cairns and Van der Schalie, 1980; Cairns, 1981; Herricks and Cairns, 1982; Matthews et al., 1982). These reviews followed on the earlier work of Hellawell (1977). These authors have emphasized the need for prolonged surveillance programs. Recent guidance has also been offered (NAS, 1990).

Some of the best examples of biological monitoring programs are associated with assessing the recovery of chemically polluted ecosystems. For example, Rosenberg (1976) presented a comprehensive report on benthic faunal dynamics following abatement of sulfite pulp mill effluent into the Staltkallefjord in Sweden. His study examined chemical and biological data during two years of increasing effluent pollution and then through eight years of decreasing effluent pollution. He measured species richness, organism density and biomass, and percentage composition, as well as calculated diversity and similarity indices for samples from five distinct regions of the fjord over the multi-year monitoring period. The Rosenberg study provided the following insights:

- density and biomass increased rapidly during the initial recovery phase but levels fluctuated with species recruitment.
- the amplitude changes in relative abundance were larger and took longer to approach equilibrium in more polluted systems.
- diversity peaked early in the recovery period and then declined bringing into question the value of this index.
- similarity measures showed an unambiguous trend in ecosystem recovery.
- the reduced fluctuations in species numbers near the end of the recovery period indicated that the system had returned to a state of 'constancy'.
- the relatively rapid recolonization of the same group of species recorded before effluent pollution indicated that the ecosystem was highly elastic.

Legge and colleagues provide a good example of a monitoring program to define the effects of toxic chemicals released into terrestrial ecosystems (Legge, 1980; Legge et al., 1981; Legge, 1982). Sulfur gas emissions were continuously monitored at both the incineration and the flare stacks at the West Whitecourt Gas Plant in Alberta for more than ten years. Intensive on-site air quality monitoring was undertaken to measure the varying concentrations of SO_2 reaching the pine forest at chosen experimental sites. Analogous sampling locations were chosen based on ecological variables such as slope aspect, soil type, species density and diversity, and environmental variables other than chemical pollutant concentrations. Statistical analysis of tree growth data revealed that distance from the sulfur source, time (in years) and their interaction had a significant effect on the woody production of pine trees. This example shows that long-term studies may be necessary to quantify subtle, sub-lethal changes in terrestrial ecosystems.

It is obvious from the examples that successful monitoring programs require considerable planning and sampling efforts. Green (1979) provides some excellent guidance in undertaking biological monitoring studies. These principles are presented in Table 4.5.

Table 4.5   Ten principles to guide field environmental studies

1. Be able to state concisely to someone else what question you are asking. Your results will be as coherent and as comprehensible as your initial conception of the problem.

2. Take replicate samples within each combination of time, location, and any other controlled variable. Differences among can only be demonstrated by comparison to differences within.

3. Take an equal number of randomly allocated replicate samples for each combination of controlled variables. Putting samples in 'representative' or 'typical' places is not random sampling.

4. To test whether a condition has an effect, collect samples both where the condition is present and where the condition is absent but all else is the same. An effect can be demonstrated by comparison with a control.

5. Carry out some preliminary sampling to provide a basis for evaluation of sampling design and statistical analysis options. Those who skip this step because they do not have enough time usually end up losing time.

6. Verify that your sampling device or method is sampling the population you think you are sampling, and with equal and adequate efficiency over the entire range of sampling conditions to be encountered. Variation in efficiency of sampling from area to area biases among-area comparisons.

7. If the area to be sampled has a large-scale environmental pattern, break the area up into relatively homogeneous sub-areas and allocate samples to each in proportion to the size of the sub-area. If it is an estimate of total abundance over the entire area that is desired, make the allocation proportional to the number of organisms in the sub-area.

8. Verify that your sample unit size is appropriate to the sizes, densities, and spatial distributions of the organisms you are sampling. Then estimate the number of replicate samples required to obtain the precision you want.

9. Test your data to determine whether the error variation is homogenous, normally distributed, and independent of the mean. If it is not, as will be the case for most field data, then (a) appropriately transform the data, (b) use a distribution-free (nonparametric) procedure, (c) use an appropriate sequential sampling design, or (d) test against simulated H_o data.

10. Having chosen the best statistical methods to test your hypothesis, stick with the result. An unexpected or undesired result is not a valid reason for rejecting the method and hunting for a 'better' one.

Source: Green (1979)

Manipulative studies

Although biological monitoring may be employed to quantify the degeneration and recovery of ecosystems put under and relieved of chemical stress, these studies often do not provide insights into the chemical–biological interactions leading to the altered structure or function of the system. On the other hand, field studies where manipulation techniques are employed may be used to test specific hypotheses related to ecological responses to perturbations. In stream systems, Peckarsky has used cage studies to examine effects of invertebrate density on colonization rate and predator–prey interactions (Peckarsky, 1979, 1980) and Benke (1978) investigated competition among dragonfly larvae in a small pond using cages. Leaf bags have been used to assess heterotrophic processes and estimate rates of energy and material transfer in streams (Peterson and Cummins, 1974). Similar manipulative techniques have been applied in terrestrial studies (Suter, 1982). Although these techniques have not been widely applied in studies of chemical pollution, their utility as tools to investigate system and community-level responses to stress would appear to be good.

Community and ecosystem toxicity testing

Toxicity testing is a predictive tool used to provide data on the potential hazards of chemicals. The test system is generally a laboratory model. There is a great deal of disparity between the ecological complexity of the common single-species toxicological test systems and the natural environment. A small number of species is generally tested in isolation to predict safe exposure levels of chemicals or effluents for the vast number of species in natural systems. Single-species toxicity tests do not take into account biological interactions which may affect a population response to toxic chemicals. Cairns (1985) argues that if we are merely endeavoring to protect certain species that are important to the general public from direct injury from waste discharges, single-species tests are adequate if validated in natural systems. If, however, single-species tests are to protect the ecosystem, Cairns (1983) argues that there is insufficient evidence available to determine the accuracy of predictions of toxicological responses from one level of biological organization to another. All these facts suggest that predictions of impact cannot be made solely on the basis of data generated by single-species toxicity tests.

There is an essential need for community-level tests of chemical toxicity not only to predict safe environmental concentrations but also to predict impacts on ecosystem structure and function. Several advantages may be realized using community-level toxicity testing models (Cairns, 1986):

1   Validation in complex natural systems is less difficult because one will be carrying out the prediction and the validation at more comparable levels of biological organization than is the case when single-species tests are used and the results extrapolated to the response of a complex natural system.
2   Critical response thresholds can be measured directly instead of using extrapolations from single-species tests that are not sufficiently complex or high enough in environmental realism to make direct measurement possible.
3   Community-level testing is less expensive than was once thought.

When the science develops further, these tests will probably be only a little more expensive than some of the elaborate single-species tests now in vogue.

4   Because of the larger number of organisms involved, errors caused by the extreme sensitivity or tolerance of a single species to the test material are eliminated or markedly reduced.

5   Although the conventional wisdom is that community-level tests are more sensitive than single-species tests because a larger array of species would include some with greater sensitivity, the functional redundancy built into complex systems may well prove this assumption false.

Arguments for inclusion of community-level toxicity testing in hazard assessment programs are not intended to be arguments for the replacement of single-species toxicity tests. Single-species testing has been and shall remain an integral part of hazard assessment. Single-species tests are the best source of information on effects on growth, reproductive success, behavior and other endpoints and their implications at the population level. Rather, arguments are for a mixture of toxicity tests at different levels of biological organization to improve the efficacy of the hazard assessment process.

This section now examines the technical approaches to community and ecosystem-level toxicity testing.

Multispecies testing

An initial review of laboratory multispecies tests for aquatic and terrestrial systems was provided by Hammons (1981). Two recent books have come from workshops examining the efficacy of multispecies toxicity test models in assessing the hazards of chemicals (Cairns, 1985, 1986). The first book, *Multispecies Toxicity Testing*, presents perspectives from representatives of industry, regulatory agencies and research groups. These papers identify expectations and perceived problems with multispecies testing from the varied perspectives and present case studies identifying applications. The papers presented in *Community Toxicity Testing* (Cairns, 1986) describe and evaluate the application of a variety of aquatic community-level toxicity test models. The three compendia provide an overview of the applicability and utility of multispecies toxicity testing.

A variety of multispecies test models have been proposed. The test models range in size and complexity from simple laboratory flask systems to large-scale complex field enclosure systems and whole-ecosystem studies. Multispecies tests have been proposed for freshwater and marine planktonic communities, freshwater and marine benthic invertebrate communities, stream communities, and terrestrial plant and soil faunal communities.

The common link between multispecies test models is that these tests have a level of organization higher than that of the single species and can be used to evaluate endpoints outside the general scope of single species test procedures.

Dickson and colleagues (1985) listed the potential uses of multispecies tests:

- identification of species interactions or compensation that would be missed in single species tests
- illustration of physical–chemical pathways
- calibration and verification of mathematical models of ecosystems
- assessment of population, community, and ecosystem-level responses to toxicants (i.e. structure, persistence, productivity, energy flow, nutrient cycling)
- elucidation of particular mechanisms and interactions
- enabling the study of functioning ecosystems with cybernetic or negative feedback loops in place
- identification of sensitive or critical species that should be used in single-species toxicity screening tests
- testing of hypotheses

In the following sections aquatic multispecies toxicity tests are examined in greater detail at three levels of complexity: generic microcosm test models, mesocosm test models and ecosystem-level field studies.

Generic aquatic microcosm toxicity tests

The suggestion of a community-level 'white rat' screening test for the ecological effects of chemicals in aquatic ecosystems has prompted the development and evaluation of a number of protocols for microcosm toxicity tests. A generic test system is based on the premise that small aquatic microcosms containing communities of algae, protozoa, bacteria and zooplankton exhibit typical ecosystem properties or processes but simulate no particular natural aquatic system.

A response model for generic microcosm toxicity test systems was proposed by Leffler (1980) and modified by Sheehan et al. (1986). This response model is presented in Figure 4.3. This model graphically describes indices of change for chemically treated systems as measured against the normal operating range of the control systems. The control operating range is defined as its response mean ± 1 Standard Error (SE). The response of treatment replicates may be assessed in terms of: (1) maximum change in treatment response range (mean ± 1 SE) from the control response range; (2) system resilience defined as the time required for the treatment range once displaced from the normal range to return; and (3) the relative impact measured as the integral difference between the perturbed trajectory of the treatment range and that of the control range. This model shows that unlike single species toxicity test data, community toxicity test data are best analyzed as a time–response curve.

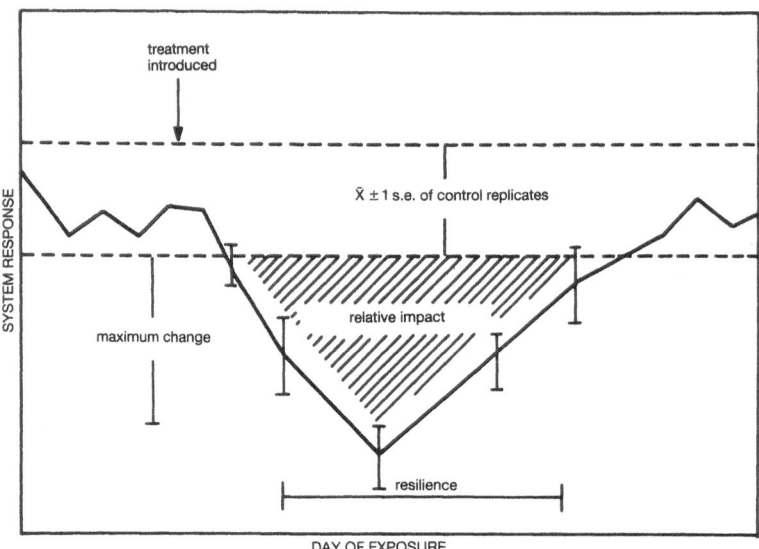

Figure 4.3 Hypothetical response model based on assessing impact in terms of changes for chemically treated systems as measured from the normal operating range of control systems

Taub and her colleagues have evaluated a gnotobiotic model which establishes experimental communities by controlled additions of organisms from monocultures (Taub *et al.*, 1982, 1986; Conquest and Taub, 1989). These standardized aquatic microcosms (SAM) have been successfully tested in four laboratories under similar conditions with copper as a reference toxicant. All experiments exhibited the same sequence of events, but the timing of these varied between experiments. Experiments provided similar statistical differences between controls and treatments within the same experiment and gave the same rank order of variables. The Taub protocol has been shown to provide data on producer and consumer communities under chemical stress and to be reproducible in interlaboratory tests. However, the gnotobiotic model has been criticized because the artificial community may not represent a co-adapted species assemblage and may not be reliable for studies of ecosystem-level properties (Hammons, 1981).

The mixed flask culture (MFC) toxicity test model was developed by Leffler (1981) and has been evaluated by Shannon *et al.* (1986). This protocol uses microcosms containing a community of organisms allowed to develop from a variety of sources over a period of time (co-adapted assemblage). Shannon and colleagues reported that the response to toxicants was similar in the SAM and MFC microcosms. The SAM protocol provided considerably more insight into changes in population density but was much more labor intensive than the MFC protocol.

Sheehan (1989) evaluated the 'generic' nature of a simple microcosm model using co-adapted species assemblages from lakes and ponds. His studies suggested that the metabolic response of these generic systems to exposures to toxic metals and pesticides was consistent in ranking hazards to the microcosm communities. Metabolic response variables such as ecosystem production, primary productivity, and bacterial activity consistently ranked the potential hazards of chemical exposure. Tests of the influence of taxonomic composition on hazard ranks indicated that community inocula from different natural sources produced consistent ranking of chemical hazards of a selected group of pesticides.

Evaluations of simple aquatic microcosm test models have demonstrated the following:

● structural and functional response variables can be effectively measured in small microcosm models.
● microcosm response is best measured as response curves reflecting changes in community variables with time rather than as a single response value at a specific time.
● there are no appropriate statistical techniques for comparing response curves, therefore the utility of these generic models is in demonstrating ecotoxicological effects and qualitatively ranking the potential hazard.
● rankings of hazard based on microcosm data appear to be reproducible in repeated tests, although the sensitivity to chemical stress and the shape of response curves may differ substantially between repeated tests with the same chemical or effluent.
● different generic models appear to provide similar insights into community-level changes induced by chemical exposure.

Generic aquatic microcosm toxicity tests will be most useful as methods to screen the potential ecological effects of toxic chemicals. The simplicity of these models should make their application cost effective but their lack of realism reduces their utility as quantitative predictors of change.

Aquatic mesocosm toxicity tests

A number of larger laboratory and field enclosure toxicity test models have been evaluated. These models incorporate more complex species assemblages including organisms of larger size, and more natural environmental conditions than do microcosm models. The greater realism of these models is expected to provide a more acceptable data base for the quantitative prediction of chemical hazard.

The use of *in situ* aquatic enclosures or 'limnocorrals' as toxicity test models has been extensively evaluated by scientists at the University of Guelph, Ontario, Canada. This group has examined the response of the enclosed aquatic ecosystems to several pesticides. With the limnocorral model, the dissipation of the applied chemicals can be monitored allowing budgets to be developed for pesticides in the system (Solomon et al., 1985). This process allows for the field validation of laboratory and theoretical models on the environmental behavior of toxic chemicals. The biological results of these limnocorral toxicity tests demonstrated the following (Kaushik et al., 1985; Herman et al., 1986; Kaushik et al., 1986; Stephenson et al., 1986):

- effects on various groups of plankton are exposure-dependent and sensitive species can be identified.
- effects on populations and communities can be demonstrated at environmentally relevant levels of pesticide exposure.
- effects on trophic interaction can be demonstrated.
- lagtime in recovery of populations and system-level functional response variables (primary productivity) from single or multiple pesticide exposures can be quantified.

Mesocosm studies of the type described appear to have utility in quantitatively predicting the damage likely from toxic chemical exposure. They may also be used to estimate the ecological buffering capacity and elasticity of aquatic systems. Mesocosm tests are obviously more complex and costly than simple microcosm or single-species toxicity tests. This model however offers a means to test chemicals in a 'natural' set of conditions without damaging a larger natural system in the test process.

Whole ecosystem toxicity tests

A number of toxicity tests have been conducted in natural terrestrial and aquatic ecosystems. Obviously the exposure of an ecosystem to toxic chemicals for the purposes of testing should be well justified, as protection of such systems is often the ultimate goal. Ecosystem tests have generally been carried out in ponds or small lakes for the purposes of documenting the effects of chemicals on ecological functions not amenable to assessment in microcosm or mesocosm models. Similarly, many field scale experiments in the coastal marine environment have been conducted to assess the impacts of oil spills (NAS, 1989).

An example of circumstances requiring a whole-ecosystem toxicity test model is the problem of defining the hazard of pesticides to ducklings previously mentioned (Sheehan et al., 1987). Aside from direct toxic effects of insecticides on ducklings, they may be harmed indirectly if aquatic exposures reduce aquatic macroinvertebrate populations, a necessary food source. In whole-lake studies, Hunter et al. (1984) implicated carbaryl as an indirect cause of reduced duckling growth (and survival) due to high mortality of macroinvertebrates. This important chemical disruption of normal duckling feeding on macroinvertebrates could only have been demonstrated in a controlled field study at the scale of the lake or pond ecosystem.

Whole-ecosystem tests are expensive and difficult to conduct, and may damage the natural systems used for testing over long periods. However, this level of testing would appear justified to assess chemical impacts on community interactions or ecosystem properties not amenable to testing with smaller, less complex models.

Conclusions on multispecies testing

There has been some general agreement on the utility of multispecies testing. This consensus may be summarized as follows:

- multispecies tests can provide useful information in assessing the hazards of chemicals.
- the response variables measured in multispecies tests are not yet as decisive as the endpoints measured in single-species tests, but with further development they may ultimately be more related to endpoints needed to make decisions about protection of the biological integrity of ecosystems.
- multispecies tests, when designed appropriately relative to the hypothesis to be tested, can be powerful analytical tools. The ability to isolate and study mechanisms while retaining a certain amount of complexity is their strongest attribute.
- multispecies tests can indicate the transformation of a test chemical into products which may be of greater or lesser biological activity.
- multispecies tests can be used as a conceptual bridge between more complex systems and less complex systems.
- the 'state of the art' in developing multispecies test systems is adequately advanced to allow test systems to simulate and track specific processes in freshwater, marine, and terrestrial ecosystems.
- multispecies tests may be less sensitive to stress caused by a chemical when compared to the acute or chronic responses of specific test organisms. However, a new 'sensitivity' may be created because ecologically relevant processes can be measured; in this sense, the test system may be more sensitive.
- multispecies tests are more complex than those conducted with single species and may exhibit more variability in structure and function. This is not necessarily detrimental to their use. Multispecies test systems often emphasize realism more than reproducibility or replicability.
- the cost of conducting multispecies tests is highly variable depending upon the design of the test. However, some multispecies test systems designed to provide information on fate and effects of chemicals are

cost-competitive with more traditional laboratory fate experiments and single-species effects tests used in hazard evaluations.

● as with single-species ecotoxicity tests, the selection of the response parameters used to assess the effects of a chemical or effluent in a multispecies test can influence the interpretation of the results.

Ecological risk assessment

The process of ecological hazard evaluation has been largely unstructured and heavily dependent on the knowledge, experience, and intuition of the evaluator to draw together exposure and effects data into a qualitative estimate of potential hazard. An effective evaluation scheme must involve a co-ordinated analysis of both environmental fate and laboratory (and field) toxicity data. It also requires a decision-making process based on acceptable criteria as well as the experience of the evaluator to provide confidence in hazard rankings and risk predictions.

The objective is an evaluation framework with efficient tools that provide consistent and accurate predictions of hazard. The model should have general utility in predicting direct impacts. A suitable framework should facilitate an analysis of indirect impacts and specific exposure situations with proper modification.

Evaluation frameworks involving computer modeling techniques are emerging in the fields of environmental hazard/risk analysis. Within these approaches simulated environmental concentrations are compared to concentrations producing toxic responses in laboratory tests. These models are a systematic approach to extrapolation which permits explicit consideration of all the assumptions that go into the extrapolation. In the initial stages of development, these models of ecological hazard can only be utilized as screening tools. It is important to realize that such hazard assessment models must, of necessity, be a compromise of realism (complexity) to provide accuracy of prediction, simplicity of use and generality of results. At the screening level, these models should require a minimum of essential input data. In addition, they must be considered as qualitative predictors of relative hazard, dependent upon the development of benchmark data substantiated by field studies to provide boundaries for rankings of relative risk and an estimate of the extent of hazard. A preliminary hazard evaluation should provide a basis for further calculation of the probability of risk for important species, communities or ecological processes. Ultimately, a hazard evaluation framework will be judged more acceptable than current procedures if predictions based on its output are more consistent and accurate than current forecasts.

For the purpose of evaluating chemical impacts, hazard modeling involves identifying the segment of the environment and taxa or ecological processes most likely to be affected by the toxic chemical or effluent. The magnitude and persistence of expected chemical concentrations in the environment are compared with concentrations toxic to representative taxa or community response variables to rank the relative risk of adverse impact (Kenaga, 1982). Ecological risk analysis is defined as the process of identifying and quantifying the probabilities of adverse changes in a population, community or ecosystem resulting from the release of the chemical. The objectives of risk analysis are thus both identification of problem situations, including relevant interactions, and the prediction of risk in probability terms (see also Chapter 5).

Hazard assessment models have taken two forms: (1) subjective scoring systems quantifying the consensus of expert evaluators (e.g. scoring, Welch and Ross, 1982; analytical hierarchy method, Barnthouse et al., 1982), and (2) applied tools for the evaluation employing exposure simulation and extrapolation of toxicity benchmarks (e.g. relative hazard evaluation, Zitko and McLeese 1980; chemical migration and risk assessment (CMRA), Parkhurst et al., 1981; analysis of error extrapolation, Suter et al., 1983; relative hazard index, Sheehan et al., 1987). Of particular interest are those techniques that fall within the latter category. They provide the evaluator with a conceptual framework as well as modeling and mathematical tools for estimating the level of hazard and for analyzing some of the sources of uncertainty inherent in these estimates. The analytical tools take the form of computer models to simulate exposure in various receiving systems (e.g. PERSISTENCE, NRCC 1981b; EXAMS, Burns et al., 1982; SERATRA, Onishi and Wise 1982; FUGACITY, Mackay and Paterson 1982; TOXIWASP, Ambrose et al., 1983) and mathematical and statistical techniques to facilitate comparison of environmental concentration and toxicological benchmarks to categorize the severity of potential damage. Complex system models incorporating ecological and toxicological interaction are also in the initial stages of development and evaluation (e.g. O'Neill et al., 1982, 1983).

Although hazard modeling appears promising, there is, not surprisingly, no single accepted model and the few currently proposed models have not been fully tested. Therefore the selection of an appropriate ecological risk assessment approach will depend on first, defining the criteria for the model and second, gathering appropriate data for the selected model. Solving the problem of gathering accurate and sufficient data and establishing relevant benchmarks will be paramount to maximizing the utility of mathematical models in the ecological risk assessment process.

Current trends

The US Environmental Protection Agency (EPA) is moving towards the development of guidelines for ecological risk assessment. In 1988 the EPA commissioned reports on the endpoints for ecological toxicology (TRI, 1988) and the trends and limitations of ecological assessment methods as currently employed by state and federal agencies (ICF, 1988). These reports provide a baseline understanding of the processes and general applications of assessment methodologies but do not go as far as to provide a step-by-step approach for conducting ecological risk assessments. Agency recognition of unacceptable population, community and ecosystem impacts from chemical exposure is the next step towards developing a regulatory framework incorporating ecological assessments.

Conclusions

A substantial amount of scientific information has been gained on the response of various components of ecological systems to chemical stress. There have been recent scientific efforts to design hazard assessment programs that integrate exposure information with effects information at the population, community and ecosystem levels. The term ecotoxicology correctly identifies this hazard assessment approach. Although there has been little indication as to how readily community and ecosystem endpoints will be incorporated into current regulations in order to better protect the environment, the efficacy of the hazard assessment process will be improved with the incorporation of well designed biological monitoring programs and multispecies toxicity tests to supplement single-species toxicity tests and chemical fate data in the regulatory process.

References

Abbott, W. (1966), 'Microcosm studies on estuarine waters: 1. The replicability of microcosms', *Journal of the Water Pollution Control Federation* 38, pp. 258–70.

Allan, D. J. (1984), 'Hypothesis testing in ecological studies of aquatic insects', in V. H. Resh and D. M. Rosenberg (eds), *The Ecology of Aquatic Insects*, (New York: Praeger), pp. 484–507.

Ambrose, R. B. Jr, Hill, S. I. and Mulkey, L. A. (1983), *User Manual for the Chemical Transport and Fate Model (TOXIWASP)*, version 1, EPA 600/3–83–005 (Athens, Georgia: US Environmental Protection Agency).

Baker, J. M. (1973), 'Recovery of salt marsh vegetation from successive oil spillages', *Environmental Pollution* 4, pp. 223–30.

Barnthouse, L. W., DeAngelis, D. L., Gardner, R. H., O'Neill, R. V., Powers, C. D., Suter II, G. W. and Vaughn, D. S. (1982), *Methodology for Environmental Risk Analysis*, Environmental Science Division Publication No. 2023, ORNL/TM8167 (Oak Ridge, Tennessee: Oak Ridge National Laboratory).

Benke, A. C. (1978), 'Interactions among coexisting predators – A field experiment with dragonfly larvae', *Journal of Animal Ecology* 47, pp. 335–50.

Burns, L. A., Cline, D. M. and Lassiter, R. R. (1982), *Exposure Analysis Modeling System (EXAMS), User Manual and System Documentation*, EPA 600/3–82–023 (Environmental Research Laboratory, Athens, Georgia: US Environmental Protection Agency).

Cairns, J. Jr (1981), 'Biological monitoring Part VI – Future needs', *Water Research* 15, pp. 941–52.

Cairns, J. Jr (1983), 'Are single-species toxicity tests alone adequate for estimating environmental hazard?', *Hydrobiologia* 100, pp. 47–57.

Cairns, J. Jr (ed.) (1985), *Multispecies Toxicity Testing* (New York: Pergamon Press).

Cairns J. Jr (ed.) (1986) *Community Toxicity Testing*, ASTM STP 920 (Philadelphia: American Society for Testing and Materials).

Cairns, J. Jr, Lanza, G. R. and Parker, B. C. (1972), 'Pollution related structural and functional changes in aquatic communities with emphasis on freshwater algae and protozoa', *Proceedings of the Academy of Natural Sciences Philadelphia* 124, pp. 79–127.

Cairns, J. Jr and van der Schalie, W. H. (1980), 'Biological monitoring Part I – Early warning systems', *Water Research* 14, pp. 1179–96.

Carney, R. S. (1987), 'A review of study designs for the detection of long-term environmental effects of offshore petroleum activities', in D. F. Boesch and N. N. Rabalais (eds), *The Long-Term Effects of Offshore Oil and Gas Development: an Assessment and a Research Strategy* (Barking, Essex: Applied Science), pp. 651–96.

Chutter, F. M. and Noble, R. G. (1966), 'The reliability of a method of sampling stream invertebrates', *Archiv. für Hydrobiologie* 62, pp. 95–103.

Colebrook, J. M. (1978), 'Continuous plankton records: zooplankton and environment, northeast Atlantic and North Sea, 1948–1975', *Oceanol. Acta* 1, pp. 9–22.

Conquest, L. L. and Taub, F. B. (1989), 'Repeatability and reproducibility of the standardized aquatic microcosm: statistical properties', in *Aquatic Toxicology and Hazard Assessment*, Vol. 12, ASTM STP 1027 (Philadelphia: American Society of Testing and Materials), pp. 159–77.

Cummins, K. W. (1974), 'Structure and function of stream ecosystems', *Bioscience* 24, pp. 631–41.

deNoyelles, F., Kettle, W. D. and Sinn, D. E. (1982), 'The responses of phytoplankton communities in experimental ponds to atrazine, the most heavily used pesticide in the United States', *Ecology* 63, pp. 1285–93.

Dickson, K. L., Duke, T. and Loewengart, G. (1985), 'A synopsis: Workshop on multispecies toxicity tests', in Cairns 1985, *op. cit.*, pp. 248–53.

Ford, J. (1989). 'The effects of chemical stress on aquatic species composition and community structure', in S. Levin, M. Harwell, J. Kelly and K. Kimball (eds), *Ecotoxicology: Problems and Approaches* (New York: Springer-Verlag), pp. 99–144.

Giddings, J. M., and Eddlemon, G. K. (1978), 'Photosynthesis/respiration ratios in aquatic microcosms under arsenic stress', *Journal of Water, Air and Soil Pollution* 9, pp. 207–12.

Giesy, J. P., and Allred, P. M. (1985), 'Replicability of aquatic multispecies test systems', in Cairns 1985, *op. cit.*, pp. 187–243.

Gilbert, R. O. (1987), *Statistical Methods for Environmental Pollution Monitoring* (New York: Van Nostrand Reinhold).

Grassle, J. F. and Grassle, J. P. (1974), 'Opportunistic life histories and genetic systems in marine benthic polychaetes', *Journal Marine Research* 32, pp. 253–84.

Green, R. H. (1979), *Sampling Design and Statistical Methods for Environmental Biologists* (New York: John Wiley and Sons).

Green, R. H. (1984), 'Statistical and nonstatistical considerations for environmental monitoring studies', *Environmental Monitoring and Assessment* 4, pp. 293–301.

Hammons, A. S. (1981), *Methods for Ecological Toxicology – A Critical Review of Laboratory Multispecies Tests* (Michigan: Ann Arbor Science Publishers, Inc).

Harte, J., Levy, D., Rees, J. and Saegebarth, E. (1980), 'Making microcosms an effective assessment tool', in J. P. Giesy Jr (ed.), *Microcosms in Ecological Research*, CONF–781101, (NTIS, Springfield, Virginia: Technical Information Center, US Department of Energy), pp. 105–37.

Hellawell, J. M. (1977), 'Change in natural and managed ecosystems: detection, measurement, and assessment', Proceedings Royal Society London, *Biological Science* 197, pp. 31–56.

Herman, D., Kaushik N. K. and Solomon, K. R. (1986), 'Impact of atrazine on periphyton in freshwater enclosures and some ecological consequences', *Canadian Journal of Fisheries and Aquatic Sciences* 43, pp. 1917–25.

Herricks, E. E. and Cairns Jr, J. (1982), 'Biological monitoring Part III – Receiving system methodology based on community structure', *Water Research* 16, pp. 141–53.

Holling, C. S. (1973), 'Resilience and stability of ecological systems', *Annual Review of Ecology and Systematics* 4, pp. 1–24.

Hunter, M. L. Jr, Witham, J. W. and Dow, H. (1984), 'Effects of carbaryl-induced depression in invertebrate abundance on the growth and behavior of American black duck and mallard ducklings', *Canadian Journal of Zoology* 2, pp. 452–6.

Hurlbert, S. H. (1984), 'Pseudo-replication and the design of ecological field studies', *Ecological Monographs* 54, pp. 187–211.

Hutton, M. (1984), 'Impact of airborne metal contamination on a deciduous woodland system', in Sheehan *et al.* (eds) 1984a, *op. cit*, pp. 365–75.

Hynes, H. B. N., Williams, D. D. and Williams, N. E. (1976), 'Distribution of the benthos within the substratum of a Welsh mountain stream', *Oikos* 27, pp. 307–10.

ICF (1988), US Environmental Protection Agency, *Review of Ecological Assessment Methods*. Prepared by ICF Incorporated for *Office of Planning and Evaluation*. EPA 10230–10–88–041.

Kaushik, N. K., Solomon, K. R., Stephenson, G. L. and Day, K. E. (1986), 'Use of limnocorrals in evaluating the effects of pesticides on zooplankton communities', in Cairns 1986, *op. cit*, pp. 269–90.

Kaushik, N. K., Stephenson, G. L., Solomon, K. R. and Day, K. E. (1985), 'Evaluation of the impacts of permethrin on zooplankton communities using limnocorrals II. Impact on zooplankton', *Canadian Journal of Fisheries and Aquatic Sciences* 42, pp. 77–85.

Kelly J. R. and Harwell, M. A. (1989), 'Indicators of ecosystem response and recovery', in S. Levin, M. Harwell, J. Kelly and K. Kimball (eds), *Ecotoxicology: Problems and Approaches* (New York: Springer-Verlag), pp. 8–35.

Kenaga, E. E. (1982), 'The use of environmental toxicology and chemistry data in hazard assessment: Progress, needs, challenges', *Environmental Toxicology and Chemistry* 1, pp. 69–79.

Kerfoot, W. C. and DeMott, W. R. (1980), 'Foundations for evaluating community interactions: the use of enclosures to investigate co-existence of *Daphnia* and *Bosmina*', in W. C. Kerfoot (ed.), *Evolution and Ecology of Zooplankton Communities*, Spec. Symp. No. 3, American Society of Limnology and Oceanography. (Hanover: University Press of New England).

Krebs, C. J. (1972), *Ecology: The Experimental Analysis of Distribution and Abundance* (New York: Harper and Row).

Kuiper, J. (1977), 'An experimental approach in studying the influence of mercury on a North Sea coastal plankton community', *Helgoländer Wiss. Meeresunter* 30, pp. 652–65.

Lawson, T. J. and Grice, G. D. (1977), 'Zooplankton sampling variability: Controlled ecosystem pollution experiment', *Bulletin of Marine Science* 27, pp. 80–4.

Leffler, J. W. (1980), 'Microcosmology: Theoretical applications of biological models', in J. P. Giesy (ed.), *Microcosms in Ecological Research*, DOE Symposium Series (Technical Information Center, US Department of Energy), pp. 14–29.

Leffler, J. S. (1981), *Aquatic Microcosms and Stress Criteria for Assessing Environmental Impact of Organic Chemicals*, EPA/68–01–5043 (Washington, DC: Office of Toxic Substances, US Environmental Protection Agency).

Legge, A. H. (1980), *Primary Productivity, Sulfur Dioxide and the Forest Ecosystem: An Overview of a Case Study.* In Proceedings of Symposium on Effects of Air Pollutants on Mediterranean and Temperate Forest Ecosystems, United States Department of Agriculture, Forest Services, Pacific Southwest Forest and Range Experimental Station, General Technical Report PSW 43, Berkeley, California, pp. 51–62.

Legge, A. H. (1982), 'Sulphur gas emissions in the Boreal Forest: The West Whitecourt case study Part II: Research Plan and Background', *Journal of Water, Air, and Soil Pollution* 17, pp.379–98.

Legge, A. H., Jaques, D. R., Harvey, G. W., Krouse, H. R., Brown, H. M., Rhodes, E. C., Nosal, M., Schellhase, H. V., Mayo, J., Hartgerink, A. P., Lester, P. F., Amundson, R. G. and Walker, R. B. (1981), 'Sulphur gas emissions in the Boreal Forest: The West Whitecourt case study Part I: Executive Summary', *Journal of Water, Air and Soil Pollution* 15, pp. 77–85.

Leppakoski, E. J. and Lindstrom, L. S. (1978), 'Recovery of benthic macrofauna from chronic pollution in the sea area off a refinery plant, southwest Finland', *Journal of the Fisheries Research Board of Canada* 35, pp. 766–75.

Lorz, H. W. and McPherson, B. P. (1977), *Effects of Copper and Zinc on Smoltification of Coho Salmon*, EPA–600 13–77–032 (Environmental Research Laboratory, Corvallis, Oregon: US Environmental Protection Agency).

Luoma, S. N. (1977), 'Detection of trace contaminant effects in aquatic ecosystems', *Journal of the Fisheries Research Board of Canada* 34, pp. 436–9.

Mackay, D. and Paterson, S. (1982), *Fugacity Models for Predicting Environmental Behavior of Chemicals*. Report to Environment Canada under Contract OSU 81–00163.

Matthews, R. A., Buikema, A. L., Cairns, J. and Rodgers, J. H. (1982), 'Biological monitoring Part IIA - Receiving system functional methods, relationships and indices', *Water Research* 16, pp. 129–39.

Matthews, R. A., Elliot, P. D., Anderlini, V. C., Girvin, D., Jacobs, S. A., Risebrough, R. W., Delong, R. L. and Gilmartin, N. G. (1976), 'Mercury-selenium-bromine imbalance on premature parturient California sea lions', *Marine Biology* 35, pp. 91–104.

Muniz, I. P. and Leivestad, H. (1980), 'Acidification – effects on freshwater fish', in A. Tollan and D. Drablos (eds), *Ecological Impact of Acid Precipitation: Proceedings of an International Conference* (Sandefjord, Norway: SNSF Project), pp. 84–92.

Naiman, R. J. and Sibert, J. (1977), 'Annual and diel variations in a small marine bay: Interpretation of monitoring data', *Journal of Experimental Marine Biology and Ecology* 26, pp. 27–40.

NAS (National Academy of Sciences) (1989), *Using Oil Spill Dispersants on the Sea* (Washington, DC: National Academy Press).

NAS (National Academy of Sciences) (1990), *Managing Troubled Waters. The Role of Marine Environmental Monitoring* (Washington, DC: National Academy Press).

Needham, P. R. and Usinger, R. L. (1956), 'Variability in the macrofauna of a single riffle in Prosser Creek, California, as indicated by the Surber Sampler', *Hilgardia* 24, pp. 383–409.

Notini, M. (1978), 'Long-term effects of an oil spill on *Fucus* macrofauna in a small Baltic bay', *Journal of the Fisheries Research Board of Canada* 35, pp. 745–53.

NRCC (1981a), *Pesticide–Pollinator Interactions*, NRCC No. 18471, Associate Committee on Scientific Criteria for Environmental Quality (Ottawa, Canada: National Research Council Canada).

NRCC (1981b), *A Screen for the Relative Persistence of Lipophilic Organic Chemicals in*

Aquatic Ecosystems – An Analysis of the Role of a Simple Computer Model in Screening, NRCC No. 18570 (Ottawa, Canada: National Research Council Canada).

O'Neill, R. V., Gardner, R. H., Barnthouse, L. W., Suter, G. W., Hildebrand, S. G. and Gehrs, C. W. (1982), 'Ecosystem risk analysis: A new methodology', *Environmental Toxicology and Chemistry* 1, pp. 167–77.

O'Neill, R. V., Gartell, S. M. and Gardner, R. H. (1983), 'Patterns of toxicological effects in ecosystems: A modeling study', *Environmental Toxicology and Chemistry* 2, pp. 451–61.

Onishi, Y. and Wise, S. E. (1982), *User Manual for the Instream Sediment Contaminant Transport Model, SERTRA,* EPA 600/3–82–055 (Athens, Georgia: US Environmental Protection Agency).

Oviatt, C. A., Pilson, M. E. Q., Nixon, S. W., Frithsen, J. B., Rudnick, D. T., Kelley, J. R., Grassle, J. F. and Grassle, J. P. (1984), *Recovery of a Polluted Estuarine System: A Mesocosm Experiment.* Marine Ecology Progress Series 16, pp. 203–17.

Parkhurst, M. A., Onishi, Y. and Olsen, A. R. (1981), 'A risk assessment of toxicants to aquatic life using environmental exposure estimates and laboratory toxicity data', in D. R. Branson and K. L. Dickson (eds), *Aquatic Toxicology and Hazard Assessment,* ASTM STP 737 (American Society for Testing and Materials), pp. 59–71.

Peckarsky, B. L. (1979), 'Biological interactions as determinants of distributions of benthic invertebrates within the substrate of stony streams', *Limnology and Oceanography* 24, pp. 59–68.

Peckarsky, B. L. (1980), 'Predator–prey interactions between stoneflies and may-flies: Behavioral observations', *Ecology* 61, pp. 932–43.

Peterson, R. C. and Cummins, K. W. (1974), 'Leaf processing in a woodland stream', *Freshwater Biology* 4, pp. 343–68.

Pimentel, D. (1972), 'Ecological impact of pesticides', *Environmental Biology* 72, pp. 1–27.

Regier, H. A. and Cowell, E. B. (1972), 'Application of ecosystem theory, succession, diversity, stability, stress, and conservation', *Biological Conservation* 4, pp. 83–8.

Resh, V. H. (1979), 'Sampling variability and life history features: Basic consider-ations on the design of aquatic insect studies', *Journal of the Fisheries Research Board of Canada* 36, pp. 290–311.

Resh, V. H. and Unzicker, J. D. (1975), 'Water quality monitoring and aquatic organisms: The importance of species identification', *Journal of the Water Pollu-tion Control Federation* 47, pp. 9–19.

Rosenberg, R. (1976), 'Benthic faunal dynamics during succession following pollu-tion abatement in a Swedish estuary', *Oikos* 27, pp. 414–27.

Rosenberg, R. (1977), 'Benthic macrofauna dynamics, production, and dispersion in an oxygen deficient estuary of West Sweden', *Journal of Experimental Marine Biology and Ecology* 26, pp. 107–33.

Saldarriga, J. G. (1987), 'Recovery following shifting cultivation', in C. E. Jordan (ed), *Amazonian Rain Forests: Ecosystem Disturbance and Recovery,* Ecological Studies 60 (New York: Springer-Verlag), pp. 24–33.

Sanders, F. S. (1985), 'Use of large enclosures for perturbation experiments in lentic ecosystems: A review', *Environmental Monitoring and Assessment* 5, pp. 55–99.

Shannon, L. J., Harrass, M. C., Yount, J. D. and Walbridge, C. T. (1986), 'A comparison of mixed flask culture and standardized laboratory model eco-systems for toxicity testing', in Cairns 1986, *op. cit.,* pp. 135–57.

Sheehan, P. J. (1984a), 'Effects on individuals and populations', in P. J. Sheehan, D. R. Miller, G. C. Butler and P. H. Bordeau (eds), *Effects of Pollutants at the Ecosystem Level* (Chichester: John Wiley & Sons), pp. 23–50.

Sheehan, P. J. (1984b), 'Effects on community and ecosystem structure and dynamics', in Sheehan et al. 1984a, op. cit., pp. 51–99.

Sheehan, P. J. (1984c), 'Functional changes in the ecosystem', in Sheehan et al., op. cit., pp. 101–45.

Sheehan, P. J. (1989), 'Statistical and non-statistical considerations in quantifying pollutant-induced changes in microcosms', in Aquatic Toxicology and Hazard Assessment, Vol. 12, ASTM STP 1027 (Philadelphia: American Society of Testing and Materials), pp. 178–88.

Sheehan, P. J., Axler, R. P. and Newhook, R. C. (1986), 'Evaluation of simple generic aquatic ecosystem tests to screen the ecological impacts of pesticides', in Cairns 1986, op. cit., pp. 158–79.

Sheehan, P. J., Baril, A., Mineau, P., Smith, D. K., Harfenist, A. and Marshall, W. K. (1987), The Impact of Pesticides on the Ecology of Prairie-Nesting Ducks, Technical Report Series, No. 19 (Ottawa, Canada: Canadian Wildlife Service).

Sheehan, P. J. and Winner, R. W. (1984), 'Comparison of gradient studies in heavy-metal-polluted streams', in Sheehan et al. 1984a, op. cit., pp. 255–71.

Solomon, K. R., Yoo, J. Y., Lean, D., Kaushik, N. K., Day, K. E. and Stephenson, G. L. (1985), 'Evaluation of the impact of permethrin on zooplankton communities using limnocorrals I. Permethrin dissipation', Canadian Journal of Fisheries and Aquatic Sciences 42, pp. 70–6.

Sprague, J. B., Elson, P. F. and Saunders, R. L. (1965), 'Sublethal copper-zinc pollution in a salmon river, a field and laboratory study', in O. Jaag (ed.), Advances in Water Pollution Research (New York and London: Pergamon Press).

Stephenson, G. L., Kaushik, N. K., Solomon, K. R. and Day, K. (1986), 'Impact of methoxychlor on freshwater communities of plankton in limnocorrals', Environmental Toxicology and Chemistry 5, pp. 587–603.

Stockner, J. G. and Antia, N. J. (1976), 'Phytoplankton adaptation to environmental stresses from toxicants, nutrients, and pollutants – a warning', Journal of Fisheries Research Board of Canada 33, pp. 2089–96.

Suter, G. W. II (1982), 'Terrestrial perturbation experiments for environmental assessment', Environmental Management 6, pp. 43–54.

Suter, G. W. II, Vaughn, D. S. and Gardner, R. H. (1983), 'Risk assessment by analysis of extrapolation error: A demonstration for effects of pollutants on fish', Environmental Toxicology and Chemistry 2, pp. 369–78.

Taub, F. B., Kindig, A. C. and Conquest, L. L. (1986), 'Preliminary results of interlaboratory testing of a standardized aquatic microcosm', in Cairns 1986, op. cit., pp. 93–120.

Taub, F. B., Kindig, A. C., Munro, P. T. and Conquest, L. L. (1982), Model Ecosystems: Effects of Chelate on Copper Toxicity, Experiment Report ME 57, FDA Contract 223–80–2352.

TRI (1987), Ecological Issues: Ecosystem Stability and Recovery Report (Rockville, Maryland: Technical Resources. Inc.).

TRI (1988), End points for Ecological Toxicity. Prepared by Technical Resources, Inc. for Office of Research and Development. US Environmental Protection Agency Contract No. 68–024199.

Welch, J. L. and Ross, R. H. (1982), 'An approach to scoring of toxic chemicals for environmental effects', Environmental Toxicology and Chemistry 1, pp. 95–102.

Westman, W. E. (1978), 'Measuring the inertia and resilience of ecosystems', Bioscience 28, pp. 705–10.

Westman, W. E. (1985), Ecology, Impact Assessment and Environmental Planning (New York: John Wiley & Sons).

Wiebe, P .H. and Holland, W. R. (1968), 'Plankton patchiness: Effects on repeated net tows', Limnology and Oceanography 13, pp. 315–22.

Winner, R. W., Boesel, M. W. and Farrell, M. P. (1980), 'Insect community structure and an index of heavy-metal pollution in lotic ecosystems', *Canadian Journal of Fisheries and Aquatic Sciences* 37, pp. 647–55.

Winner, R. W., Van Dyke, J. S., Caris, N. and Farrell, M. P. (1975), 'Response of the macroinvertebrate fauna to a copper gradient in an experimentally-polluted stream', *Verh. Internat. Verein. Limnol.* 19, pp. 2121–7.

Zitko, V. and McLeese, D. W. (1980), *Evaluation of Hazards Used in Forest Spraying to the Aquatic Environment*, Canadian Tech. Rep. Fisheries Aquatic Science No. 985.

Suggested readings

Cairns, J. Jr and Mount, D.I. (1990), 'Aquatic toxicology. Part two of an ecotoxicology series', *Environmental Science and Technology* 24(2), pp. 154–61.

Freedman, B. (1989), *Environmental Ecology. The Impacts of Pollution and Other Stresses on Ecosystem Structure and Function* (San Diego: Academic Press).

Levin, S. A., Harwell, M. A., Kelly, J. R. and Kimball, R. D. (eds) (1989), *Ecotoxicology: Problems and Approaches* (New York: Springer-Verlag).

Moriarty, F. (1988), *Ecotoxicology. The Study of Pollutants in Ecosystems*, 2nd edn (London: Academic Press).

Munawar, M., Dixon, G, Mayfield, C. I., Reynoldson, T. and Sadar, M. H. (eds) (1989) 'Environmental bioassay techniques and their application', *Hydrobiologia* 188/189.

Sheehan, P. J., Miller, D. R., Butler, G. C. and Bourdeau, P. (eds) (1984), *Effects of Pollutants at the Ecosystem Level. Scope 22.* (Chichester: John Wiley and Sons).

5 Integrating effect and exposure information: an industrial viewpoint*

S. L. DANIELS & C. N. PARK

Introduction

Rhabdomancy is the use of a forked stick to divine water, gold, oil or buried treasure. Specific applications have included locating well sites, tracing underground streams, determining amounts and depths of water and even analyzing water. Ancient rhabdomancers also used forked sticks to affix blame, predict the future and cure disease (Ellis, 1917). While the tools have changed considerably, the intent of modern rhabdomancers or risk assessors remains the same.

Risk assessment, which some regard as the modern 'art' of rhabdomancy, has similar goals. These have now been translated into effect analysis, exposure analysis and risk evaluation (Daniels, 1983b). As in the case of divining for water, it is not enough to know that the material, in this case a chemical, has been found. Risk assessors also need to know its quantity, its concentration, its form and whether it is mobile and persistent.

Without any criteria followed by priority setting, a series of dilemmas is created. How much exposure, which may lead to an adverse effect, may pose a significant hazard? What is the probability of that effect occurring? Can it be mitigated safely? Risk management, as shall be discussed in Chapter 12, involves making choices of appropriate technical and regulatory controls to balance assessed risks, costs and benefits. The overall objectives are the safe manufacture, transport and use of materials as well as the proper storage, treatment and disposal of wastes.

Quantitative risk assessment may not be the answer for everyone. Some disdain risk assessment because of misunderstanding, basic mistrust, lack of hard data, need to assign relative risks, or simple reluctance to accept finite non-zero risks. A cursory risk assessment based upon poor data or invalid models also may create more harm than good. The selection of hazards for

* adapted from an earlier presentation on multimedia risk assessment/risk management by Daniels and Park (1984).

consideration are often taken for granted and assumed to be intuitively obvious, when they may be very uncertain (Kletz, 1977; Douglas and Wildavsky, 1982) as described in Chapter 4. Recognition and quantification of acceptable risks are preferred, however, over blind assumptions and irrational perceptions of risk.

Social, political and scientific awareness of trace environmental constituents has increased for several reasons. These include: new uses and increased demands for products, more sensitive analyses, more frequent and widespread sampling, and improved detection of possible long-term biological effects. Unfortunately, assessments of new data have not kept pace with data collection as toxicological studies and their interpretation have lagged far behind the tremendous improvements in analytical capability in chemistry and physics. The danger exists that with the limited human resources available, more stringent controls may be imposed to minimize perceived risks despite consensus that mere presence of some constituent at some trace level in some environmental medium is not necessarily reflective of true risk. Quite correctly, where contamination is quantitatively determined to affect specific health and environmental qualities, sources of particular constituents must be found and appropriate management options identified and implemented.

This chapter will distinguish risk assessment from risk management, consider flexible approaches to assessment, describe extrapolation of dose–responses and environmental assessment principles, briefly discuss ranking and modeling systems, and conclude with several hypothetical examples.

Risk management follows logically from risk assessment. Neither are mutually exclusive. Various options are applicable during production, distribution, and use of materials and energies, and during treatment and disposal of wastes. Risk management as conducted by industry and government can be made more economically affordable and legally enforceable if it is founded upon good risk assessment.

No natural phenomenon or anthropogenic activity can ever be guaranteed to be 100 per cent risk-free. Hopefully, insistence on living in a zero-risk environment will lessen with time. But public confidence will only be restored when significant and controllable risks can be managed and facilities properly sited (Conservation Foundation, 1983). The public must understand fundamental environmental and health issues. The commercial user must properly manage products and wastes. The governmental regulator must protect environmental resources. There may be individual differences of opinion but the common goal is to minimize risk. If all chemical constituents are assumed to comprise equal risks in all environmental media, and all hazardous wastes are perceived to be managed at no visible cost, then nothing is prioritized, and resources are ineffectively utilized.

History of risk assessment

Safe levels or acceptable daily intakes (ADIs) were historically based on safety or 'uncertainty' factors. No observed effect levels (NOEL) or no observed adverse effect levels (NAEL) have been experimentally derived while safety factors have been derived on the basis of differences in sensitivity between (1) test animals and humans and (2) among humans themselves. A typical safety factor is 100 (ADI + 0.01 NOEL). Arbitrariness of safety factor selection and dependence, to some extent, of the NOEL upon a limited number of species tested have been criticized. Conservatism in the safety factor is evident, however, and specific magnitudes of safety factors are seen as empirical and open to change (Dourson, 1983). Model results may appear more precise than real. For several reasons, ranging from conservative desires to scientific hypotheses, the safety factor approach was rejected for the regulation of animal carcinogens. A modified probit method for setting safe levels for animal carcinogens was proposed as early as 1961 (Mantel and Bryan, 1961). In the early to mid-1970s, the regulatory agencies proposed the use of another version of the probit method. A number of scientific and procedural deficiencies resulted in its subsequent withdrawal. It is often overlooked that it was essentially a safety factor approach.

The next method proposed for regulatory use was the 'one-hit' model (Regulatory Council, 1979) in which age-specific tumor rates are assumed proportional to dose. It is often referred to as the 'linear' model, but in complete form it is exponential. At low doses it approximates a linear increase of 'risk' with dose. This model is justified by 'biological' assumptions stated in 'mathematical terms' which have been interpreted incorrectly as 'proving' low-dose linearity (Guess et al., 1977). Any exposure is assumed to produce an additive risk onto an identical ongoing background mechanism of carcinogenesis. In conjunction with low projected risk levels, the model results in an extremely conservative risk assessment. The one-hit model has been strongly criticized and has been largely replaced by a linearized multi-stage model (Crump, 1981) currently used by most federal and state agencies. Calculations for the latter are more complex, but concepts are simple. A polynomial series instead of a single linear dose term, is fitted to the data. The original multistage model (Doll, 1971) has been further modified into the linearized multistage (LMS) model. The linear term is adjusted until the latter model just marginally fits the data. The most important feature of the LMS model is that results for most data sets are very similar to the one-hit model even for extremely nonlinear data. The one-hit, LMS, and Mantel–Bryan models, approaches are similar. 'Safe' levels are approximated by simple functions of effect or no-effect levels.

There are limitations to all models. None provide completely valid estimates of human risk due to uncertainties and the lack of realistic animal-to-man extrapolations. Models do have a place in regulation, but they do not quantify risk. 'Best estimates' are of limited value since they are very sensitive to even minor biological variability which can produce a thousandfold variation in estimated risk (Krewski, 1981). Most models incorporate no biological factors except observed tumors. Pharmacokinetics, metabolism, excretion, mechanisms of attack, genotoxicity, tumor type, and time to tumor are ignored (Park and Snee, 1983). Unrealistic risk assessments and inconsistent resource expenditures are then applied to solve sometimes real, but often imagined problems.

Risk assessment systems

The public and private sectors, and various regulatory agencies, have expressed interest in risk assessment (Ruckelshaus, 1983). Weighing of risks, costs, and benefits is anathema to some and panacea to others (Anderson and Dower, 1980). However, with the large number of chemicals in use and our dependence on them, systematic processes for assessing risks, costs and benefits are necessary. Many systems have been described for assessing and ranking both constituents and facilities (Tables 5.1 and 5.2).

Present Environmental Protection Agency policy is to improve the scientific basis for risk assessment, and to separate risk assessment from risk management (Ruckelshaus, 1983). Components of risk assessment have included identification, dose–effect, transport, exposure pathways, and sensitivity/uncertainty analyses. Several systems, for example, the EPA Hazard Ranking System (US EPA, 1982), the US Air Force Hazard System (Hawley and Napier, 1985), and the Syracuse Research Corporation Chemical Hazard Assessment System (Howard et al., 1982) have been reviewed (Chu et al., 1986). The complexity of the ranking system in being able to differentiate priority chemicals or groups will vary depending on the responsibility of the management agency. A manufacturer of specialty chemicals may have an inventory of a few dozen chemicals while an environmental or public health regulatory agency may be concerned with the full spectrum of 100,000 commercial chemicals. In our view, it is impractical and unnecessary to assess each chemical constituent individually.

Table 5.1   Generalized assessment and ranking schemes

Blair, E. H. and Bowman, C. M., 1983	Howard, P. H. et al., 1982
Cairns, J. Jr et al., 1978	National Academy of Sciences, 1981
Cairns, J. Jr and Maki, A. W., 1979	Neely, W. B. and Blau, G. E., 1985
Chemical Manufacturers Assoc., 1984	Stanford Research Institute, 1978
Daniels, S. L., 1981	Stephenson, M. E., 1977
Daniels, S. L. et al., 1985	Venezian, E. C., 1977
Harris, R. H. et al., 1984	Versar Inc., 1986

Table 5.2 Specialized assessment and ranking schemes

Consumer and bulk products	Becker, D. S., 1978
	Beckman, R. B., 1974
Pesticides	Kenaga, E. E., 1980
	Kenaga, E .E., 1979
	Olsen, A. R., and Wise, S. E., 1982
Nuclear	Hawley, K. A. and Napier, B. A., 1985
	Poston, J.W., 1978
Hazardous waste	Chu, M. S. Y. *et al.*, 1986
	Daniels, S. L., 1981(a)
	Daniels, S. L., 1981(b)
	Koerner, R. M. *et al.*, 1986
	Lee, G. F. and Jones, R. A., 1982
Synfuels	Travis, C. C. *et al.*, 1983
Gas storage	Kunreuther, H.C. *et al.*, 1983
Water	Daniels, S. L., 1983(b)
	Hedden, K. F. *et al.*, 1982
	Smith, J. H. *et al.*, 1978
Waste water	Dickson, K. L. *et al.*, (eds) 1982
	Falco, J. W. *et al.*, 1981
	Fiksel, J., 1981
Multimedia pollutants	Cleland, J. G. and Kingsbury, G. L., 1977
	Conservation Foundation, 1985
	Daniels, S. L., 1987
	Daniels, S. L., 1986

Grouping according to physico-chemical properties, toxicological properties and other significant parameters are often used and are recommended (Daniels *et al.* 1979). The relative importance of different groupings or parameters within groups may also depend on the particular stage in the life cycle of a chemical product. As an example, the Hazard Response System has been used to prioritize uncontrolled hazardous waste sites for remedial action (Harris *et al.*, 1984).

The benefits and detriments of models have been debated at international, national and state levels (Cobler and Hoerger, 1983; Cumming, 1983). A number of models have been constructed to assist in the assessment process especially by predicting and quantifying the potential for exposure (Table 5.3). One of the better known models is the fugacity model (Mackay, 1979; Mackay *et al.*, 1985; Mackay and Paterson, 1981; Neely and Mackay, 1982). This model calculates the escaping tendency of a chemical from one medium into another.

The National Academy of Sciences of the United States supported appropriate use of risk assessment (National Academy of Sciences, 1983). Scientific risk assessment and risk management were distinguished as integral parts of the regulatory process. In the review, the academy stressed that the adequacy of data bases and validity of evaluation procedures should

Table 3   Models designed to support the assessment process

General use	Onishi, Y. *et al.*, 1980
	Ott, W. R., 1978
	Ott, W. R. (ed.), 1976
Wastewater effluents	Baughman, G. and Lassiter, R., 1978
	Bergman, H. L. *et al.*, 1982
	Burns, L. A. *et al.*, 1982
	Cairns, J. Jr, 1983
Solid wastes	Daniels, S. L., 1980
	Klee, A. J., 1976
Aquatic environments	Connolly, J. P. and Winfield, R. P., 1984
Groundwater	Javandel, I. *et al.*, 1984
Agricultural runoff	US Dept. of Agriculture 1980
Site evaluations	Chang, S. *et al.*, 1981
	Chu, M. S. Y. *et al.*, 1986
	ICF Inc., 1984
	JRB Associates Inc. 1980
	Long, F. A. and Schweitzer, G. E., (eds) 1982
	US EPA, 1980
Multimedia pollutants	Cohen, Y., 1986
	Cohen, Y. and Ryan, P. A., 1985
	Mackay, D., 1979
	Mackay, D. *et al.*, 1985
	Mackay, D. and Paterson, S., 1981
	Neely, W. B., 1980
	Neely, W. B., 1979
	Neely, W. B. and Blau, G. E., 1985
	Neely, W. B. and Mackay, D., 1982
	Urban, D. J., and Cook, N., 1986

be carefully reviewed. The 'nature and quality of the relevant components of risk assessment' should be recognized; 'gaps and uncertainties in available information' should be acknowledged. The development of uniform inference guidelines for risk assessment and the establishment of a board to evaluate risk assessment methodology also were recommended. The Interdisciplinary Panel on Carcinogenicity convened by the American Industrial Health Council reviewed criteria for chemical carcinogenicity in light of new bioassay and epidemiological data, improved estimates of dose–response, continued research on mechanisms of action, impacts of biometrics on model development, and continued controversy over quantification of risk by carcinogens. The panel noted that evaluations are dependent on properties, metabolism, uses, availability of data, and other factors. Animal studies augmented by studies of mechanisms, metabolism, and pharmacodynamics were recommended for better risk assessment (Interdisciplinary Panel on Carcinogenicity, 1984). Sincere efforts are underway to improve data bases in many sectors of industry and government for risk asssessment. However, the degrees of sophistication of the models which are becoming increasingly available should be proportional to the availability of data: that is to say that if the data are not available, model users should not expect that sophisticated models will provide better answers.

Risk assessment and risk management

The relationship of risk assessment to risk management (National Academy of Sciences, 1983; Park and Snee, 1983) can be paraphrased in a five-part query (Daniels, 1984) (Table 5.4).

Table 5.4   Risk assessment and risk management

Query	Step	Process
What effect is possible?	Effect analysis	Risk identification
Upon whom is the effect possible?	Exposure analysis	
How much of the chemical is present where?		
What chance is there of the effect occurring?	Probability analysis	Risk evaluation
What degree of effect demands what action?	Regulatory/technical responses	Risk management

All steps in the risk assessment and management process are not always fully addressed. Distinctions between toxicity, hazard and risk are sometimes muddled by regulators and the media. This can occur when apparently insignificant chemical constituents are reported and when worst case scenarios are assumed. In the latter case, conditions influencing exposure may not be included in the risk identification process and managers or regulators may move directly from laboratory effects studies to control action. There may be instances when such direct action is warranted but for many chemicals in use today, the risk can be mitigated.

Distinctions between toxicity, hazard and risk

Distinctions between toxicity, hazard and risk are important. Toxicity is an adverse effect upon a living species by a constituent administered at a toxic concentration in a biologically available form by a particular route and during a fixed exposure time. Hazard is a possibility of a toxic effect being manifested upon exposure. Risk is a probability of hazard occurring. Distinctions should be made between what does happen under controlled laboratory conditions, what might happen, and what is likely to happen in

Table 5.5   Kinds of risk and interpretations of hazard

Kind of Risk	Interpretation of hazard	Example
Public	Separation of technical, economic, social and political components	Nuclear *v.* solar power
Private	Restriction of personal freedoms in pursuit of safety	Seat belts; detergent substitutes; sidewalks
Controllable	Application of new or improved technology	Occupational protection; waste minimization and waste treatment
Uncontrollable	Occurrence of natural disasters	Earthquakes; volcanoes; hurricanes; tidal waves
Unknown	Caution in absence of precise information	Chronic exposure to very low levels of chemicals

the real world. Exposure analysis comes to the fore in these last two questions.

Risks can be classified as complex or simple, technical or political, controllable or uncontrollable, voluntary or involuntary, or as combinations of various kinds (Table 5.5). Individuals appear to differentiate between unavoidable and unnecessary risk (Cohen and Lee, 1979). Unfortunately, the application of new technologies and improvements of old technologies continue to create some new risks, eliminate some old risks, and decrease unavoidable risk (Starr, 1969).

Distinctions between hazards

Hazardous wastes are perceived by some to be reflective of all kinds of risks. They are 'new', 'complex', and 'unknown' in comparison to 'old', 'simple', and 'familiar' risks. In the latter case, gasoline as a motor fuel and chlorine gas as a disinfectant for water are routine and accepted uses, yet they are a flammable liquid and a toxic gas, respectively. In general, they are beneficial when used under carefully controlled conditions.

Distinctions between a 'waste' and a 'nonwaste', and between a 'solid' waste and a 'hazardous' waste are not always evident. In industry, a waste and a nonwaste are distinguished primarily by economic and legal factors, and not by technical factors. Interpretations of degrees of hazard (or non-hazard) (Branson, 1980; Daniels, 1980; Poston, 1978), therefore, are more often policy matters and not scientific decisions which establish numerical rankings (Petak and Atkisson, 1982). This does not preclude the quantification of risks, but legal and technical definitions can clash. For example, 'hazardous waste' has been broadly defined to include solids, liquids, semi-solids, and contained gases, posing a variety of hazards of differing extents (Daniels *et al.*, 1982). Technically, standard methods are needed to characterize exposure (e.g. volatility, solubility, and extractability), and effect (e.g. corrosivity, ignitability, reactivity, and toxicity). Standards are also needed for specifications of materials and wastes, performances of waste management processes, and terminology. Some of these standards are being developed by voluntary consensus standard setting groups such as the American Society for Testing and Materials (ASTM).

Hazard is 'how much', 'where', and 'when for how long', may do 'what' to 'whom'. Effects are the 'how much' may do 'what' to 'whom'. Exposures are the 'how much', but particularly the 'where and when for how long'. Hazard requires analysis of both an exposure and an effect. Effect is often emphasized to the detriment of exposure. A waste containing a highly toxic constituent at a low concentration may pose a relatively low degree of hazard. Conversely, a waste containing a slightly toxic constituent at a higher concentration may pose a higher degree of hazard because of its exposure to people, plants or animals (Daniels, 1985).

Natural and synthetic constituents are sometimes difficult to distinguish and there is often a perception that natural compounds are more acceptable (Daniels, 1983a). For example, petroleum hydrocarbons are naturally derived, refined as fuels, recombined as polymers and other useful products, and discarded in various forms. Each constituent of petroleum from whatever source may be associated with a variety of health and environmental concerns of varying degrees. Thus crude oil, gasoline, benzene, solvent waste, and scrap plastic are each perceived differently. The dilemma facing many regulatory agencies is in establishing relative hazards posed by diverse constituents from various sources in multiple environmental media rather than the hazards associated with individual chemicals or groups of chemicals. The unfocused social perception of hazards posed by generic groupings of chemicals needs revision. Some families of chemicals are perceived collectively to be hazardous, mobile and persistent in the environment, and therefore recalcitrant to treatment. But it must be recognized that environmental properties vary significantly, and associated risk may also vary by many orders of magnitude.

The integration of effects and exposure analyses

Risk assessment requires quantitative measurements and/or predictions of both effects and exposures (Conway, 1982; Lee and Jones, 1982; Stern and Walker, 1978). While environmental effects have traditionally been determined for individual constituents in pure forms under laboratory conditions (Daniels, 1981a; Daniels, 1981c; Freitag et al., 1985; Korte et al., 1978; Kotzias et al., 1981; National Academy of Sciences, 1981b), environmental exposures, by comparison, are determined under site-specific conditions (Daniels et al., 1985; Neely and Blau, 1985). The problem faced by managers is that a chemical may be very toxic, even lethal, under laboratory conditions but exposure in the workplace or the community is not likely to occur due to physical or chemical properties of the chemical itself or of the medium into which the chemical is released. Some constituents are relatively nonpersistent and immobile and pose only acute hazards. Other constituents may be more persistent and mobile and pose chronic hazards. Chemically indistinguishable constituents are sometimes perceived to pose different hazards in various environments irrespective of their properties. A waste with trace constituents in a landfill is perceived as more hazardous than an article in commerce having the same constituents at much higher concentrations. Mere presence of a 'toxic' constituent does not constitute hazard. For a hazard to be posed, a constituent must be capable of exerting some adverse effect, such as toxicity, involving some dose or concentration upon an exposed organism (Freitag et al., 1985; Korte et al., 1978; US Air Force, 1983). Thus, exposures are necessary for toxic effects to result and these may be acute or chronic, reversible or irreversible (Daniels, 1981b). As described in Chapter 3, acute toxicity is determined by relatively short and inexpensive tests. Chronic toxicity is determined in much longer and more expensive tests which may span several years and cost millions of dollars in animal care, analytical support, quality assurance, and data assessment. Neither of these types of laboratory tests may reflect the actual impact of the chemical in the environment. This is not to suggest that toxicity testing should be eliminated but rather that its limitations should be recognized.

Risk is a function of duration, frequency, and probability of exposure. Exposure is related to how a constituent moves about and how long it persists in the environment. How fast such things happen (i.e. the rate) (Blau et al., 1975) and how completely they occur (i.e. the equilibrium) (Mackay et al., 1985; Paterson, 1985) are both important (Haque, 1980; Hedden, 1984; Thibodeaux, 1979; Tinsley, 1979). Movement in the air, water and soil and to and from living organisms depends upon intrinsic properties of individual constituents. Evaporation from water to air, solubility from solid to water, and sorption from water to soil are commonly measured (Lyman et al., 1982; McCall et al., 1983; Thibodeaux, 1979; Verschueren, 1983). Water-soluble constituents will obviously prefer the water phase; volatile ones will prefer the air phase; ones with high sorptivity to soil will prefer a solid phase; and ones with high affinity for lipids will bioconcentrate. Chemical and biological degradations can occur in water solution or on soil particles, and photochemical degradations can occur in both air and water (Hedden, 1984; Howard et al., 1982; Neely and Mackay, 1982; Swann and Eschenroeder, 1983). Hazards are reduced through degradation and attenuation both in the environment and in organisms. Multimedia considerations must be reflected in the assessments (Cohen and Ryan, 1985; Daniels, 1987; Daniels, 1986; Hedden, 1984; Hedden et al., 1982; Neely, 1980). Constituents from both natural and anthropogenic sources occur at different levels in air, drinking water, foods, and wastes.

Some degree of finite risk is accepted in all human endeavors including hazardous waste management. Everything in a biologically available form is toxic at some concentration, even those necessary to sustain life. Exposures to most highly toxic constituents are usually limited to nontoxic concentrations. Uncertainty arises, however, over determining real and significant hazards, and safe and acceptable exposures. The vast infrastructure being created to address remedial actions at abandoned sites which may constitute imminent hazards is a case in point. Decisions must be made of how dirty is 'dirty' in initial assessments and how clean is 'clean' in final assessments. Unfortunately, existing data bases are insufficient for proper risk assessment (Office of Technology Assessment, 1983). In many instances, management decisions are being made from assumed, non-existent, incomplete or erroneous data. Some of this is due to a lack of monitoring data whether ecological or epidemiological though this should not be surprising due to the large number of chemicals in commerce compared to the number of chemicals which have undergone ecotoxicological assessments. As a result, possible hazards posed by various constituents may not be placed in proper perspective. Consequently, risk assessments are needed to help prioritize both problems and solutions.

Risk assessment and human health

There has been considerable discussion in recent years concerning the assessment of the carcinogenic potential of chemicals. Attention has focused on rigid mathematical approaches used by various federal agencies, particularly the Environmental Protection Agency and the Food and Drug Administration, and state agencies, e.g. California and Michigan (California Department of Health Services, 1982; Michigan Department of Natural Resources, 1983). Much controversy has arisen because the regulatory approach to carcinogenesis is so different from that used for other toxic effects. For most toxicological phenomena, the concept of a safety factor or margin of safety is accepted for evaluating potential hazards. Choices of safety factors depend on judgemental, subjective, and qualitative factors underlying scientific, regulatory, and societal considerations. When tumorigenesis is observed in animals, the regulatory approach has been either to attempt an impossible zero exposure, or to use a very rigid and conservative risk assessment with little or no scientific judgement. Typical exposure limits are the order of one thousand times more conservative than those which would have been proposed for an endpoint other than carcinogenesis. The dichotomy has fueled much of the controversy surrounding the relevance of animal testing to potential human results and reflects the public's aversion to a dreaded disease.

If animal tumorigenicity were conceptually treated like other toxicological phenomena, then the public and industry might view the decisions as more logical. Observed animal carcinogenicity is often only a secondary result of other observable toxicity manifested at extremely high doses. The use of a modified safety factor approach to establish acceptable human exposure limits for animal carcinogenesis has been proposed (Park and Snee, 1983). The strengths and weaknesses of current mathematical approaches have been summarized and a simplistic procedure with some scientific judgement was described. It is accepted that different factors may be used for different regulatory scenarios and modified depending upon available data. This approach directly addresses multimedia concerns for a particular constituent that may manifest its effects in considerably different forms and pathways. The 'modifier' to the safety factor is estimated separately for each of the following factors depending upon availability of data:

1 Bioassay results: numbers of negative and positive responses.
2 Species sensitivity: comparative metabolism and pharmacokinetics; specific tumor type as ultra-sensitive tumorigenic endpoints.
3 Mechanism of action: evidence of genotoxic potential; evidence of carcinogenesis secondary to other toxicity; use of a safety factor based on NOEL for primary toxicity.
4 Time of tumor appearance; spontaneous rate in animals and man: increased incidence of lesions with age and high spontaneous rate are less indicative of human hazard than early appearance of tumors with low background rate.
5 Associated toxicity: observed incidence of tumors in animals are a very crude measure of inherent hazard. Tumor counts are somewhat subjective, expected to vary with time, and strain and species. Occurrence of tumors is often directly correlated to a more easily observable form of toxicity. Observed threshold is often used as a NOEL and a safety factor applied.
6 Epidemiology: other relevant data likely on case-by-case basis.

Using this interpretive approach, exposure limits can be more logically set for diverse chemical substances. Potential carcinogenicity can also be integrated with other forms of toxicity. Although there have been strong arguments for the case-by-case approach, there is nothing inherently inconsistent with accepting 'generic' guidelines that incorporate the above approach. Another aspect of the proposal is that the reliance on 'risk' numbers, such as 10^{-6}, are avoided. Such numbers have no real meaning but often are incorrectly interpreted as equivalent to an event affecting a certain segment of the population, such as one person in a million.

A related approach with a slightly different emphasis has also been proposed (Squire, 1981). This is essentially a scoring scheme dealing specifically with animal data to the extent to which they imply human hazard. The classifications include:

1   number of species affected;
2   number of histogenetically different neoplasms observed;
3   spontaneous tumor incidence in control animals;
4   percentage of observed neoplasms that are malignant;
5   genotoxic potential of the compound;
6   dose at which the compound is tumorigenic in animals.

Each of these categories except the last reflect attempts to identify compounds whose outcomes are likely to be somewhat species dependent with a dose–response that probably does not extend to low doses compared to those compounds which imply greater human hazards. None of these schemes specifically addresses the scientific quality of the reported data. The International Registry of Potentially Toxic Chemicals (IRPTC) and the Ontario Government in Canada have addressed this problem. Both groups have guidelines for judging the validity and usefulness of published studies. The Ontario approach derives a score representing the quality of any given study while the IRPTC uses a peer review approach. Recent multidisciplinary workshops have also addressed quality of data collection procedures.

Risk assessment of chemicals in the environment

There have been parallel activities in classification of chemicals that reflect inherent differences in levels of effects and/or exposures (Table 5.6). Six principles of environmental risk assessment can be elaborated (Daniels *et al.*, 1985):

1   Assessment begins with determination of effect and exposure: relative significance of various effects and their relations to direct and indirect exposures.
2   Appraisal of production/use patterns indicates extent of data: closed or open systems; consumptive or dispersive uses; single or multiple sites; limited or extensive multimedia transport; limited to extensive degradation; voluntary or regulated controls.
3   Assessment is an interactive and interdisciplinary process: involvement of various scientific disciplines; application of various models and field verifications.
4   Assessment is an iterative (staged) process: preliminary screening versus detailed tiered assessments; reassessment of data as necessary.
5   Necessary data are highly variable and chemically specific: anticipated release rates; intrinsic properties of chemical constituents; extrinsic properties of the environment.

Table 5.6   Classification systems for chemicals

Aquatic environments	Astill, B. D. *et al.*, 1982
	Bergman, H. L. *et al.*, 1982
	Branson, D. R., 1980
	Howard, P. H. *et al.*, 1982
	Kotsias, D. *et al.*, 1981
	Neely, W. B., 1980
	Neely, W. B. and Blau, G. E., 1985
	Saxena, J., 1983
	Stephenson, M. E., 1977
	Stern, A. M. and Walker, C. R., 1978
Solid and hazardous wastes	Daniels, S. L., 1981(b)
	Daniels, S. L., 1980
	Howard, P. H. *et al.*, 1982
	Lee, G. F. and Jones, R. A., 1982
	Long, F. A. and Schweitzer, G. E., (eds) 1982
	Michigan Dept. of Natural Resources, 1983

6   Chemical similarities are few, but evaluations are common: estimation of release rates; predictions of multimedia partitioning; evaluations of persistence and degradation; and evaluations of ecotoxicity.

These principles are relevant to both scientific and regulatory approaches to risk assessment. They encourage flexibility in risk assessment and risk management, while devoting scarce resources for assessments on a tiered basis toward prioritized concerns. Examples of exposure analysis as applied to both 'new' and 'old' chemicals under the Toxic Substances Control Act (US) are described elsewhere (Daniels *et al.*, 1985).

Examples of risk assessment

While there is a definite need to become more quantitative, it is recognized that decisions regarding the manufacture, use and disposal of many commercial chemicals will have to depend on informal judgement supported where possible by tools and techniques such as quantitative structure activity relationships and fugacity models. Qualitative environmental assessments can be illustrated by simplified examples which comprise a cross-section of typical eventualities that may be encountered. They are not substance or location-specific, but are derived in part from real-world experiences. Actual case histories for specific substances or locations may comprise hundreds of pages compiled over periods of years after considerable expenditures. Detailed quantitative assessments have been made of chemicals in commerce (Blair and Bowman, 1983; Haque, 1980; National Academy of Sciences, 1981(a)) and wastes (Daniels and McElroy, 1985; Daniels and Park, 1984), at specific sites (Chang *et al.*, 1981; Howard *et al.*,

1982; JRB Associates, Inc., 1980; Michigan Department of Natural Resources, 1983) or in various environmental media (Baughman and Lassiter, 1978; Bergman *et al.*, 1982; Hedden *et al.*, 1982; Lyman *et al.*, 1982; Ott, 1976; Smith *et al.*, 1978; Tinsley, 1979). The general approaches are applicable to various constituents and facilities depending upon classification by use (Daniels and McElroy, 1985) or type of waste (Daniels, 1987; Daniels, 1981c).

The examples are presented in two parts: (1) the assessment of the properties of five hypothetical chemicals and their potential effects, and (2) the assessment of the same five chemicals and their potential exposure under five types of release to the environment.

Assessment of effects: The form, partition, fate, and potential effects of each of five hypothetical chemicals are categorized in Table 5.7. It is important to recognize that pure constituents in solid, liquid, or gaseous forms may undergo phase changes in the environment (see Chapter 2). They may partition into various environmental media. They may be dispersed and diluted or undergo chemical changes and degradation. They may exert a myriad of effects at various stages in the environment.

Table 5.7 Potential effects of five hypothetical chemicals

Chemical	Pure form	Partition	Fate	Effect or condition
A	Gas	Volatile	Photodegradable	Inhalation Ignition
B	Solid	Soluble	Chemodegradable	Dermal burn Corrosion
C	Liquid	Soluble	Biodegradable	Fish toxicity Taste/odor
D	Solid	Insoluble nonvolatile	Inert, persistent	Constituents in drinking water
E	Liquid	Immiscible, sorptive	Biodegradable	Accumulation to toxic levels

Table 5.8 Potential exposures to five hypothetical releases

Chemical	Release	Env. medium	Activity	Situation	Exposure
A	Emission	Atmosphere	Production	Controlled	Short
B	Spill	Land	Transportation	Remedial	Short
C	Discharge	Surface water	Treatment	Controlled	Long
D	Leachate	Groundwater	Disposal	Controlled	Long
E	Sorbate	Sediment	Use	Dispersal	Short

Note: The hypothetical chemicals are all of natural extraction: A – carbon monoxide, B – lye, C – fuel oil, D – coal ash, and E – creosote.

Table 5.9   Possible management strategies for hypothetical events

Example A: Emission to the atmosphere during production
This example is intended to illustrate environmental assessment of a short-term controlled event consisting of a limited quantity of a known substance being released as a single emission to the atmosphere during production of another substance. The response might include an evacuation alert and routine reporting depending upon site-specific factors.

Example B: Accidental spill during transportation
This example is intended to illustrate environmental assessment of a short-term uncontrolled event consisting of a limited quantity of a known substance being released as a single spill to the land during transportation. The response might include dilution with water with limited site monitoring and routine reporting.

Example C: Discharge to surface water after treatment
This example is intended to illustrate environmental assessment of a long-term controlled event consisting of a known substance being released continuously at a low concentration as a discharge to surface water after treatment of a wastewater. The response might include periodic monitoring and reporting as specified by permit.

Example D: Leachate to groundwater after disposal
This example is intended to illustrate environmental assessment of a long-term uncontrolled event consisting of a known substance being released continuously at a low concentration as a leachate to groundwater after disposal of a waste. The responses might range from short-term monitoring and remedial action to long-term collection and treatment.

Example E: Sorption to sediment after use
This example is intended to illustrate environmental assessment of a long-term controlled event consisting of a known substance being released continuously at a low concentration as a dispersal to the environment after use. The response might range from precautionary notice to avoid release into sensitive environments to remedial action dependent upon site-specific factors.

In summary, these brief examples only touch upon the details of environmental assessments that might be conducted for specific chemicals associated with potential exposures.

Assessment of exposures: Five hypothetical releases, their degree of control, the receiving medium, the source activity, and the extent of exposure, involving the five hypothetical chemicals are presented (Table 5.8). These examples include release of a chemical to the environment through:

1   controlled emission to the atmosphere during production;
2   accidental spill on the ground during transportation;
3   controlled discharge to surface water after waste treatment;
4   controlled leachate to groundwater after waste disposal;
5   dispersal to the environment after use.

Each example involves a different type of exposure, environmental medium, and activity. Each example can involve a short or long-term concern, or a controlled or uncontrolled situation. Responses are site specific and can range from limited to detailed action steps.

It should be emphasized that while attempts are usually made to mini-

mize releases of constituents that may exert the greatest effects in the·
shortest times or at the lowest concentrations, there are always finite
constraints. No releases are absolutely 'zero' in the practical sense. Releases
are also somewhere between 'all at once' and 'continuous'. Balances are
struck where the releases are below detectable limits or below thresholds of
effect that are associated with adequate margins of safely. In some jurisdictions the assimilative capacity of the natural environment is also considered
in setting release limits while in other countries, a best practicable or
available technology limit is imposed. In the case of certain chemicals
background levels of compounds and in some instances, naturally occurring compounds, must also be evaluated.

Conclusions

Once the scientific findings and policy judgements embodied in risk
assessment have been established, then various political, economic and
technical factors are integrated into risk management (Daniels and
McElroy, 1985). The options available to manage chemical risks include:
manufacturing guidelines, labelling and packaging standards, transportation guidelines, recommended uses and use levels, consensus guidelines for recycling, standards for treatment and disposal, and various
regulatory guidelines and standards or discharges (Daniels, 1987). Many of
these are oriented to reduction of exposures. Options depend upon the
relative magnitudes of the risks involved, the degree of voluntary control
by private parties, the extent of public concern, and the nature of governmental policies. Clearly with the number of chemicals involved each
having certain risks and benefits, the environmental policies of the future
will require the application of multiple options.

Simple assumptions of high risks should be avoided. Most quantitated
involuntary risks have been shown to be much smaller than voluntary
risks. Listings and 'action' levels are not needed for most environmental
contaminants. Regulatory agencies have historically believed, however,
that it is impossible to quantify risks associated with trace constituents over
long exposures. Conventional regulatory approaches have emphasized
source control or prevention, rather than specific treatment or remedial
action (Berthouex and Rudd, 1977). The relative benefits of controls
'before' rather than 'after' the fact have not been well quantified though
there is a general view as expressed by the World Commission on Environment and Development (1987) that 'anticipate and prevent' strategies are
more appropriate than 'react and cure' approaches. Suitable methodologies
are available or under development for characterization; technologies for
risk management are being devised which are considered adequate to
protect health and the environment.

Regulation is in a state of flux, however, with controls of additional contaminants in various environmental media under consideration. Many lists of hazardous contaminants have been developed, but attempts to set *de minimus* levels have been largely superficial and not based on formalized, quantitative risk assessment. The most appropriate regulatory options for risk management are very often chemical and site specific. In some cases, maximum contaminant levels may be appropriate; in others, designated uses may suffice; in still others, treatment performance standards may be best.

Ratios of preventive to remedial costs are difficult to substantiate. It is simpler to quantitate recovery versus treatment versus disposal costs. Some corrective actions can be costly, especially if an excessively stringent treatment level is imposed, i.e. best available technology to the limit of detection. This appears to be the case with dioxins at the present time where the chemicals are being measured at the part per quadrillion level. Imposition of unrealistic or unnecessary regulations in areas where potential hazards are negligible can be burdensome. The regulatory agencies, the manufacturers, the commercial establishments, and the consumer public are faced with decisions over 'what extent' demands 'what action'? Technical, economic, legal, and social debates will continue over the benefits, feasibilities, and general selection of risk management options (Anzenberger, 1985; Kleindorfer and Kunreuther, 1987).

Recognition of the values and limitations of risk assessment will assist in the decision-making process. The options of risk management must be scientifically sound, health and environmentally defensible, technically feasible, economically affordable, and legally enforceable. Better risk assessment, with a greater emphasis on the determinations of both effects and exposures, will lead to a sound scientific foundation to ensure that regulations, treatment facilities, and remedial actions are truly effective in controlling actual hazards and minimizing actual risks. The linkage of assessment and management systems is exemplified by the Risk–Cost Analysis Model of the Resource Conservation and Recovery Act of the United States (ICF, Inc., 1984) which considers tradeoffs of costs and risks among wastes, environments and technologies to arrive at feasible management alternatives.

Maidens' Prayers and Tinkers' Damns

We've finally learned no maiden's prayer
Can help assess all risks.

The multimedium abounds:
Air–water–solid exists.

Risk assessment: how much of what
Is where, how long persists.

And what it might affect, and how
And who is hit or missed.

But risk is more than simple terms,
Or names placed on some lists.

Risk management: how much, and when,
Of what to do, consists

For risks don't mean a tinker's damn,
If we sat around and guessed.

S. L. Daniels

References

Anderson, R. C., and Dower R. C. (1980), 'The use of cost-benefit analysis for hazardous waste management', in *Disposal of Hazardous Waste*, EPA 6th Annual Research Symposium, pp. 145–66.

Anzenberger, L. (1985), 'The Environmental Protection Agency's Regulatory Practices: The impact of a holistic approach', *Env. Prog.* 4 (3), pp. 155–60.

Astill, B. D., Lockhard Jr, H. B., Moses, J. B., Nasr, A. N. M., Raleigh, R. L. and Terhaar, C. J. (1982), 'Sequential testing for chemical risk assessment', in Conway, *op. cit.*, pp. 412–33.

Baughman, G., and Lassiter, R. (1978), 'Prediction of environmental pollutant concentration', in Cairns *et al.*, op. cit., Philadelphia pp. 35–54.

Becker, D. S. (ITTRI) (1978), *Algorithmic Ranking of Consumer Chemicals by Health Hazard*, prepared for Consumer Products Safety Commission (US), US CPSC, 35 pp.

Beckman, R. B. (1974), *System for Evaluation of the Hazards of Bulk Water Transportation of Industrial Chemicals* (Washington, DC; National Academy of Science [US]), 42 pp.

Bergman, H. L., Kimerle, R. A., and Maki, A. W. (1982), *Environmental Hazard Assessment of Effluents*, proc. 5th Pellston Environmental Workshop, Valley Ranch, Wyo., pp. 22–7 .

Berthouex, P. M., and Rudd, D. F. (1977), *Strategy of Pollution Control* (New York: John Wiley), p. 579.

Blair, E. H., and Bowman, C. M. (1983), *Control of Existing Chemicals*, ACS Symposium Series no. 213, pp. 67–79.

Blau, G. E., Neeley, W. B., and Branson, D. R. (1975), 'Ecokinetics: A study of the fate and distribution of chemicals in laboratory ecosystems', *AIChE J.* 21, pp. 854–62.

Branson, D. R. (1980), 'Prioritization of chemicals according to the degree of hazard in the aquatic environment', *Env. Health Perspectives* 34, pp. 133–8.

Burns, L. A., Cline, D. M., and Lassiter, P. R. (1982), *Exposure Analysis Modeling System (EXAMS): User Manual and System Documentation*, EPA–600/3–82–j023, pp. 443.

Cairns, J. Jr. (1983), 'Regulating hazardous chemicals in aquatic environments', *Boston College Env. Affairs Law Review* 11 (1), pp. 1–10.

Cairns, J. Jr., Dickson, K. L., and Maki, A. W. (eds) (1978), *Estimating the Hazard of Chemical Substances to Aquatic Life*, ASTM STP 657 (Philadelphia, Pa: American Society for Testing Materials).

Cairns, J. Jr., and Maki, A. W. (1979), 'Hazard analysis in toxic materials evaluation', *J. Water Poll. Con. Fed.* 51, pp. 666–71.

California Department of Health Services (1982), *Regulatory Criteria for Identification of Hazardous and Extremely Hazardous Wastes*, Summary draft.

Chang, S., Barrett, K., Haus, S., and Platt, A. (Mitre Corp.) (1981), *Site Ranking Model for Determining Remedial Action Priorities Among Uncontrolled Hazardous Substances Facilities*, EPA contract no. 68–01–6278, Working Paper WP–81woo141.

Chemical Manufacturers Association (1984), *Risk Management of Existing Chemicals*, Proceedings of a Seminar (Washington, DC: CMA) 184 pp.

Chu, M. S. Y., Rodricks, J. V., St. Hilaire, C, and Bras, R. L. (1986), Risk Assessment and Ranking Methodologies for Hazardous Chemical Defense Waste: A State-of-the-Art Review and Evaluation, Task 1 Report SAND86–0530, Waste Management Systems Division, Sandia National Laboratories, Albuquerque,. N. Mex., 74 pp.

Cleland, J. G., and Kingsbury, G. L. (1977), *Multimedia Environmental Goals for Environmental Assessment*, Vols. 1 and 2, EPA–600/70–77–136a and b.

Cobler, J. G., and Hoerger, F. D. (1983), *Analysis of Agency Estimates of Risk for Carcinogenic Agents*, presented at a meeting of the Society of Risk Analysis, New York City.

Cohen, B. L., and Lee, I. S. (1979), 'A catalog of risks', *Health Physics* 36, 707–22. Philadelphia.

Cohen, Y. (1986), 'Organic pollutant transport', *Env. Sci. Tech.* 20, pp. 538–44.

Cohen, Y. and Ryan, P. A. (1985), 'Multimedia modeling of environmental transport, trichloroethylene test case', *Env. Sci. Tech.* 19, pp. 412–21.

Connolly, J. P., and Winfield, R. P. (1984), *A user's guide for WASTOX, A Framework for Modeling the Fate of Toxic Chemicals in Aquatic Environments, Part 1: Exposure Concentration*, EPA–600/3–84–077, 126 pp.

Conservation Foundation (1983), *Siting Hazardous Waste Management Facilities – A Handbook*, prepared by members of the Hazardous Waste Dialogue Group under sponsorship of the Program for Environmental Dispute Resolution, The Conservation Foundation, published jointly by The Conservation Foundation, Chemical Manufacturers Association, and the National Audubon Society, Washington, DC, 71 pp.

Conservation Foundation (1985), *New Perspectives on Pollution Control – Cross-Media Problems*, The Conservation Foundation, 1717 Massachusetts Ave., NW, Washington, DC 20036, 88 pp.

Conway, R. A. (1982), 'Introduction to environmental risk analysis', in Conway, R. A. (ed.) *Environmental Risk Analysis for Chemicals*, (New York: Van Nostrand Reinhold) pp. 1–30.

Crump, K. (1981), 'An improved procedure for low-dose carcinogenic risk assessment from animal data', to be published in *J. Env. Pathol. Toxicol.*

Cumming, R. B. (1983), 'Risk assessment and the government policy-making analysis process', *Risk Analysis* 3 (1), pp. 1–3.

Daniels, S. L. (1980), 'Characterization of solid waste by degree of hazard, hazardous waste characterization', in *Proceedings of National Conference on Hazardous and Toxic Waste Management*, Vol. II, (Newark, NJ: New Jersey Institute of Technology), pp. 571–600.

Daniels, S. L. (1981a), *Development of Realistic Tests for Effects and Exposures of Solid Wastes, Hazardous Solid Waste Testing, First Conference* (eds) Conway and Malloy, STP760 (Ft Lauderdale, Fla: American Society for Testing and Materials)., pp. 345–65.

Daniels, S. L. (1981b), 'Characterization of wastes: short- and long-term toxic effects', in Worm, B. C., Dantin, E. J., and Seals, R. K. (eds), *Proceedings of the Symposium on Hazardous Waste Management: Protection of Water Resources* (Baton Rouge, LA: Louisiana State University), November, pp. 5–28.

Daniels, S. L. (1981c), Assessment of Risk for Hazardous Waste, presented at the Symposium on *Research and Decision Criteria in Environmental Risk Assessment – Perspective and Retrospective Hazard Assessment*, 2nd Annual Meeting, Society of Environmental Toxicology and Chemistry, Arlington, Va.

Daniels, S. L. (1983a), Applications of Chemical Engineering Fundamentals for Assessment of Environmental Effects and Exposures, presented at the Symposium *Chemical Engineering Fundamentals Applied to the Environment*, AICHE Diamond Jubilee Meeting, Washington, DC.

Daniels, S. L. (1983b), Groundwater – Contamination or Consternation? (Perspectives of Chemical Constituents), presented at the closing general session *Ground Water – Hidden Resource or Lurking Danger?*, California Water Pollution Control Association, Annual Meeting, Palm Springs, Calif.

Daniels, S. L. (1984), 'Risk assessment and risk management of hazardous wastes', *ASTM Standardization News* 12 (3), pp. 16–19.

Daniels, S. L. (1985), Selection of Recovery, Treatment, and Disposal Alternatives for Hazardous Waste Management, presented at the Symposium: *Active Mitigation of Industrial Wastes for Repository Characteristics*, AAAS Annual Meeting *Science and Engineering: Diversity and Convergence*, Los Angeles, Calif.

Daniels, S. L. (1986), Integrated Multimedia Approaches Applied to the Management of Industrial Hazardous Wastes, presented at the Fifth Conference on *Environmental Engineering Education*, Michigan Technological University, Houghton, Mich., July 22, 1986.

Daniels, S. L. (1987), Multimedia Approaches to Waste Mitigation, presented at the Symposium on *Multimedia Approaches to Pollution Control*, sponsored by National Research Council, Board on Environmental Studies and Toxicology, Committee on Multimedia Approaches to Pollution Control, Washington, DC.

Daniels, S. L. Hoerger, F. D. and Moolenaar, R. J. (1985), 'Environmental exposure assessment: experience under the Toxic Substances Control Act', *Env. Toxicol. Chem.* 4, pp. 107–17.

Daniels, S. L., Kuszaj, J. M., and Martin, J. B. (1982), 'Technical and legal implications of the regulatory definition of hazardous waste', *Env. Prog.* 1 (4), pp. 231–6; *Ibid.*, 'Hazardous material and waste management – perspectives of a large integrated chemical manufacturing complex', *Env. Prog.* 1 (4), pp. 236–41.

Daniels, S. L., and McElroy, F. T. R. (1985), 'Innovations in the proper management of hazard', in E. A. Glysson, D. E. Swan, and E. J. Way (eds), *Innovations in the Water and Wastewater Fields* (Boston: Butterworth), pp. 360–88.

Daniels, S. L., Neely, W. B. and Bailey, R. E. (1979), 'Priority pollutant perspectives – regulatory assessment, environmental evaluation, and waste treatment', *Proc. Purdue Ind. Waste Conf.* 34, pp. 287–308.

Daniels, S. L., and Park, C. N. (1984), Multimedia Risk Assessment/Risk Management of Hazardous Waste Constituents, Symposium: *Toxic and Hazardous Wastes: Multimedia Considerations in Risk Assessment*, AIChE Summer National Meeting, Philadelphia, Pa.

Dickson, K. L., Maki, A. W., and Cairns, J., Jr. (eds) (1982), *Modeling the Fate of Chemicals in the Aquatic Environment*, (Ann Arbor, Mich: Ann Arbor Science).

Doll, R. (1971), 'Age distribution of cancer', *J. Royal Statistical Soc.*, Ser. A, pp. 133–66.

Douglas, M. and Wildavsky, A. (1982), 'Risk and Culture: An Essay on the Selection of Technical and Environmental Dangers' (Berkeley and Los Angeles: Univ. Calif. Press).

Dourson, M. L. (1983), 'Regulatory history and experimental support of uncertainty (safety) factors', *Regulatory Toxicology and Pharmacology* 3, pp. 224–38.

Ellis, A. J. (1917), *The Divining Rod, A History of Water Witching*. US Geological Survey, Water Supply Paper 416, 59 pp.

Falco, J. W., Mulkey, L. A., Swank, P. R., Lipcsei, R. E. and Brown, S. M. (1981) (SRI Int.), *A Screening Procedure for Assessing the Transport and Degradation of Solid Waste Constituents in Subsurface and Surface Waters*, Draft.

Fiksel, J. (1981) (ADL Inc.), *Categorization of Priority Pollutants by Exposure Pathways and Effects*, EPA Contract no. 658–01–5949, Draft, Final Report.

Freitag, D., Ballhorn, L., Geyer, H., and Korte, F. (1985), 'Environmental hazard profile of organic chemicals', *Chemosphere* 14, pp. 1589–1616.

Guess, H., Crump, K., and Peto, R. (1977), 'Uncertainty estimates for low-dose-rate extrapolations of animal carcinogenicity data', *Cancer Research* 37, pp. 3475–83.

Haque, R. (ed.) (1980), *Dynamics, Exposure, and Hazard Assessment of Toxic Chemicals* (Ann Arbor, Mich: Ann Arbor Science).

Harris, R. H., Highland, J. H., Humphreys, K., and Rodricks J. V. (1984), 'Comparative risk assessment: Tools for remedial action planning', *Hazardous Waste* 1 (1), pp. 19–33.

Hawley, K. A., and Napier, B. A. (1985), *A Ranking System for Sites with Mixed Radioactive and Hazardous Wastes*, Pacific Northwest Laboratories, Draft Report.

Hedden, K. F. (1984), 'Multimedia fate and transport models: An overview', EPA 600/5k–84–060; *J. Toxicol.-Clin. Toxicol.* 21, 65–95; NTIS PB84–229871.

Hedden, K. F., Mulkey, L. A., and Tucker, W. A. (1982), 'Application of multimedia exposure assessment to drinking water', *Env. Monitoring and Assessment* 2, pp. 57–69.

Howard, P. H., Santodonato, J., and Durkin, P. R. (1982), 'Syracuse Research Corporation's approach to chemical hazard assessment', in *Environmental Risk Analysis for Chemicals 1982*, pp. 379–98.

ICF, Inc. (1984), *The RCRA Risk-Cost Analysis Model, Phase III Report*, prepared for US EPA (Washington, DC: Office of Solid Waste).

Interdisciplinary Panel on Carcinogenicity (American Industrial Health Council) (1984), 'Criteria for evidence of chemical carcinogenicity', *Science* 225, pp. 682–7.

Javandel, I., Doughty, C. and Tsang, C. F. (1984), *Groundwater Transport: Handbook of Mathematical Models*, Water Resources Monograph Series 10 (Washington, DC: American Geophysical Union), 228 pp.

JRB Associates Inc. (1980), Methodology for Rating the Hazard Potential of Waste Disposal Sites, Draft Report, McLean, Va.

Kenaga, E. E. (1979), *Risk Analysis of Pesticides – Environmental Effects Testing, presented at the 178th National Meeting*, ACS, Washington, DC.

Kenaga, E. E. (1980), 'Use of environmental toxicology and chemistry data in hazard assessment: Progress, needs, challenges', in Proceedings of First Annual Meeting, Soc. Env. Toxicol. Chem., *Environmental Risk Assessment – An Integrated Approach* (Rockville, Md: SETC).

Klee, A. J. (1976), 'Models for Evaluation of Hazardous Waste', *J. Env. Eng. Div.*, ASCE 102(EE1), pp. 111–25.

Kleindorfer, P. R., and Kunreuther, H. C. (eds) (1987), *Insuring and Managing Hazardous Risks – From Seveso to Bhopal and Beyond* (New York: Springer-Verlag).

Kletz, T. A. (1977), 'What risks should we run?', *New Scientist* 74 (1051), pp. 320–2.

Koerner, R. M., Parkin, G. F., and Pipes, W. O. (eds) (1986), *Hazardous Wastes Treatment and Disposal*, Research Needs Workshop, sponsored by the National Science Foundation, Drexel University, Philadelphia, Pa.

Korte, F., Freitag, D., Geyer, H., Klein, W., Karus, A. G., and Lahaniatis, E. (1978), 'Ecotoxicological profile analysis', *Chemosphere* 1, pp. 79–102.

Kotzias, D., Geyer, H., Viswanatha, R., Kraus, A., Freitag, D., Klein, W., and Korte, F. (1981), *An Approach to the Ecotoxicological Evaluation of Environmental Chemicals, Toxicol. Aspects* (9th Int. Congr. Eur. Assoc. Poison Control Centers), 1980, pp. 257–67: Chem. Abst. 95, 91829 (1981).

Krewski, D. (1981), Simulation of Low-Dose Extrapolation Procedures. Presented at the workshop on *Biological and Statistical Implications of the ED Study and Related Data Bases*, sponsored for the Society for Toxicology and the National Center for Toxicological Research, Deer Creek State Park, Mt. Sterling, Oh.

Kunreuther, H. C., Linnerooth, J., Lathrop, J., Atz, H., Macgill, S., Mandl, C., Schwarz, M., and Thompson, M. (1983), *Risk Analysis and Decision Processes – The Siting of Liquefied Energy Gas Facilities in Four Countries* (New York: Springer-Verlag).

Lee, G. F., and Jones, R. A. (1982), 'A risk assessment approach for evaluating the environmental significance of chemical contaminants in solid wastes', in Conway, *op. cit.*, pp. 529–49.

Long, F. A., and Schweitzer, G. E. (eds) (1982), *Risk Assessment at Hazardous Waste Sites*, SCS Symposium Series No. 204 (Washington, DC: American Chemical Society).

Lowrance, W. W. (1976), *Of Acceptable Risk – Science and the Determination of Safety* (Los Alsos, Ca: William Kaufman Inc.).

Lyman, W. J., Reehl, W. F. and Rosenblatt, D. H. (Arthur D. Little, Inc.) (1982), *Handbook of Chemical Property Estimation Methods, Environmental Behavior of Organic Compounds* (New York: McGraw-Hill).

Mackay, D. (1979), 'Finding fugacity feasible', *Env. Sci. Tech.* 13, pp. 1218–23.

Mackay, D. and Paterson, S. (1981), 'Calculating fugacity', *Env. Sci. Tech.* 15, pp. 1006–14 .

Mackay, D., Paterson, S., Cheung, B., and Neely, W. B. (1985), 'Evaluating the environmental behavior of chemicals with a level III fugacity model', *Chemosphere* 14, pp. 335–74.

McCall, P. J., Laskowski, D. A., Swann, R. L., and Dishberger, H. J. (1983), 'Estimation of environmental partitioning in model ecosystems', *Res. Rev.* 85, pp. 231–44.

Mantel, N., and Bryan, W. R. (1961), ' "Safety" testing of carcinogenic agents', *J. National Cancer Institute* 27, pp. 455–70.

Michigan Department of Natural Resources, Site Assessment System (AS), June 1983 *et seq.*

National Academy of Sciences, National Research Council (1981a), Assembly of Life Sciences, Board on Toxicology and Environmental Health Hazards, Steering Committee on Identification of Toxic and Potentially Toxic Chemicals for Consideration by the National Toxicology Program, *Strategies to Determine Needs and Priorities in Toxicity Testing* (Washington, DC: National Academy Press).

National Academy of Sciences, National Research Council (1981b), Committee to Review Methods for Ecotoxicology, working papers prepared for background for *Testing for Effects of Chemicals on Ecosystems* (Washington, DC: National Academy Press).

National Academy of Sciences, National Research Council (1983), Committee on the Institutional Means for Assessment of Risks to Public Health, *Risk Assessment in the Federal Government: Managing the Process* (Washington, DC: National Academy Press).

Neely, W. B. (1979), 'An integrated approach to assessing the potential contact of organic chemicals in the environment', in *Workshop Anal. Hazard Eval.*, 1978 (Bethesda, Md: Amer. Fish. Soc.), pp. 74–82.

Neely, W. B. (1980), *Chemicals in the Environment – Distribution, Transport, Fate, and Analysis* (New York: Marcel Dekker).

Neely, W. B., and Blau, G. E. (1985), *Environmental Exposure from Chemicals*, Vols. I and II (Boca Raton, Fla: CRC Press).

Neely, W. B., and Mackay, D. (1982), 'Evaluative model for estimating environmental fate', in Dickson *et al.*, *op. cit.*, pp. 127–43

Office of Technology Assessment (1983), *Technologies and Management Strategies for Hazardous Waste Control*, particularly Chapter 6, pp. 221–61, 'Managing the risks of hazardous waste', and Appendices 7A and 7B, 'Hazard ranking system and risk/cost policy model', pp. 386–97, resp., Report No. OTA-M–196, (Washington, DC: US GPO).

Olsen, A. R., and Wise, S. E. (1982), *Frequency Analysis of Pesticide Concentrations for Risk Assessment (FRANCO Model)*, EPA–600/3–82–044.

Onishi, Y., Brown, S. M., Olsen, A. R., and Parkhurst, M. A., (1980), (Battelle PNL), *The Chemical Migration and Risk Assessment (CMRA) Methodology Summary*.

Ott, W. R., (ed.) (1976), *Proceedings of the EPA Conference on Environmental Modeling and Simulation*, (300 authors), EPA–600/9–76–016.

Ott, W. R. (1978), *Environmental Indices, Theory and Practice*, (Ann Arbor, Mich: Ann Arbor Science).

Park, C. N. and Snee, R. D. (1983), 'Quantitative risk assessment: State-of-the-art for carcinogenesis', *Fund. Appl. Toxicol.* 3, pp. 320–33.

Paterson, S. (1985),'Equilibrium models for the initial integration of physical and chemical properties', Chapter 9, pp 217–31, in Neely and Blau, *op. cit.*

Petak, W. J., and Atkisson, A. A. (1982), *Natural Hazard Risk Assessment and Public Policy – Anticipating the Unexpected*, (New York: Springer-Verlag).

Poston, J. W. (1978), *Measures of Equivalent Hazard of Radionuclides: A Survey*, Report no. Y/OWI/SUB–7278/2, Subcontract 7278, School of Nuclear Engineering, Georgia Institute of Technology, subcontract to Union Carbide Corp., Contract W–7405 eng. 26, Office of Waste Isolation, US Department of Energy.

The Regulatory Council (US) (1979), *Regulation of Chemical Carcinogens 1979* (Washington, DC).

Ruckelshaus, W. D. (1983), 'Science, risk, and public policy', *Science* 221, pp. 1026–8.

Saxena, J. (1981), (1983), (1984), (1985). *Hazard Assessment of Chemicals – Current Developments*, Vols. 1–4, (New York: Academic Press).

Smith, J. H., Mabey, W. R., Bohonos, N., Holt, B. R., Lee, S. S., Chous, T. W., Bamberger, D. C., and Mill, T. (1978), (SRI), *Environmental Pathways of Selected Chemicals in Freshwater Systems*, Parts I and II, EPA–600/7–77–113 and EPA–600/7–78–074.

Squire, R. A. (1981), 'Limitations of histopathology analysis', presented at the workshop on *Biological and Statistical Implications of the ED Study and Related Data Bases*, sponsored by the Society for Toxicology and the National Center for Toxicological Research, Deer Creek State Park, Mt. Sterling, Ohio.

Stanford Research Institute (1978), *Systems for Rapid Ranking of Environmental Pollutants*, EPA–600/5–78–012.

Starr, C. (1969), 'Social benefit versus technological risk', *Science* 165, pp. 1232–8.

Stephenson, M. E. (1977), 'An approach to the identification of organic compounds hazardous to the environment and human health', *Ecotoxicol. Env. Safety* 1, pp. 39–48.

Stern, A. M., and Walker, C. R. (1978), 'Hazard assessment of toxic substances: Environmental fate testing of organic chemicals and ecological effects testing', in Cairns *et al.* (1978) *op. cit.*, pp. 81–131.

Swann, R. L., and Eschenroeder, A. (eds) (1983), *Fate of Chemicals in the Environment*, ACS Symposium Series 225, (Washington, DC: American Chemical Society).

Thibodeaux, L. J. (1979), *Chemodynamics – Environmental Movement of Chemicals in Air, Water, and Soil*, (New York: Wiley-Interscience).

Tinsley, I. J. (1979), *Chemical Concepts in Pollutant Behavior*, (New York: Wiley-Interscience).

Travis, C. C., (1983), *Exposure Assessment Methodology and Reference Environments for Synfuel Risk Analysis*, Oak Ridge National Laboratory, ORNL/TM–8672.

Urban, D. J., and Cook, N. (1986), 'Hazard Evaluation Division Standard Evaluation Procedure', *Ecological Risk Assessment*, EPA 540/9–86/167, NTIS PB 86–247657.

US Air Force (1983), *Installation Restoration Program, Phase I. – Records Search*, Lowry AFB, Colorado, US Air Force AFESC/DEV, Tyndall AFD, Fla, Final Report AD-A133469.

US Department of Agriculture (1980), CREAMS: *A Field Scale Model for Chemicals, Runoff, and Erosion from Agricultural Management Systems*, 3 vols., Conservation Report No. 26.

US EPA, (1982), *Uncontrolled Hazardous Waste Site Ranking System*, 40 CFR Part 300, Appendix A.

US EPA, OTS (1980), *Proceedings of the EPA Workshop on the Environmental Scoring of Chemicals*, compiled by Ross, R. M., and Welch, J., Report ORNL/EIS–158, EPA–560/11–80–010.

Venezian, E. C. (1977), (A. D. Little Inc.), *Pre-Screening for Environmental Hazards – A System for Selection and Prioritizing Chemicals*, EPA–560/1–77–002, 117 pp.

Versar Inc. (1986), *Draft Superfund Exposure Assessment Manual*, EPA Contract No. 68–01–7090.

Verschueren, K. (1983), *Handbook of Environmental Data on Organic Compounds*, 2nd edn, (New York: Van Nostrand Reinhold).

World Commission on Environment and Development (1987), *Our Common Future* (Oxford: Oxford University Press).

6 Control strategies and technologies

D. F. BISHOP

Introduction

This chapter describes control strategies and technologies used to reduce environmental risks from toxics. As an example of needed legislative support, the chapter includes an overview of the supporting legislative infrastructure for the US Environmental Protection Agency (EPA). It also includes as background, discussion of the factors contributing to the complexity of the toxic control problem and five regulatory methods for establishing control limits. Using the strategies evolving from the EPA's legislative infrastructure as comprehensive examples, the chapter examines important individual strategies in five functional control areas used for overall control. The chapter also describes four important types of treatment approach used and examines the advantages and disadvantages of each type. The chapter presents representative technologies within each type of treatment and considers current limitations of the treatment state-of-the-art.

The widely varying types and amounts of organic and inorganic chemicals in our modern society produce inherently complex control problems for the reduction of environmental risks. Toxicity from chemicals is derived both from the molecular and structural properties of compounds and from the elements themselves, such as metals and radioactive isotopes, both as pure substances and within compounds. The toxic effects primarily related to elemental characteristics are usually conserved during treatment while those associated specifically with the molecule and its structure can be destroyed. Consequently, three overall environmental control goals exist: one is to achieve molecular destruction of appropriate toxic substances in wastes and the environment; another is to contain conservative toxicants permanently; and the third is to prevent pollution by source control of toxic chemicals.

The inherent complexity of the control problem requires multiple control strategies for comprehensive risk management of toxic chemicals. Thus, the existing and evolving control strategies described in this chapter have been conveniently classified into five functional areas: regulation of production and use; waste minimization; control of release; waste treatment; and clean-up of contamination. The first two functional areas and portions of the third can be collectively called pollution prevention.

The order in which the areas are listed is a desirable applications sequence for the control of toxics. Appropriate regulation of production or use, waste minimization and control of release should precede waste treatment and clean-up of contamination. This sequence reduces the compositional complexity and cost of waste treatment and remediation. Control strategy areas that are not addressed in this chapter include those used to ensure food and drug quality, strategies for worker health and safety, and community strategies for evacuation and/or emergency medical support in response to catastrophic releases of toxics.

The development and implementation of the control strategies to achieve the overall environmental goals in the selected five functional areas requires co-operative interaction between the public and private sectors and an appropriate governmental legislative infrastructure. As the most comprehensive example, a diverse legislative infrastructure, driven by the inherent control complexity, has evolved to provide control authority for the US EPA, although there are other agencies with responsibility for control of chemicals, especially in the workplace. The principal legislative acts of the infrastructure in the United States are presented in Table 6.1. They provide a framework for control and clean-up of the multimedia dispersal of toxic chemicals.

While this extensive legislative infrastructure produces regulatory overlap, its diverse elements are needed for effective control of air, water and land releases and for clean-up of contamination. In the United States inconsistencies from overlap of the diverse regulatory programs are minimized by the EPA through the use of a Risk Assessment Forum. The Forum's goals are to promote consensus on scientific issues across regulatory offices and to ensure that the consensus is incorporated into appropriate guidance.

Before considering the individual control strategies, it is important to identify the factors contributing to the complexity of toxic control problems and regulatory methods. Important factors include:

- diversity of physical, chemical and biochemical properties of the toxic compounds;
- interacting multimedia (air, land, and water) transport pathways for toxics;
- diversity in chemical producers and users;
- past disposal or use practices requiring site clean-up;
- uncontrolled or accidental releases of toxics during production, storage, transport, use and disposal;
- diversity of regulatory responsibility.

Table 6.1 Types of control strategies, as related to regulatory authority of the US Environmental Protection Agency

Type of control strategy	Applicable acts (major functions)
Regulation of producion and use	Federal Insecticide, Fungicide and Rodenticide Act (FIFRA) of 1947 and its amendments Toxic Substances Control Act (TSCA) of 1976 Asbestos Hazardous Emergency Response Act (AHERA) of 1986 Clean Air Act (CAA) of 1970 and its amendments
Waste minimization	Resource Conservation and Recovery Act (RCRA) of 1976 Hazardous and Solid Waste Amendments (HSWA) of 1984
Control of release	Clean Water Act (CWA) of 1972 and its amendments CAA and its amendments Marine Protection, Research and Sanctuaries Act (MPRSA) of 1972 and its amendments Comprehensive Environmental Response, Compensation and Liability Act (CERCLA) of 1980 The Superfund Amendments and Reauthorization Act (SARA) of 1986 Safe Drinking Water Act (SDWA) of 1986 TSCA; RCRA; and HSWA Uranium Mill Tailings Radiation Control Act (UMTRCA)
Waste treatment	CWA and its amendments CAA and its amendments RCRA and HSWA CERCLA and SARA
Clean-up of contamination	CWA and its amendments CAA and its amendments CERCLA; SARA; and AHERA SDWA

The selection of control technologies or management practices in a control strategy to reduce risk, like the fate of toxics in the environment, is substantially determined by the first two factors, the diversity of properties and the multimedia transport pathways. Detailed consideration of the effects of the properties of toxics on the multimedia transport pathways and on the treatment (toxic removal) mechanisms occurring in control technologies is inappropriate here. Discussion of the effects of the widely variable properties of toxics and their interacting multimedia transport pathways on a widely used sedimentation/activated sludge (biological) wastewater treatment system is presented later and provides some perspective of control complexity.

The third factor, the diversity in chemical producers and users, ranging from the bulk producers of agricultural and industrial chemicals to manu-facturers of the sophisticated formulations of the pharmaceutical industry, produces major challenges for devising effective toxic control strategies. Since the toxicity and use of compounds vary widely, production levels alone are not an indicator of probable toxic problems. Relatively high levels of production or use of a compound with known substantial toxicity, however, help establish control priorities for addressing the wide variety of individual compounds produced by industry.

The manner in which toxics are produced, used and discharged into end-of-pipe waste treatment systems or directly into the environment significantly contributes to the complexity of control. Widespread pro-duction or use generate mixtures of wastes or wastewaters of widely variable concentrations and content at many sites. The wastes discharged with intermittent and irregular frequencies into treatment systems thus often contain unidentified toxic chemicals from many different sources. The intermittent discharges of complex mixtures of chemicals with vari-able constituents and concentrations can produce significant treatment difficulties.

As examples, slug discharges can inhibit widely used biological treatment processes and even produce catastrophic health and safety inci-dents (us Environmental Protection Agency, 1986). Further, the uni-dentified complex mixtures of toxic chemicals, even if highly diluted as in municipal wastewater effluents, can produce substantial environmental impacts, such as rivers devoid of fish (Figure 6.1). Indeed, substantial amounts of unknown genotoxic substrates (Figure 6.2) with unknown health or environmental impacts (Meier et al., 1987) have been observed passing through biological wastewater treatment systems, even while the plants meet their permit requirements for conventional pollutants. Thus, with diversity in production and use, the tracing to source, identifying the specific toxicants and characterizing appropriate treatability or control techniques for these complex mixtures force development of innovative control strategies, sophisticated chemical and biological monitoring tools and multiple control technology approaches.

With the environmental health and ecosystem risks associated with modern chemicals not widely appreciated until recent decades, careless use and poor disposal practices, especially into uncontrolled landfills, pits, ponds and lagoons and directly onto land, produced many cases of locally severe contamination. This contamination, often consisting of complex toxic mixtures, involves geological and hydrological variability, and requires increasing innovation in control technologies to contain and clean up.

The impacts of the last two factors, uncontrolled release of chemicals and the diversity of regulatory responsibility, are best illustrated by uncontrolled air emissions of toxics. Air emissions contribute to major environmental problems including acid rain, deterioration of the upper atmospheric

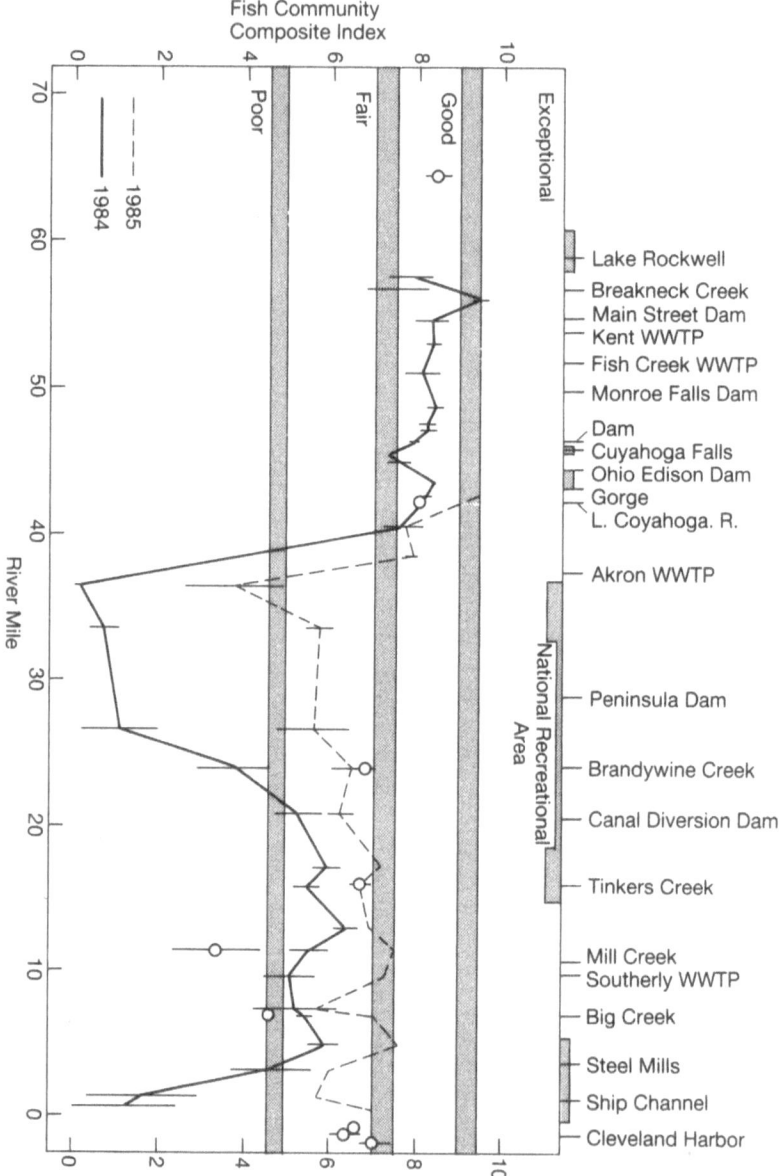

Figure 6.1 Cuyahoga River fish survey data. *Source*: Neiheisel *et al.* 1988

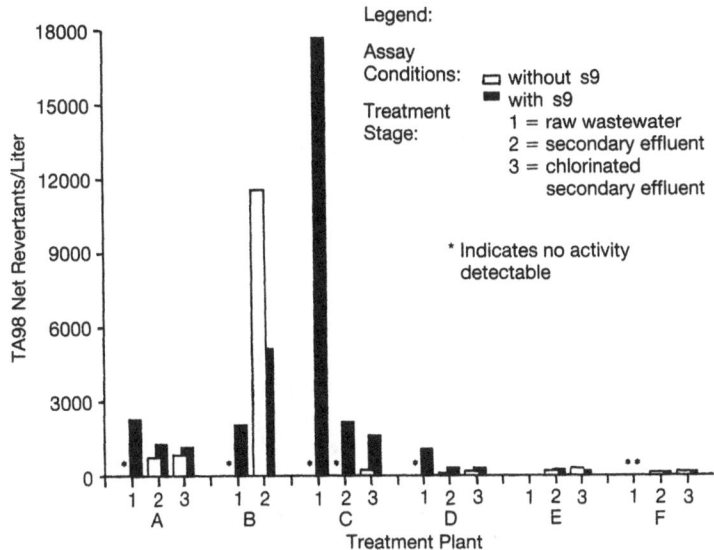

Figure 6.2 TA98 mutagenic activities of second series of municipal wastewater samples at different stages of treatment

ozone layer, production   smog with high ozone concentrations at ground level and even global warming. Uncontrolled air emissions of volatile or air dispersable toxics involve large numbers of emission sources ranging from mobile sources to large industrial standing sources. These sources include waste treaters and abandoned waste disposal sites. With rapid and uncontrollable dispersal after release, control of toxic air emissions is essential and one of the most important challenges in toxics management. It requires strategy elements in all five functional areas presented in this chapter.

Currently in the United States, air emissions are regulated by a variety of environmental legislation. The responsibility for developing regulatory strategies and the knowledge of environmental impacts is distributed over various federal agencies. The diverse regulatory structure unfortunately is not integrated and awaits the development of new Clean Air Act amendments to ensure effective co-ordination of the required multiple control strategies. Regulatory co-ordination for such multiple control strategy approaches requires tiered programs with co-operation between international, national, regional (State, Province, etc.) and local authorities.

In developing toxics control strategies, five important regulatory methods are used to establish control limits. One regulatory method uses chemical-by-chemical concentrations or waste application limits. Here permitted amounts of chemicals or substances are established to prevent damage to the ecosystem or to human health. The chemical-by-chemical method, based on control criteria or standards, is most effective when detailed knowledge concerning the toxicological and ecotoxicological nature and the multimedia fate of each chemical is available.

The large financial resources required for the fully definitive chemical-by-chemical approach and the practical limitations of technology in achieving selected media quality standards has led to an alternative chemical-by-chemical method called technology-based effluent or discharge limitations. In technology-based limitations, specific chemicals in representative industrial wastewater categories, in listed hazardous waste classifications and at contaminated sites are regulated by chemical-by-chemical concentration limits on the treated discharges from applicable control technologies. These limits are established by evaluating the capability of the control technologies to remove, contain or destroy the important toxics in various wastewaters and hazardous wastes and at contaminated sites.

A third method for establishing control limits is used at sites contaminated by past disposal practices and by accidental or unavoidable releases. The objective is to leave concentration residuals at the site after clean-up equal to the background concentrations found in adjacent uncontaminated environments.

A fourth regulatory method, the establishment of the permissible levels of toxicity from complex mixtures discharged into receiving media, uses a modified media, quality-based approach. This media, quality-based method includes, when available, chemical-by-chemical limits established by media quality criteria or standards. The approach in water also features aquatic toxicity testing (US Environmental Protection Agency, 1985) on waste effluents to establish effluent limits for toxicity from unidentified complex mixtures of toxics. Dilution of the discharge by the receiving water is usually considered in establishing the toxicity limits for the discharge. The approach is now being applied to control toxicity impacts on aquatic ecosystems.

The fifth regulatory method, the mass balance, is used to examine the fate of toxic chemicals during application of management practices and control technologies and to aid the development of regulatory discharge limits (waste load allocation) for chemicals. It is also used as a regulatory limit for selected control approaches such as the 99.99 per cent destruction and removal efficiency for thermal destruction methods applied in the United States under the Resource Conservation and Recovery Act.

Control strategies

The US legislative structure has successfully generated detailed governmental strategies in all control areas presented here. It also is generating major research and development on the methods used. Because of the broad range of US legislation to regulate toxics, the described control strategies are generally those evolving in response to the US legislative structure. Although supporting legislative structures in other countries are less extensive, control strategies, similar to those arising from the US regulatory framework, are evolving worldwide with supporting development of control tools. In the future, the application of control strategies generated by such structures will be governed by risk assessment/risk management decision systems, such as described in Chapter 12, to permit more effective environmental protection.

The scope of control problems for toxics requires presentation of generalized control strategy approaches, supplemented with selected examples to illustrate individual control problem complexity and adopted strategy. The control tools used in the strategies, for convenience, are divided into control technologies and management practices. In this discussion, control technologies are treatment systems used to remove or destroy the toxic chemicals as, for example, in the waste treatment control area. Management practices are techniques to reduce or control the release of chemical wastes entering treatment systems and the environment. Completely distinct separation of control technologies and management practices does not occur and combinations of the tools are applied in all control strategy areas. In this discussion, the diverse control strategies will be presented by principal functional area with recognition that individual strategies often overlap functional areas. Control approaches for most toxic environmental problems usually involve integration of control strategies from more than one of the described five functional areas.

Regulation of production and use

Government regulation of production and use of toxic chemicals can be divided into two general approaches. The first is the review or re-review of the environmental impacts of chemicals or substances currently in production. Control options include bans on the production or limits on use of a substance. The second approach is the review of proposed new chemicals or substances to establish requirements for production and use and, if appropriate, to prevent the future production of high-risk substances. To reach regulatory decisions, both overall approaches in this control area use chemical-by-chemical evaluation methods. They consider technical data (known or estimated) on toxicology, ecotoxicology, persistence, biomagnification, environmental fate and transport, production and use levels or projections, and treatability or control by appropriate technology and management practices.

In the United States, pesticide regulation was recognized in the 1940s as needing a legislative support structure (Table 6.1) and evolved as one of the first legislative acts to regulate production and use of chemicals. Later, the Toxic Substances Control Act was enacted to regulate other toxic chemicals. With separate legislation, an important difference evolved in US strategies for regulation of pesticides and other toxic chemicals. The difference is that cost/benefit assessments are used in regulatory decisions on pesticides and are not used in the control strategies on other toxics. In the United States, pesticide regulation features registration (licensing) for new products and re-registration of existing products. Extensive labelling requirements also are employed to encourage proper use of products. In the United States, the mechanism for control of new substances, other than pesticides, is called a pre-manufacturing notification (PMN) review.

Regulation of production and use, when resulting in a complete ban on use of existing chemicals, should include procedures for management of existing unused stocks of the chemical. The US legislative structure requires that the EPA purchase residual inventories and assume responsibility for the costs of disposal of unused inventories. Control technologies such as thermal destruction are considered for managing unused stocks of banned pesticides and other chemicals.

Important early examples of the strategies for regulation of production and use resulted from various legislative acts. One strategy evolved under the Clean Air Act to reduce the use and eventually to eliminate lead 'anti-knock' compounds in gasoline. The strategy first eliminated lead compounds from gasoline used in new engines in the automotive fleet. It then produced schedules for gradual reduction of the use and concentration of lead in gasoline, consistent with the gasoline octane requirements for the changing automotive fleet. As an adjunct to the control of lead, a parallel strategy was developed to reduce consumption of gasoline and engine emissions of unburned hydrocarbons. One of the hydrocarbons, benzene, is a carcinogen regulated by the US Occupational Safety and Health Administration (OSHA). High levels of the aggregated hydrocarbon emissions also produce smog and high ozone levels in cities. The governmental strategy gradually increased the average mile-per-gallon performance level for the automotive fleet and thus forced reduction in engine size and power. The reduced gasoline consumption, coupled with a mandated use of catalytic converters to destroy the unburnt hydrocarbons in the engine emissions, led to fewer released volatile hydrocarbons.

A second example of the regulation of production and use evolved in the United States for the banning of selected asbestos products such as insulation for buildings. The strategy also includes clean-up of asbestos-contaminated sites such as in schools. The control level in the asbestos clean-up component of the overall strategy uses background air concentration levels from adjacent uncontaminated areas as the regulatory target as well as health established particle concentrations in air. Other important examples of these types of control strategies are restrictions on pesticide use and the banning of selected pesticides, and the world-wide agreement on production limits for the volatile halogenated hydrocarbons that reduce upper atmospheric ozone.

Waste minimization

Because of increasing technical complexity, costs and, for the near term, limited available capacity in 'end-of-pipe' treatment technologies and in technologies for clean-up of contamination, waste minimization is receiving world-wide emphasis. Control strategies in this area are applied to industrial production and use of toxic chemicals to reduce waste and the subsequent need for treatment, storage and disposal of hazardous wastes. The strategies feature four management approaches as shown in Table 6.2 (us Environmental Protection Agency, 1987a). Treatment of segregated wastes may also be employed.

Since these approaches involve management and operation of chemical production or use facilities, governmental practices in Western Europe, Japan and North America to date generally rely on co-operative voluntary efforts (us Environmental Protection Agency, 1987a). The government practices (Table 6.3) include incentives and information exchange. They also try to identify and overcome obstacles to successful waste minimization, but usually do not employ mandatory performance or other regulatory methods.

In addition to governmental encouragement, waste minimization will be essential for competitiveness in the chemical industry. Obstacles to the use of waste minimization (Freeman and Eby, 1987) include industry fear of

Table 6.2   Waste minimization approaches and techniques

Inventory management and improved operations	*Production process changes*
Inventory and trace all raw materials	Substitute nonhazardous for hazardous raw
Purchase fewer toxic and more nontoxic production materials	materials
	Segregate wastes by type for recovery
Implement employee training and management feedback	Eliminate sources of leaks and spills
	Separate hazardous from non-hazardous
Improve material-receiving, storage and handling practices	wastes
	Redesign or reformulate end products to be
	less hazardous
Modification of equipment	Optimize reactions and raw material use
Install equipment that produces minimal or no waste	
Modify equipment to enhance recovery or recycling options	*Recycling and re-use*
	Install closed-loop systems
Redesign equipment or production lines to produce less waste	Recycle on-site for re-use
	Recycle off-site for re-use
Improve operating efficiency of equipment	Exchange wastes
Maintain strict preventive maintenance program	

Table 6.3 Waste minimization practices in various countries

	Japan	Canada	Germany	Sweden	Netherlands	Denmark
Tax incentives						
Waste end taxes	●		●		●	●
Tax incentives		●	●			
Economics						
Price support system for recycling	●	●	●			
Government grants as subsidies	●		●	●	●	●
Low interest loans			●		●	
Technical assistance						
Information and referral service	●	●	●		●	●
Site consultation	●	●				●
Training seminars			●			
R & D assistance						
Technical development labs	●	●	●	●	●	●
Demonstration projects			●	●	●	
Industrial research				●		
Permits and plans						
National waste management plans					●	
Waste reduction agreements	●			●		
Waste reduction as a part of permits						
Waste exchange						
Regional waste exchanges	●	●	●		●	●
Public information						
Focus on corporate image	●					
Focus on consumer practices			●			

Source: Center for Environmental Assessment (1986).

reducing the quality of products, technical barriers in process design and lack of engineering information on source reduction and recycling techniques.

An evolving tool in the promotion of waste minimization is called the waste minimization assessment or 'audit' procedure (US Environmental Protection Agency, 1987a). This procedure (Table 6.4) produces a step-by-step list of waste reduction opportunities and ranking guidance for selecting the optimum option for any industry.

Table 6.4   Suggested steps of a waste minimization assessment

1. Prepare background material for the assessment
2. Conduct a preassessment visit to identify candidate waste streams
3. Select waste streams for detailed analysis
4. Conduct a detailed site visit to collect data on selected waste streams and controls and related process data
5. Develop a series of potential waste minimization options
6. Undertake preliminary option evaluations (including development of preliminary cost estimates)
7. Rank options by: waste reduction effectiveness; extent of current use in the industry; potential for future application at the facility
8. Present preliminary results to plant personnel along with a ranking of options
9. Prepare a final report, including recommendations to plant management
10. Develop an implementation plan and schedule
11. Conduct periodic reviews and updates of assessments

Successful examples of waste minimization (US Environmental Protection Agency, 1987a) include:

- One of the largest chemical manufacturers in the Netherlands uses waste segregation, removal of solvents in water solutions by distillation, and other source reduction measures to reduce the company's annual wastewater output by 80 per cent.
- In Sweden, a major pharmaceutical producer initiated a program to recycle approximately 10,000 tons of hazardous waste solvents per year through the company's onsite distillation plant, thereby reducing by 60 per cent the amount of solvent waste that was shipped off site for disposal.

Examples of the potential for waste minimization are given by the waste minimization audit (Drabkin and Rissman, 1987) performed at an electric arc steelmaking complex. The procedure was applied to a stainless steel pickling facility in the complex and identified as a technically and economically feasible recycling option for the recovery of fluorspar (calcium fluoride) from a corrosive waste stream. The recycling approach produces a projected annual saving of $168,000 US and a 30 per cent reduction in final waste disposal volume.

Control of release

In this control approach, strategies for regulation of production and use and waste minimization are linked with the strategies in waste treatment and clean-up of contamination. The transition strategies in control of release are often placed in one of the other strategy areas since these strategies usually overlap or strongly interact with control actions in the other four areas. Important strategies, often still actively evolving, are:

- pretreatment of industrial wastewaters and hazardous wastes;
- regulation of the disposal of municipal sewage sludges;
- regulation of ocean disposal of wastes and sludges;
- control of disposal of hazardous wastes;
- air emissions control;
- indoor air pollution control;
- minimization of accidental releases.

The US strategy for pretreatment of industrial wastewaters before discharge to municipal wastewater collection and treatment systems established, in a tiered regulatory approach under the CWA, national control limits on selected toxics called 'priority pollutants' for selected 'categorical' industries (Table 6.5). The strategy used local control limits for other industries and commercial waste generators. The EPA was authorized under the Act to prescribe 'best management practices' (BMPs) to strengthen toxics control in wastewater treatment. These BMPs were selected to prevent the release of toxic and hazardous pollutants from plant site runoff, spillage or leaks, sludge or waste disposal and drainage from raw material storage when these events are associated with or are ancillary to the manufacturing or treatment process.

In-place elements of the pretreatment strategy include use of waste discharge control limits established by chemical-by-chemical methods (technology based) for existing sources and for new sources of chemicals. The evolving elements of the strategy will use material balance approaches (wasteload allocations) to establish allowable discharges for individual industrial users of the central treatment system. Toxicity testing with control limits for complex mixtures of chemicals also are likely to be included. Since discharges of toxic chemicals into central municipal treatment systems may lead to health and safety hazards, the evolving strategy is also likely to include national regulations containing elements to manage slug (intentional batch and accidental) discharges.

Industrial dischargers, under local regulations, often already are required to report accidental spills into sewers and to develop improved management practices to minimize spills and to control releases. An example of a simple practice is to install holding capacity to permit controlled intentional release of waste batches and thus to minimize impact and improve

Table 6.5 Industrial categories subject to United States National Effluent Limitations and Standards

Aluminum forming	Metal finishing[b]
Asbestos mfg	Metal molding and casting
Battery mfg	Mineral mining
Builder's paper	Nonferrous metals forming
Carbon black mfg	Nonferrous metals mfg
Cement mfg	Oil and gas extraction
Coal mining	Organic chemicals and plastics and
Coil coating	synthetic fibers[c]
Copper forming	Paint formulating
Dairy products processing	Paving and roofing (tars and asphalt)
Electroplating[a]	Pesticides
Electrical and electronic components	Petroleum refining
Explosives mfg	Pharmaceuticals
Feedlots	Phosphate mfg
Ferroalloy mfg	Photographic
Fertilizer mfg	Plastics molding and forming[d]
Fruits and veg. processing mfg	Porcelain enameling
Glass mfg	Pulp and paper
Grain mills mfg	Rubber processing
Gum and wood chemicals	Seafood processing mfg
Hospitals	Soaps and detergents mfg
Ink formulating	Steam electric
Inorganic chemicals	Sugar processing mfg
Iron and steel mfg	Timber products mfg
Leather tanning and finishing	Textiles[d]
Meat processing	

[a] Cross reference to metal finishing
[b] Cross reference to electroplating
[c] 'Organic chemicals and manufacturing' has been combined with the 'plastics and synthetics' point source category; pretreatment standards for new sources are still in effect
[d] Category is regulated only by the general pretreatment standards

treatability at the central plant. The central treatment system receiving the discharges also should develop response programs to monitor for spills and to establish emergency response procedures. Finally, health and safety controls such as levels of exposure Immediately Dangerous to Life or Health (IDLH) and Lower Explosive Limits (LEL) and emergency response action plans are likely to be incorporated in the evolving pretreatment control strategy.

A component of the pretreatment approach uses toxicity testing of complex waste mixtures in Toxicity Reduction Evaluation (TRE) procedures. The TRE methods such as those developed by Botts et al. (1989) for municipal wastewater treatment systems provide systematic procedures for tracing toxicity to its sources and for identifying, if possible, the individual toxics causing the toxicity in the wastewater effluent. The TRE also provides procedures for evaluating treatability of the toxicity, either at the central treatment plant or through pretreatment or waste minimization techniques. The TRE methods, with sensitive toxicity testing tools, will play an increasingly important role in the management of discharges of industrial and hazardous wastes into central treatment systems, both as part of the pretreatment strategy, and as part of the waste treatment strategies.

The early US regulation for the utilization and the disposal of municipal sewage sludges, excluding marine disposal (Code of Federal Regulations, 1979), emphasized pathogen control and regulation of cadmium and polychlorinated biphenyls (PCBs). In this strategy, the regulation established total loading levels on agricultural land to prevent excessive uptake of cadium into the food supply.

Currently, a sludge utilization and disposal strategy (Federal Register, 1989) is evolving under the CWA to address additional specific toxic chemicals and to control groundwater contamination from land disposal of sludges using the maximum contaminated level (MCL) of specific chemicals in drinking water as control targets. In the strategy, municipal sewage sludges which exhibit leachable toxic chemical concentrations exceeding limits established through the toxicity characteristic leaching procedure (TCLP) (Federal Register, 1986) would be managed under RCRA hazardous waste regulations. Limited surveys (Walker, 1987) reveal that most municipal sludges should pass the TCLP (all eighteen diverse sludges in the surveys passed) and thus be regulated by the evolving municipal sludge strategy.

The municipal sewage sludge strategy addresses the principal sludge utilization and disposal practices; these are application to land, distribution and marketing (sludges as fertilizers, compost, etc.), sludge-only landfills and incineration. Regulation of sludge and refuse co-disposal landfills and co-incineration are evolving separately under solid waste regulations.

The municipal sewage sludge strategy will be the first comprehensive regulatory method to use risk assessment methodologies (Bruins et al., 1987) to establish control limits for selected toxic pollutants. The method employs algorithms which permit calculation of health and ecosystem exposure and risk for the important exposure pathways in the various disposal practices. The method sets sludge pollutant concentrations, loading rates (kilograms/hectare) or emission concentrations, as appropriate, depending upon the disposal practice or technology. The evolving sludge disposal strategies propose incineration control requirements for air release of toxic chemicals, especially the conservative toxicants associated with metallic residuals and the release of toxic organic products of combustion.

An interim strategy for ocean disposal of wastes and sludges in the United States has considered two basic control limits, chemical-by-chemical control limits and toxicity testing control limits. The interim ocean strategy uses the controlled release of chemicals in the wastes and sludges in marine areas outside the continental shelves with dilution to concentrations not exceeding marine water quality criteria or whole waste bioassay limits (Walker et al., 1987). The bioassay limits are established from toxicity testing of the undefined complex mixtures in the waste. The strategy considers biomagnification of selected toxics such as mercury and highly chlorinated organics in establishing water quality criteria.

The disposal of municipal sewage sludge in the United States, by far the largest waste-category in ocean disposal, has been permitted at one site, the ocean dump site 106 miles off the New York/New Jersey coast. Ocean disposal of municipal sludge is currently practised by other nations such as the United Kingdom with discharges into the North Sea. Key elements in the interim US ocean disposal strategy include monitoring of toxic contaminant distributions at the 106-mile ocean disposal site (O'Connor et al., 1985) and developing modeling methods to establish wasteload allocations for disposal of sludges at this site (Walker et al., 1987).

A final legislative strategy in the United States bans ocean disposal of municipal sludges and industrial wastes (Statutes at Large, 1988). The strategy features a gradual phasing out of ocean disposal with fees charged for disposal during the phase-out period.

The strategy for controlled disposal of hazardous wastes (RCRA and HSWA) features far-reaching concepts for control of wastes containing toxic chemicals. The strategy includes the following major elements:

- the listing (classification) of hazardous wastes and cradle-to-grave manifesting of listed wastes to prevent uncontrolled release;
- banning of liquid hazardous wastes in land disposals;
- mandated storage of hazardous waste residuals in secured landfills;
- groundwater monitoring and control requirements;
- controls on air emissions from hazardous waste landfills, surface impoundments, waste piles and waste treatment storage and disposal facilities;
- control of blending and burning of hazardous wastes;
- corrective action for release of hazardous wastes at permitted hazardous waste facilities, beyond facilities boundaries and from underground storage tanks.

The criteria for listing hazardous waste classes are broadly defined in the regulations (Code of Federal Regulations, 1987) such that the EPA Administrator may list a class of wastes as hazardous if he has reason to believe that individual wastes within the class typically or frequently are hazardous as defined in the RCRA. Each waste (Table 6.6) is then assigned an EPA Hazardous Waste Number. This number is used in record keeping, reporting and for compliance controls in cradle-to-grave manifesting of the listed wastes. This manifesting produces major administrative and liability requirements for the control of disposal of generated hazardous wastes.

In addition to cradle-to-grave control of the listed hazardous wastes, two key strategy elements produce major impacts on the control of disposal of hazardous waste. These are banning of liquid hazardous wastes in land disposal and mandated storage of immobilized hazardous waste

Table 6.6   Resource Conservation and Recovery Act (RCRA) hazardous wastes

Hazard codes:	Ignitable (I), corrosive (C), reactive (R), EP toxic (E), acute hazardous (H) and toxic (T)
Listed wastes:	
D wastes	Characteristic wastes ignitable (I), corrosive (C), reactive (R) and EP toxic (E)
F. wastes	Non-specific sources of hazardous wastes such as spent halogenated solvents, spent cyanide baths, quenching bath residues, etc.
K wastes	Specific sources of hazardous waste such as distillation bottoms, tars and ends and process residues and waste streams from specific chemical production from organic, inorganic, pesticides and other chemical manufacturing
U wastes	Discarded commercial chemical products off-specification species, container residues and spill residues that are classified as toxic (T) wastes
P wastes	Discarded commercial chemical products, off-specification species, container residues and spill residues that are classified as acute hazardous (H) wastes

Source: Code of Federal Regulations (1987).

residuals in secured landfills. The two elements should reduce substantial environmental releases of hazardous substances from current land disposal (secured landfills) and eliminate the costly clean-up of these sites.

Strategies to manage air emissions of toxics include a range of pretreatment approaches and management practices to minimize release from large numbers of small sources as well as from large sources. In the evolving overall approach, the selection of important sources employs mass balance estimations and risk assessments to establish control priorities. Unfortunately the large number of small point or standing sources (gas stations, refrigeration systems, dry cleaners, etc.) and mobile sources in aggregate contribute substantial amounts of air emissions and require management practices that are intrusive to the entire community. As indicated earlier, the overall integration of the complex issues and resulting multiple control strategies from various environmental legislative acts into a comprehensive control approach has not yet been accomplished. Specific components in the overall approach to minimize uncontrolled air releases of toxics include or are likely to include:

- pretreatment of wastes for removal of volatile toxics before discharge to central industrial and municipal wastewater treatment facilities and control of air emissions in some wastewater treatment facilities;
- control of emissions during treatment in hazardous waste treatment storage and disposal facilities;
- control of emissions through recycle and re-use, during handling and

refilling of chemical and fuel storage tanks, and maintenance and recharging of refrigeration systems;

- control of emissions during solvent production or use in industrial and commercial applications including improved housekeeping and maintenance programs;
- control of emissions during incineration of wastes;
- control of emissions of volatile and dispersable toxics in Superfund site clean-up (see p. 165);
- reduction of hazardous emissions especially oxides of sulfur in power plant stack gases, including cleaning of dirty fuels to reduce emissions;
- improved application methods to minimize air dispersal of pesticides;
- reductions of oxides of nitrogen and carbon monoxide emissions in the transportation industry.

Indoor air pollution (us Environmental Protection Agency, 1988) causes significant human exposure through toxics build-up in buildings designed with low air exchange rates for efficient heating and cooling. In contrast to uncontrolled releases into open air, the containment of toxics such as radon gas seeping into buildings or formaldehyde from building materials requires increased air exchange to minimize the build-up. Major elements of the indoor air pollution control strategy related to toxic chemicals include: control of dispersed emissions of asbestos and lead in contaminated buildings; increased exchange of air and sealing of radon entry routes in buildings contaminated with radon gas; and exchange of air and control of organic compounds such as pesticides, benzene, cleaning solvents, formaldehyde, tobacco smoke, and other combustion products.

The final control of release strategy involves management and engineering practices to minimize accidental releases. Such practices are incorporated as components of most chemical production, storage, use and disposal facilities. They include the use of appropriate materials of construction, the provision of containment capacity in the event of a spill or emission and the use of effective maintenance, housekeeping and operations control practices to ensure proper functioning and safety within the facility.

Waste treatment

Waste treatment strategies refer to control of toxic chemicals through application of the treatment technology to the 'end-of-the-pipe' waste discharges and to specific waste streams in various industrial and chemical production processes. These wastes often are in liquid form but may include solid wastes from which selected toxics leach into water or volatilize into air. As described in the other control areas, treatment technologies also are extensively applied in control of release (pretreatment), in waste minimization and in clean-up of contamination.

The first comprehensive regulation of toxic chemicals in waste discharges occurred under the United States' CWA in response to a court order consent decree (NRDC v. Train, 1976) for a group of 129 'priority pollutants'. The resulting wastewater treatment strategy established integrated technology based limitations for control of both conventional and priority pollutants in industrial and municipal wastewaters. The wastewater treatment strategy included the following major elements:

- Developing national industrial categories and subcategories for twenty-one major groups of industries classified by their Standard Industrial Classification (SIC) code. (These groups were subsequently divided into thirty eight industrial categories (Table 6.5) known as 'priority' projects.)
- Establishing effluent limitations in mg/l and new source and pretreatment standards from the characterization studies. (The regulations using the effluent limitations did not require any particular control technology but they did require treatment plants to achieve effluent limitations through proper operations of any equivalent technology.)

The overall control concept under the CWA recognized two general approaches to pollutant reduction, pretreatment or in-plant process control (waste minimization) and end-of-the-pipe treatment. The end-of-the-pipe treatment included preliminary treatment, primary (sedimentation) treatment, secondary (usually biological) treatment and advanced waste treatment sometimes called tertiary treatment. The integrated pollution control regulations were based on three levels of treatment technology for both conventional (biological oxygen demand, suspended solids and nutrients) and toxic pollutants and included a control application schedule.

The first level was best practicable technology currently available (BPT). Existing industrial direct discharges in the earliest version of the strategy were to achieve effluent limitations for conventional pollutants based on the application of the BPT by July 1, 1977. The BPT level represented the average of the best existing performances of plants within the industry of various age, sizes, processes, or other common characteristics. When existing performance was uniformly inadequate, BPT could be transferred from a different industrial subcategory or category. BPT also focused on end-of-pipe treatment rather than process changes or internal controls, except when these technologies were common industry practice.

Second, existing industrial direct discharges later were to achieve effluent limitations, including toxic priority pollutants, based on the application of the best available technology (BAT) economically achievable by July 1, 1984. This level represented the best economically achievable performance of plants varying in age, size, processes, or other shared characteristics. Factors considered in defining BAT included the age of the equipment and facilities involved, the processes used, engineering aspects of the control technology, process changes, non-water-quality environmental impacts (including energy) and the costs of applying such technology. Uniformly inadequate performance within a category or subcategory required transfer of BAT from a different subcategory or category. Many BATs included process changes or internal controls, even when these technologies were not common industry practice.

Third the level of technology control in the final integrated strategy was not an additional limitation but replaced BAT for the control of conventional pollutants and was also to be achieved by July 1, 1984. The factors considered in defining the best conventional pollutant control technology (BCT) for an industry were the relationship between the cost of attaining a particular improvement in effluent and the resultant benefits attained, plus a comparison of the cost and the level of reduction of such pollutants by publicly-owned treatment works (POTWs) and industrial plants. The Act required that BCT limitations be assessed according to a two-part 'cost–reasonableness' test. The EPA must find that limitations are 'reasonable' under both tests before establishing them as BCT. In no case can BCT be less stringent than BPT.

The specific control limits (as mg/l) of the important priority pollutants found in each industrial subcategory were incorporated into the industrial wastewater treatment plant's National Pollutant Discharge Elimination System permit. The overall strategy also included the in-place elements of the pretreatment strategy as described earlier. Thus, the control of toxics not fully treatable by the POTW was achieved in the strategy through pretreatment established nationally for 'categorical' industries discharging to POTWs and through pretreatment established by local control limits for other sources of toxics.

Although the CWA waste treatment strategy was not legislatively restricted to 'priority pollutants', its application has been used chiefly to control them. Thus, to complement toxics control by the wastewater treatment strategy under the CWA, the US Congress expanded the technology-based limitations for control of toxics to include the broad range of 'listed' hazardous wastes regulated under the RCRA (Table 6.1). These hazardous wastes, in five major categories (Table 6.6), are described earlier in the control of release strategy area.

In the RCRA hazardous waste treatment, the EPA is required to promulgate regulations specifying treatment or equivalent treatment control levels so that the short-term and long-term threats of these wastes to human health and the environment are minimized. In the strategy, RCRA listed hazardous wastes are treated to treatment levels or by methods specified in regulations and the waste residuals from the treatment technologies are disposed of in a secured land disposal facility. The evolving strategy with legislated control achievement dates is based upon the level of existing technology and is called best demonstrated available technology (BDAT).

The waste treatment strategy using BDAT employs the most effective, existing, full-scale technologies which are commercially available or licensed to establish regulatory control levels. For wastes with volatile toxics, the technologies must prevent air emissions of the toxics. The treatment level achieved by the technologies selected for a hazardous waste may be established using full-scale or pilot testing. The actual treatment applied to specific regulated wastes, however, may be any technology capable of meeting the established level of treatment. The bulk of the hazardous waste managed under these regulations is treated chiefly with on-site systems by the industries generating the wastes. Wastes also are shipped off-site for treatment in RCRA-permitted treatment, storage and disposal facilities (TSDF).

The RCRA waste treatment strategy, like the CWA wastewater treatment strategy, also allows discharge of the hazardous waste into sewers for treatment at central POTWs or industrial wastewater treatment plants. Dilution by the wastewater and treatment at the plant must provide appropriate control of the specific toxics or toxic characteristics in the listed hazardous waste. Satisfactory treatment at the POTW without treatment inhibition, proscribed air emissions, or sludge contamination eliminates the extensive regulatory 'cradle-to-grave' administrative and liability provisions of the RCRA. If hazardous wastes are not satisfactorily treatable in POTWs or industrial wastewater treatment plants or by RCRA-regulated TSDFs, such wastes are controlled through pretreatment prior to final waste treatment at POTWs, industrial wastewater treatment plants or TSDFs.

The CWA wastewater treatment and RCRA waste treatment strategies do not necessarily control subsequent toxic environmental impacts from the many intermittent discharges of wastes into municipal and industrial wastewater treatment plants or into RCRA treatment, storage and disposal facilities. Indeed, as described earlier, the large numbers of chemicals and their interactions in complex mixtures suggest that regulation by chemical-by-chemical approaches will not achieve fully satisfactory control of toxics. Thus, an evolving waste treatment strategy features toxicity (bioassay) testing of the effluents from waste treatment facilities to confirm treatment effectiveness and toxicity reduction. In the strategy, toxicity (bioassay) limits are included in the treatment plant's discharge permit. Practical use of the strategy, currently under development for the water medium, requires control of the toxicity either through pretreatment, as described earlier, at the sources of the toxicity or application of suitable 'end-of-the-pipe' waste treatment alternatives at central treatment facilities.

Successful integration of the toxicity testing (ecosystem and health effects) and toxicity reduction evaluation strategies for complex mixtures (US Environmental Protection Agency, 1985; Mount and Anderson-Carnahan, 1988 and 1989; and Botts et al., 1989) into the control strategies for waste treatment (BAT, BDAT) and control of release (pretreatment, air emissions, sludge disposal, etc.) will complete approaches for control of toxics in hazardous wastes and wastewaters.

Clean-up of contamination

Locally severe contamination caused by poor or uncontrolled disposal practices for hazardous waste have created worldwide clean-up needs. The most comprehensive strategy for such costly clean up has evolved in the United States under its CERCLA and SARA Acts, and is commonly called the Superfund to clean up hazardous waste sites. The overall program also includes a research and development effort called the Superfund Innovative Technology Evaluation (SITE) Program to produce improved clean-up technology. The most important goal of the Superfund and SITE programs is to provide permanent remedies to the contamination. A second goal under CERCLA, SARA and CWA is to provide emergency response to spills, including oil spills, and a corrective action program for leaking, underground, petroleum storage tanks.

The Superfund clean-up activities (Hall *et al.*, 1987) involve two types of responses: short-term removal actions involving spills or other emergencies requiring immediate action; and longer term remedial responses aimed at providing permanent solutions to problems at uncontrolled hazardous waste sites. The short-term removal actions provide immediate response to a diversity of contaminations, ranging from highly hazardous or ecologically damaging spills such as accidental release of flammable or explosive chemicals, pesticides and crude oil, to the leaking of underground petroleum products storage tanks into groundwater supplies. Elements of the removal action provide:

- skilled personnel trained in emergency response at local, regional and national levels;
- proper protective and clean-up equipment;
- rescue teams;
- decisive and effective on-site control action;
- safety procedures to minimize hazards to emergency response personnel and to the public;
- hazard monitoring;
- emergency decontamination and clean-up services.

The EPA has listed more than 800 hazardous waste sites on its National Priority List for longer term remedial responses. Each site must be investigated and, as appropriate, dealt with under an investigative and remedial process consisting of two phases: first, site characterization to define the nature and extent of the contamination and the public health and environmental risks; and second, the post-screening field investigation to support evaluation of remedial alternatives and to recommend a solution. A feasibility study in the process consists of three phases: the development of remedial alternatives; an initial screening of the alternatives; and a detailed analysis of the alternatives. The results of the process lead to remedy selection, a record of decision, a remedial design and finally remedial action.

The research and development of innovative or alternative technology to support Superfund under the SITE Program (US Environmental Protection Agency, 1989) involves three program areas, the Demonstration Program, the Emerging Technologies Program and the Measurement and Monitoring Technologies Program. The Demonstration Program provides engineering and cost data on selected technologies from pilot or demonstration projects. The Emerging Technologies Program supports bench-scale testing of new and innovative technologies to develop candidates for the Demonstration Program. The Measurement and Monitoring Program develops and demonstrates innovative measurement and monitoring technologies needed to characterize Superfund sites, such as immunoassays for toxic substances and fiber optic sensing for *in situ* site analysis.

Control technologies

With the diversity in properties of toxic chemicals, multimedia transport pathways, and widespread contamination, a broad range of treatment technologies is needed for environmental management of toxics or hazardous wastes. These technologies include four types of treatment: biological degradation; chemical degradation; separation processes; and containment. Successful control technologies often employ combinations of the four approaches to address specific environmental control problems.

Biological degradation involves low input energy for degradation of mixtures of organic chemicals, usually to innocuous residuals. Indeed, anaerobic biosystems may produce excess energy in the form of methane gas. The biomass in biosystems requires effective contacting along with appropriate nutrients for efficient degradation. Biodegradation sometimes only transforms chemicals, including some metallic or organic metallic species, into alternative species. The transformation may increase or decrease the toxicity of the chemical. Toxic chemicals at various concentrations often inhibit the degradation process. Some chemicals also are highly resistant to biological degradation, although adaptation of natural or development of engineered organisms capable of degrading resistant chemicals is often possible. Finally, toxicity associated with elemental properties such as with metals is not fully degradable, although the level of toxicity may be altered by biotransformation.

Biosystems may be applied to contaminated aqueous wastes using suspended growth bioreactors such as activated sludge or biofilm reactors such as biofilters and rotating biological contactors, and to contaminated air using biofilters. The biosystems may also be applied to slurried soils and used as *in situ* treatment for contaminated soils and solid wastes. Biosystem limitations, such as inhibition effects, resistance to the degradation of some chemicals and the volatilization transport pathway in aerobic biosystems, require treatability assessments on wastes, contaminated wastewaters and contaminated sites. This would determine applicability of the various bio-system approaches to specific toxic control problems. In spite of possible limitations, the low energy cost associated with biosystems and their poten-tial capability to permanently destroy both toxic and non-toxic organic molecules *in situ* stimulate substantial research and development efforts.

Chemical degradation of toxics is usually achieved through thermal destruction (Oppelt, 1987). Degradation through chemical oxidation and chemical reduction (dehalogenation) also destroys toxic molecules. Chemi-cal degradation of toxics exhibits wide variability in energy requirements. The requirements depend upon the relative concentrations and nature of the toxic and the non-toxic materials in the wastes or at contaminated sites. Indeed, wastes with high concentrations of organic toxics, especially with low water content, can be burned as fuels in boilers or other devices equipped with appropriate air pollution controls. Toxics dispersed in water, soil or air may also be thermally destroyed but the energy costs of destruction increase sharply as the amounts of noncombustible substances associated with the toxic organics increase. Similarly in chemical oxidation or reduction, chemical energy (oxidant or reductant) requirements increase markedly with increasing concentrations of non-toxic but chemically active substances associated with the toxics.

Such characteristics impact the economic practicality of the chemical degradation technology. As examples, thermal destruction is preferable for high strength wastes or contaminated disposables requiring low energy input while chemical oxidation, using combinations of oxidants such as ozone, hydrogen peroxide and ultraviolet radiation, is preferable for low concentrations of oxidizable materials such as those found in groundwaters or some air releases. The high costs associated with chemical degradation also stimulate research and development of chemical degra-dation approaches that can be selectively focused on specific toxics, especially for clean-up of the persistent toxics (polychlorinated biphenyls and highly chlorinated pesticides) in soils and sediments. As an example, soil-bound polychlorinated biphenyls (pcbs) can be dechlorinated using a formulation of polyethylene glycol (peg) and KOH, termed kpeg (Kornel and Rogers, 1985). The products of the reaction below 100°c are peg-substituted chlorobiphenyls, with chlorine being displaced at the site of substitution.

The third type of treatment, separation processes, ranges from simple sorption/sedimentation/filtration systems to sophisticated processes such as steam distillation. The processes usually involve physical separation mechanisms but also can involve chemical or biochemical phenomena in achieving separation of the toxicant from its associated non-toxic materials. As with chemical degradation, the input energy required for the separation may range from none for gravity sedimentation to relatively high levels for steam distillation. Separation processes do not destroy toxics and must be incorporated with other control technologies or management practices for destruction, recycle and re-use or final disposal.

Containment, as the fourth type of toxic control technology, employs two overall approaches to prevent release of toxic wastes. The first approach attempts to establish essentially 'permanent' containment of toxics. As an example, conservative toxics such as metals and other residuals classified as RCRA hazardous wastes are immobilized or stabilized and are permanently stored in 'secured landfills'. The landfills are secured with impervious liners and caps. Immobilization involves solidification usually within cement matrices or vitrification within glassy matrices. The 'permanent' stabilization processes may also be applied as *in situ* treatment at Superfund sites to minimize environmental release of toxics.

The second approach involves temporary containment to prevent release from contaminated sites until final remedial solutions can be applied. This practice includes containment as an emergency response to spills and other uncontrolled toxic releases, and also to isolate contaminated sites such as Superfund landfills. Isolation methods include the provision of liners or impervious barriers (slurry walls) around sites. When complete isolation is not possible, leachate pumping and treatment are also used to ensure that groundwater flows only into the site. After establishing temporary containment, appropriate treatment such as thermal destruction, *in situ* vitrification or *in situ* biodegradation can sometimes be used as final remedial solutions.

One containment technology, deep well injection, has been used to manage aqueous toxic wastes, especially conservative toxics such as metal-containing wastes. Uncertainty related to the ultimate fate and impacts of the injected wastes, however, is forcing review of this technology as a 'permanent' containment approach. It is likely to be restricted in the future.

A desired perspective on the complexity of selecting and integrating control technologies is provided by considering the fate of toxics in the sedimentation/activated sludge treatment process that is widely used in industrial and municipal wastewater treatment. This technology employs chiefly biological degradation and pollutant separation processes, although the technology's sludge disposal options may also include a chemical degradation process, sludge incineration.

The principal treatment mechanisms (Bishop and Jaworski, 1986) of sedimentation/activated sludge wastewater treatment technology are sorption of organic compounds or sorption/precipitation of metal species and organics onto solids with subsequent solids removal by sedimentation; volatilization of organic toxics into air, by stripping during wastewater aeration and by surface desorption in sedimentation basins and flow channels of the collection and treatment systems; and biotransformation/biodegradation of the toxics by the aerobic biomass in the activated sludge reactors and by the anaerobic biomass in the technology's sludge digesters. Other less important treatment mechanisms, such as hydrolysis (Mabey and Mill, 1978) or photochemical decomposition (Zebb and Cline, 1977) may also contribute to the destruction of selected chemicals in wastewater during transport and treatment.

From the perspective of water pollution, the principal removal mechanisms in sedimentation/activated sludge wastewater treatment compete for toxics removal and complement each other for control of toxics entering the aqueous environment. Thus, the technology with proper operation and bioacclimation of the biomass to toxic chemicals provides efficient removal of a wide variety of toxic organics and significant reduction in the metals released into the aqueous environment (Petrasek et al., 1983; Petrasek et al., 1985; and Hannah et al., 1986).

Unfortunately, certain toxics, such as some metals and pesticides (Petrasek et al., 1983 and Hannah et al., 1986) and selected dyes (Shaul et al., 1988), possess properties of suitable water solubility, low volatility and resistance to aerobic biotransformation/biodegradation. These types of toxics substantially pass through sedimentation/activated sludge treatment into the aqueous environment. Also, since the metallic elements in compounds in the wastewater are conservative through treatment processes (i.e. they may be biotransferred or complexed but not destroyed during treatment), the metals not passing directly into the aqueous environment are transferred from the water to the waste sludges from the treatment system. The sludge disposal practices may then reintroduce these toxics into the environment (US EPA, 1986).

In addition, organic toxics, substantially removed from the water by the treatment process, may be volatile or strongly sorbed on solids in wastewater systems. These toxics may be relatively rapidly or only slowly degraded by biotransformation/biodegradation. Such toxics are partially transferred through the competing removal mechanisms or volatilization and sorption from the wastewater into the air or onto land or water during disposal of the sludges. Even with high biodegradation rates, those organics which are highly volatile in aqueous systems can also be partially transferred into the air during transport through sewers and other flow channels. Thus, with the widely varying properties of the toxics, and interacting transport pathways, releases into the environment are not fully controlled by this widely used control technology. Additional central plant treatment processes or control approaches through pretreatment or waste minimization techniques must be incorporated in the overall wastewater treatment approach.

The developmental status of technologies for treatment of toxics varies and has been conveniently classified by the EPA as: proven or demonstrated technology; alternative and innovative technology; and emerging technologies. Proven or demonstrated technologies are well documented treatment systems used successfully in field applications. A proven or demonstrated technology, as indicated by the discussion on sedimentation/activated sludge treatment systems, does not infer that a technology can be

Table 6.7 Control technologies for toxic wastes. Legend: P or D (proven or demonstrated); I and A (innovative and alternative); and E (emerging)

Treatability approach	Status	Waste application
Biological degradation		
Aerobic biosystems		
Activated sludge treatment	P or D	Aqueous
Activated sludge with powdered activated carbon	P or D	Aqueous
Biofilters	P or D	Aqueous
Rotating biological contactors (RBC)	P or D	Aqueous
Extended aeration	P or D	Aqueous
Aerated lagoon	P or D	Aqueous
Composting	P or D	Solids
In situ biotreatment	I and A	Aqueous, solids
White rot fungus on RBC	E	Aqueous, solids
Bioengineered organisms	E	Aqueous, solids
Soil slurry reactors	I and A	Solids
Anaerobic biosystems		
Sludge digestion	P or D	Sludges
Biofilters	P or D	Aqueous
Expanded bed biofilters	E	Aqueous
Expanded bed biofilters with activated carbon	E	Aqueous
Chemical degradation		
Thermal destruction		
Incineration	P or D	Liquids, solids, air
Kiln	P or D	Liquids, solids, air
Boilers	P or D	Liquids, solids, air
Infrared beds	I and A	Solids
Chemical oxidation and reduction		
Ozone and UV	I and A	Aqueous
H_2O_2 and UV	I and A	Aqueous
Polyethylene glycol and Potassium hydroxide (KPEG)	I and A	Aqueous
for dechlorination	E	Solids, liquids
Separation processes		
Sedimentation	P or D	Liquids
Chemical precipitation/ sedimentation	P or D	Liquids
Filtration	P or D	Liquids
Activated carbon adsorption	P or D	Liquids or air
Air stripping with carbon adsorption of air	P or D	Liquids, air
Soil washing with and without detergents	I and A	Soils
Extraction, solvent	I and A	Solids, liquids
Extraction, vacuum	I and A	Solids, liquids
Extraction, critical fluid	E	Solids, liquids
Eutectic freezing	I and A	Aqueous
Steam distillation	P or D	Liquids
Thermal separation	I and A	Solids

Table 6.7 continued

Containment		
Permanent		
Secured landfills	P or D	RCRA solid wastes
Vitrification	I and A	Metals, radio-active substances
Stabilization/solidification	I and A	Metals
Temporary		
Liners	P or D	Landfills
Slurry walls	P or D	Landfills, spills
Mechanical containment	P or D	Spills

used to control all toxics problems in any specific waste or contaminated site. Alternative and innovative technologies refer to technologies being evaluated for toxics control applications under the EPA SITE Program at contaminated sites in pilot and field demonstrations. Emerging technologies are technologies under development in bench-scale studies for control of toxics in RCRA-listed wastes and for the EPA SITE Program.

Representative technologies in the four treatability approaches are presented in Table 6.7. Selection of these technologies for control of a specific toxic waste or contaminated site usually requires a treatability assessment to establish applicability of the technology to the specific waste or contaminated site. The selection of appropriate treatment technologies can usually achieve technical solutions to most problems. Unfortunately, achieving molecular destruction or effective containment using current technology may not be practical or cost effective. As examples, thermal destruction (Oppelt, 1987) can be operated to achieve essentially complete molecular destruction (> 0.99999). Thermal destruction of low concentrations of potent toxicants in soils, sediments or water not only requires excessive fuel, but can produce adverse environmental impacts. These impacts occur through release of CO_2 and other products of combustion and through the release of toxicants during handling for thermal destruction. Similarly, steam distillation with effective recovery and recycling of many solvents in aqueous wastes requires increasing needs for steam and, therefore, fuel, as the concentration of the solvents in the aqueous wastes decrease.

Indeed, treatment applicability, efficiency and costs are often highly specific to the waste, site or contaminant mixture. Current field applications and research are producing a perspective on an array of technologies being developed worldwide. Contributions to the perspective are included in various proceedings and overviews (Englund and Mafrica, 1987; Oppelt, 1987; American Petroleum Institute *et al.*, 1987; US Environmental Protection Agency, 1987b; and Nunno *et al.*, 1989). The perspective unfortunately reveals wide variability in treatment efficiency and provides limited cost assessments.

A detailed overview of the array of technology, especially the emerging technology, is beyond the scope of this chapter. There are, however, key gaps in the technology. One gap is produced by a lack of more effective biosystems with broad toxicant applicability and increased degradation efficiency. A second is produced by deficiencies in methods for effective *in situ* treatment. A third gap is produced by deficiencies in practical separation processes, especially to efficiently separate toxicants from soils and sediments. A fourth gap is produced by limited development of innovative combinations of treatment types that complement strengths and overcome weaknesses of the individual treatment approaches. With rapidly advancing bioengineering technology, biosystems, especially for low toxicant concentrations, offer enormous opportunities to achieve treatment practicality and cost effectiveness. Aerobic and anaerobic treatment systems coupled with effective methods for *in situ* treatment, low energy separation processes and toxicant specific chemical degradation are under development. Success in this effort will lead to more effective technology solutions, especially for remediation of contaminated sites and accidental spills.

Conclusions

This chapter has reviewed the five functional control strategies: regulation of production and use, waste minimization, control of release, waste treatment and clean-up of contamination. These can be further classified as either preventive or mitigative, and of course, they are not mutually exclusive. In many industries, a combination of strategies is employed to satisfy regulatory standards, community expectations or company policies.

The selection of control strategies and the technologies which support them is becoming a more complex decision-making process. The diversity of producers and users along with the potential for accidental releases of highly toxic materials have led to an extensive regulatory framework, such as that in the United States. This framework must be comprehensive yet flexible enough to ensure that risks are reduced to acceptable levels.

References

American Petroleum Institute, Environmental Protection Agency, and us Coast Guard (1987), 1987 Oil Spill Conference, Proceedings – Tenth Biennial, Baltimore, Maryland, April 6–9.

Bishop, D. F. and Jaworski, N. A. (1986), 'Biological treatment of toxics in wastewater: The problems and opportunities'. *Proceedings of the International Conference on Innovative Biological Treatment of Toxic Wastewaters*, Arlington, Virginia, June 24–6.

Botts, J. A., Braswell, J. W., Zyman, J., Goodfellow, W. L. and Moore, S. B. (1989), *Toxicity Reduction Evaluation Protocol for Municipal Wastewater Treatment Plants*, us Environmental Protection Agency, EPA/600/2–88/062, NTISPB89–195218. (Cincinnati, Ohio: Risk Reduction Engineering Laboratory).

Bruins, R. J. F., Fradkin, J. and Stara, J. F. (1987), 'Summary of risk assessment methodologies for municipal sludge reuse or disposal options'. *Proceedings of the National Conference on Municipal Sewage Treatment Plant Sludge Management*, Boston, Massachusetts, May 27–9.

Center for Environmental Assessment (1986), *Foreign Practices in Hazardous Waste Minimization* (Medford, Mass.: Tufts University).

Code of Federal Regulations (1979), Title 40 EPA Part 257. Criteria for classification of solid waste disposal facilities and practices, para. 257.3–6, Disease Interim Final Regulations.

Code of Federal Regulations (1987), Title 40, Chapter 1, July 1, pp. 374–87.

Drabkin, M. and Rissman, E. (1987), *Waste Minimization Audit Report: Case studies of corrosive and heavy metals waste minimization at a specialty steel manufacturing complex*. US Environmental Protection Agency, EPA/600/2–87/055, NTISPB88–107180. (Cincinnati, Ohio: Hazardous Waste Engineering Research Laboratory).

Englund, H. M. and Mafrica, L. F. (eds) (1987), *Treatment Technologies for Hazardous Wastes*. APCA Reprint Series, RS13, from JAPCA APCA, Pittsburgh, Pennsylvania.

Federal Register (1986), *Rules and Regulations*, 51, 40643–52, November 7.

Federal Register (1989) *EPA Standards for the Disposal of Sewage Sludge*, Proposed Rule 54(23), pp. 5746–902.

Freeman, H. M. and Eby, E. (1987), 'A review of hazardous waste minimization in the United States'. Presented at the International Congress on Recent Advances in the Management of Hazardous and Toxic Wastes in the Process Industries, Vienna, Austria; US Environmental Protection Agency, Hazardous Waste Engineering Research Laboratory, Cincinnati, Ohio.

Hall, R. M., Watson, T., Davidson, J. J., Case, D. R. and Bryson, N. S. (1987), *Superfund Manual Legal and Management Strategies* 2nd edn (Rockville, Maryland: Government Institutes Inc.).

Hannah, S. A., Austern, B. M., Eralp, A. E. and Wise, R. H. (1986), 'Comparative removal of toxic pollutants by six wastewater treatment processes', *Journal Water Pollution Control Federation* 58 (1), pp. 27–34.

Kornel, A. and Rogers, C. (1985), 'PCB destruction: a novel dehalogenation reagent', *Journal of Hazardous Materials*, 12, pp. 161–76.

Mabey, W. R. and Mill, T. (1978), 'Critical review of hydrolysis of organic compounds in water under environmental conditions', *J. Phys. Chem. Ref. Data* 7, 383.

Meier, J. R., Blazak, W. F., Riccio, E. S., Stewart, B. E., Bishop, D. F. and Condie, L. W. (1987), 'Genotoxic properties of municipal wastewaters in Ohio', *Arch. Environ. Contam. Toxicol*, 16, pp. 671–80.

Mount, D. I. and Anderson-Carnahan, L. (1988), *Methods for Aquatic Toxicity Identification Evaluations: Phase I Toxicity Characterization Procedures*, US Environmental Protection Agency, EPA/600/3–88/034. (Duluth, Minnesota: Environmental Research Laboratory), September, 1988.

Mount, D. I. and Anderson-Carnahan, L. (1989), *Methods for Aquatic Toxicity Identification Evaluations: Phase II Toxicity Characterization Procedures*, US Environmental Protection Agency, EPA/600/3–88/035, NTISPB89–207005. (Duluth, Minnesota: Environmental Research Laboratory), February, 1989.

Neiheisel, T. W., Horning, W. B., Austern, B. M., Bishop, D. F., Reed, T. L. and Estenik, J. F. (1988), 'Toxicity reduction at municipal wastewater treatment plants', *Journal Water Pollution Control Federation*, 60 (1), pp. 57–67.

NRDC (Natural Resources Defense Council) *et al.* v. Train (1976), 8 ERC 2120, (DDC).

Nunno, T., Hyman, J., Spawn, P., Healy, J., Spears, C. and Brown, M. (1989), *Assessment of International Technologies for Superfund Applications — Technology*

Identification and Selection, US Environmental Protection Agency, EPA/600/2–89/
017, NTISPB89–205959 (Cincinnati, Ohio: Risk Reduction Engineering
Laboratory).

O'Connor, T. P., Walker, H. A., Paul, J. F. and Bierman, V. J. (1985), 'A strategy
for monitoring of contaminant distributions resulting from proposed sewage
sludge disposal a the 106-Mile Ocean Disposal Site', *Marine Environ. Res.* 16 (2),
pp. 127–50.

Oppelt, E. T. (1987), 'Incineration of hazardous waste, a critical review', JAPCA,
vol. 37 (5), pp. 558–86.

Petrasek, A. C., Austern, B. M. and Neiheisel, T. W. (1985), 'Removal and
partitioning of volatile organic priority pollutants in wastewater treatment',
Proceedings of the Ninth US–Japan Conference on Sewage Treatment Technology, US
Environmental Protection Agency, EPA–600/9–85/014, NTISPB85–227379
(Cincinnati, Ohio: Water Engineering Research Laboratory) pp. 557–94.

Petrasek, A. C., Kugelman, I. J., Austern, B. M. , Pressley, T. A., Winslow, L. A.
and Wise, R. H. (1983), 'Fate of toxic organic compounds in wastewater
treatment plants'. *Journal Water Pollution Control Federation*, 55 (10), pp. 1286–96.

Shaul, G.M., Dempsey, C. R. and Dostal, K. (1988), *Fate of Water Soluble Azo
Dyes in the Activated Sludge Process*, US Environmental Protection Agency,
EPA/600/2–88/030, NTISPB88–208251. (Cincinnati, Ohio: Water Engineering
Research Laboratory), February, 1988.

Statutes at Large (1988), Title 1 of Public Law 100–68, 102 Stat. 4139, November
18.

US Environmental Protection Agency (1985), *Technical Support Document for Water
Quality-Based Toxics Control*, EPA/440/4–85/032. (Washington, DC: Office of
Water), September, 1985.

US Environmental Protection Agency (1986), *Report to Congress on the Discharge of
Hazardous Wastes to Publicly Owned Treatment Works*, EPA/530/sw–86/004.
(Washington, DC: Office of Water Regulations and Standards), February, 1986.

US Environmental Protection Agency (1987a), *Waste Minimization: Environmental
Quality with Economic Benefits*, EPA/530/sw–87/026. (Washington, DC: Office of
Solid Waste and Emergency Response), October, 1987.

US Environmental Protection Agency (1987b), 'Land disposal, remedial action,
incineration and treatment of hazardous waste', EPA/600/9–87/015, NTISPB87–
233151, *Proceedings of the Thirteenth Annual Research Symposium*, Hazardous
Waste Engineering Research Laboratory, Cincinnati, Ohio, July, 1987.

US Environmental Protection Agency (1988), *Environmental Progress and Challenges*
EPA/230/7–88/033. (Washington, DC: Office of Policy Planning and Evaluation),
August.

US Environmental Protection Agency (1989), *Superfund Innovative Technology
Evaluation (SITE) Program*, EPA/540/8–89/002, EPA Research Symposium, Centre
for Environmental Research Information, Cincinnati, Ohio, April 10–12.

Walker, H. A., Paul, J. F. and Bierman, V. J. (1987), 'Methods for waste load
allocation of municipal sewage sludge at the 106-Mile Ocean Disposal Site',
Environmental Toxicology Chemistry, 6 (6), pp. 475–89.

Walker, J. (1987), 'Comparative testing of municipal sewage sludges by the toxicity
characteristic leaching procedure and compositional analysis'. *Proceedings of the
National Conference on Treatment Plant Sludge Management*, Boston, Massachusetts, May 27–9.

Zebb, R. G. and Cline, D. M. (1977), 'Rates of direct photolysis in aquatic
environments', *Environ. Sci. Technology*, 11 (4), pp. 359–6.

7 Occupational and environmental health linkages

R. C. PAEHLKE

Introduction

The health hazards to which some persons are exposed in their work are not unrelated to the environmental hazards which permeate the ecosystems within which we are all integrally bound. Yet for the most part, occupational health policies are determined as if there were no such connection. This chapter will explore some of the implications of this observation.

We will begin our inquiry with a look at the historic context of occupational health policy, noting that at least a quasi-scientific awareness of workplace health hazards has, in some cases, been established for several centuries. However, effective preventive action has followed very slowly. Some of the reasons for this slow translation from indicative evidence to explicit workplace change are briefly examined. This first section then concludes with a short discussion of 'special' workplace situations. The second section offers a brief review of the connections between occupational hazards and environmental hazards. The third sets out two general approaches to workplace health and safety regulation: the so-called 'command-and-control' approach associated with the Occupational Health and Safety Act (OSH Act) in the United States and the so-called internal responsibility system (IRS) common to several Canadian provinces. The chapter concludes with an evaluation of the relative success and failure of these two systems and the implications of the linkages between occupational and environmental health for both approaches and for occupational health policy in general.

From science to policy: accounting for historic inaction

Wagoner (1976) speaking on the 200th anniversary of the discovery of occupational carcinogenesis by Percival Pott noted that coke-oven workers in the United States steel industry were still exposed to airborne substances of the same class that caused scrotal cancer in Pott's chimney sweeps. He further observed that 'ninety years after miners in the Erz Mountains of Central Europe were observed to be dying of lung malignancy and thirty years after radioactivity within these mines was generally accepted to be the cause of these cancers, thousands of American uranium miners were still working in radon daughter concentrations of such magnitude as to triple their prospect of dying of lung cancer.'

Observations such as those of Wagoner led to widespread public concern in the 1970s, to new legislation in most industrialized countries, and to considerable improvement in many workplaces. Wagoner went on to list other hazards where there have been comparable historic gaps between scientific knowledge and effective action: arsenic, various aromatic amines, and asbestos. Others have made similar observations regarding PCBs, mercury and lead, coal dust, cotton dust, and grain dust (Commoner, 1973; Tataryn, 1979; Wallick, 1972; Epstein, 1978; Paehlke, 1979, 1980). Epstein (1978) argued that there has also been a considerable hesitation to act decisively regarding several substances of more modern origins: Bischloromethylether (BCME), benzene, and vinyl chloride monomer. I will not elaborate the details of these cases here, but rather I will try to set out some of the reasons why this pattern has been so consistent and long standing. These reasons are important as a background to the considerable policy ambivalence regarding occupational health protection which has returned in the 1980s particularly in the United States. More recently in Canada, the collaboration between federal and provincial governments in the development, promotion and implementation of the Workplace Hazardous Material Information System (WHMIS) has raised the profile of occupational health protection considerably.

The reasons for the extended history of delay and ineffectiveness are, unfortunately, numerous. One might begin with two reasons that have at least, in part, been eclipsed in recent decades. First, there was for centuries little by way of understanding that cancer, in particular, was to a significant extent both environmentally induced and preventable. Second, there was also in many cases a lack of technical capacity to prevent workplace exposures to substances known to be hazardous without eliminating or radically reducing their use. Regarding the first of these reasons one might note the observation of Sir Richard Doll (quoted in Tataryn, 1979) that in the 1930s, even after the identification of carcinogens, 'it was commonly assumed that cancers were an inevitable accompaniment of ageing and that little could be done to reduce the mortality they caused. It is now clear, however, that most if not all cancers have environmental causes and can in principle be prevented.' Doll and Peto (1981), defining environment very broadly, came to a conclusion similar to that in Doll's earlier statement. However, tobacco, alcohol and diet are found to be the dominant contributing environmental factors. Further, as Eisenbud (1978) and others have agreed, cancer rates are age-specific and removing environmental factors might well defer onset and/or mortality rather than, strictly speaking, preventing them. None the less, the debate more often now centers on which substances are most hazardous, on what are acceptable levels of exposure, and on how much money should be spent to prevent exposures beyond those levels. It is no longer doubted that environmental exposures are important.

This new debate is a vast improvement over the old. None the less, it has brought to the surface a more fundamental set of problems. Most of these problems might be summarized as the general political bias which exists in most, if not all, democratic political economies in favor of private corporate interests. One need not subscribe to elaborate leftist orthodoxes to recognize the truth of the characterization in Lindblom (1977) regarding the 'privileged position of business in government and politics'. In these systems the electoral success of political leaders is highly dependent on economic performance which is, in turn, dependent on the existence of a climate relatively conducive to capital investment. Further, those who have the most to gain financially from the profits of a business enterprise are less likely to be exposed themselves to health or safety hazards. In occupational health policy virtually all of the benefits of protection accrue to those who do not participate in capital investment decisions, and virtually all of the costs to those who do.

Thus, prior to the 1970s, governments acted autonomously and decisively regarding occupational health only when countervailing political pressures were exceptionally strong. Hamilton (1943) observed that such pressures were sometimes present before and during the Second World War and early technical strides were made by medical practitioners and the pioneering mine and factory inspectorates of Great Britain in the 19th and early 20th centuries (Hunter, 1978). None the less, as a generalization it is fair to say that the 1970s stand out as a singular period of consistent government initiatives in occupational health protection.

There are many other possible explanations for the commonplace inaction regarding occupational health protection prior to the 1970s. These reasons are indicative of the difficulties which corporate and government decision-makers have faced in dealing with occupational health controversies. Evidence regarding occupational hazards is often ambiguous; decisive evidence may take decades to generate. Even the very onset of occupationally-induced diseases is often long delayed. Further, in many cases the dollar costs of improving the workplace environment are quite high, especially if imposed on workplaces where extensive capital improvements were not otherwise on the agenda. Ison (1975, cited in Swinton, 1982) has noted that the time when health and safety advances can be most easily and inexpensively undertaken is precisely the time when plants are in the design stage and unions are often not in place to exert pressure. That is also the moment when business has enormous leverage over government, witness the recent financial supports available to very prosperous corporations from already debt-ridden governments.

Also relevant here is the long-standing tradition in labor relations to avoid contesting decision-making regarding capital investment. Specifically, Tucker (1984) observed that there has been 'a tacit acceptance of managerial rights to control the production process, which included choice of technology and direction of the work force. Rather than raising such concerns, North American unions have focused their demands on improving wages, benefits and security provisions, leaving management to manage.' With a few notable exceptions, there have been few attempts on the part of unions to influence workplace design, for whatever reason, even to protect jobs. Productivity improvements have generally been acceded to when the remaining jobs are at a higher wage level. Whatever the general consequences of this pattern, and a case can be made that it is an approach with many merits, the consequences for workplace health and safety have been unfortunate. Unless unions can be involved at the outset in workplace design, management is free to make an often, though by no means always, valid case that the retrofitted improvements are prohibitively costly. Indeed management has made such a case to workplace health and safety committees, to government and to the courts (Sagoff, 1985; Mendeloff, 1979; Noble, 1986). After-the-fact corrections are indeed much more expensive than engineering changes at the design stage. It is particularly interesting to note that the same arguments have been made about environmental improvements.

There are other important reasons why effective action has been difficult to achieve. Schrecker (1984) noted that the 'burden of proof is often assigned to those arguing for control of a particular hazard'. Further he observed that 'even where the burden of proof is nominally imposed on the proponent, the application of lenient standards of proof can turn the process into a ritual which embodies a presupposed right to market, pollute, and so forth, and in which approval is almost automatic.' The standard of proof demanded of or imposed on those seeking to limit action can be severe and, indeed, only rarely in matters of occupational and environmental health is the burden of proof even nominally assigned to proponents of chemical use.

Of particular importance are the doubts some would cast on the validity of evidence obtained in animal tests. Again, as Schrecker (1984) put it 'suspected carcinogens are tested on animals at doses several orders of magnitude higher than those to which a human being would be exposed, because this is the only way to induce a high enough number of tumors to produce statistically significant results in manageable numbers of animals.' Yet many of the public, and some working in regulatory agencies, have been induced to treat evidence based on animal testing as invalid, believing that such tests show only that those who are exposed to very high levels of hazardous substances are at risk. The linear extrapolations that scientists are prepared to make are often not made by the general public. Those whose potential profits are on the line, needless to say, do not remind the public that where a small number of generally healthy animals are exposed for a short period of time, it is usually a large number of humans, not all adult or healthy ones, who will be exposed for a long period of time.

There are three further explanations of slow and ineffective action worth noting. The first follows from the preceding discussion regarding public perceptions of the risks associated with a product such as saccharine. Generally there have been few effective means of communicating the findings of the epidemiological and toxicological studies, and the ethical implications of those sciences, to the wider public. This is particularly true regarding those of the public most likely to actually be at risk. High school health courses rarely touch on the important concepts and issues except recently perhaps in relation to smoking and alcohol and drug abuse. The media convey alarm, but far too little by way of conceptualization, thoughtful reflection, and understanding. Further there are only a limited number of widely available university-based adult education courses or programs in occupational or environmental health especially for those who are not health professionals.

Second, it should be noted that occupational health protection may well be particularly vulnerable to general economic stringency, recession, or a wartime sense of emergency. One might also observe that the continual presence of at least one of the three is normal in this century. Peace and prosperity are more the exception to the rule. This is an especially telling observation when one recognizes that, for example, even industry-specific slowdowns have, in some locations, led to reduced protection of worker health and safety. Even the postwar urgency of the 1950s was partially responsible for some of the risks incurred by uranium miners in both the United States and Canada. More generally here, Noble (1986) has argued that the 1980s have witnessed rollbacks on occupational health protection principally because of the 'liberal' (non-mobilizing) nature of the us Occupational Health and Safety Act. However, one might suspect that the problem is broader than Noble would have it. Any system of worker health protection would be vulnerable when there is a widespread perception of significant present or near-term future economic difficulties. This mood has pervaded North America in the 1980s. Capital is globally mobile, capital investment decisions are almost wholly based on yield prospects, and in a 1980s context health and safety has sometimes been perceived by those facing difficult decisions as something of a frill. Until quite recently, particularly in the more developed countries, this was also the case in environmental protection. Further, in a context of restrained capital investment there are likely to be few process or equipment changes of any sort, affording fewer opportunities to significantly abate hazards.

Third and finally here one must acknowledge, with Swinton (1983), that what is an acceptable level of safety will be 'determined by financial as well as ethical consideration' and the 'cost of ensuring a level of safety for the average worker is inevitably cheaper than protecting the most susceptible'.

However, this does not mean that either financial considerations are determining in every case or that protection of especially susceptible workers could not ultimately become the standard for many, if not all workplaces. In the environmental area, only in the case of endangered species have standards more stringent than those that would protect the average organism or a suitable indicator species been applied. Protecting the most susceptible is a possible approach which is generally underemphasized and involves questions that will be difficult to resolve. Such an approach would deny, for example, that adequate occupational health protection could be achieved through such devices as uniform pay differentials to those in hazardous occupations (Viscusi, 1983; Graham and Shakow, 1981).

Special susceptibility involves two different sorts of issues. The first involves physical characteristics which made certain individuals especially vulnerable to particular hazards. For example, pregnant women have a special vulnerability to lead exposure (Stellman, 1977) and smokers have a special vulnerability to the cancer risk associated with asbestos (Epstein, 1978). But probably even more significant in policy terms than considerations of this kind are the particular vulnerabilities associated with special workplace situations. As with the 'exceptional times' discussed above, these 'special' situations (Paehlke, 1981) are far more usual than we ordinarily acknowledge.

These special situations would include (1) workplaces where a significant percentage of the employees at risk do not speak or read the language of the workplace, because they are immigrants, or because labels are in a foreign technical language, or because they are only marginally literate, (2) workplaces which are not unionized or where existing unions are not affiliated with those larger bodies which have relevant professional staff, (3) the many smaller workplaces where there is no management expertise regarding hazardous chemicals, (4) hazardous workplaces where there is a high employee turnover and/or an extensive use of temporary, casual or part-time labor, (5) workplaces where most or all workers are temporary and/or non-resident, as in migrant farm labor situations, and (6) workers in isolated communities where there are few or no employment alternatives. This last item alone, when one considers how many workers with narrow skills are also tied to mortgages and family requirements, indeed renders doubtful the theoretical labor mobility on which protection through market differentials rests (Viscusi, 1983).

But more than this, a regulatory system, whether based on command-and-control 'paternalism' or a mandated internal responsibility system, or a combination of the two, must contend with and be effective in these difficult but all too normal situations. Indeed these are precisely the situations where the worst workplace health and safety abuses are most likely to occur. In many cases those in high-risk occupations are less educated, less mobile, and lacking even a realistic prospect of unionization.

Where such workers are unionized, they are often in what are now seen as sunset industries, where any new difficulties may tip the balance towards plant closure (Kazis and Grossman, 1982). It must also be noted that within any given workplace the more vulnerable individuals are often those with the least seniority, and thereby often the least experience and knowledge. (Those with relatively more seniority generally can opt to work elsewhere in the plant.) The most vulnerable are sometimes told by supervisory staff that something one cannot smell cannot be very harmful, or that in the old days things were much worse, or that the supervisor himself has been breathing this dust or chemical for years (Nelkin and Brown, 1984). In addition, the most vulnerable cannot read a newspaper with ease, let alone a complex label or government regulatory notice. The most vulnerable are continuously and realistically fearful of imminent unemployment. They generally do what they are told, even if it is immediately and certainly hazardous, let alone ambiguously hazardous in the long term. The opinions of the most vulnerable are rarely solicited by either management or by health experts, whether independent professionals, government employees, or company doctors. Thus, even with the best of regulatory regimes occupational health protection is not assured for all.

The historic links between occupational and environmental hazards

Commoner (1973) in his now classic discussion of the history of PCBs made clear something very important about the general relationship between occupational and environmental hazards. These two types of hazards are interconnected both historically and ecologically in complex and often unappreciated ways. Commoner noted that within a short time after their commercial introduction in 1929, PCBs were found to have resulted in a widespread incidence of chloracne among plant workers and even in a few cases their spouses and children. However, it was not until more than thirty years later, when PCB residues were found in fish and birds, that PCBs were restricted to 'contained' uses. But the so-called contained uses continued to result in human exposures particularly at the point of disposal and total use through the years has resulted in a widespread and persistent environmental presence. One could argue that in this case only prompt and effective action in the 1930s could have prevented present problems associated with PCBs. It is important to grasp that, aside from preventing additional future releases, there is precious little that can be done at this point in time. In general, one might even conclude that if prompt and comprehensive action is not taken prior to or soon after initial workplace exposures, persistent chemicals will almost inevitably find a quite permanent place in the general waterborne environment.

The case of asbestos is also still instructive despite its frequent recounting.

The Romans understood that there were some health problems associated with the use of asbestos and early in this century insurance companies assumed that it involved serious health risks. In the mid-1930s there was some epidemiological evidence of health risk and in the early 1960s Selikoff (Selikoff, Churg, and Hammond, 1964) demonstrated a statistical link to cancer. None the less asbestos was still widely used in the construction of office buildings and schools after that date and today extensive exposures still occur in the attempts to remove the substance from buildings. Widespread environmental exposures also occurred well after the early 1960s as a result of exposures to particles released from brake linings. Remarkably, Canadian asbestos exports have now returned to very near their peak levels, with a greater proportion now bound for Third World nations. One can only hope, but not expect, that airborne exposures in these new settings will be severely limited.

The general point here is, however, the more important one. Substances which are produced in large quantities which come to be used in a variety of ways in several locations will find their way into the general environment. Once there, if persistent, they generally cannot be removed. Many of the general public persist in the illusion that the environment can be 'cleaned up'. However, those who are thoughtful about these matters will recognize that this is largely a futile hope. One can only reduce or eliminate further additions and perhaps anticipate rates of dilution, leaching, deposition or precipitation.

The pathways by which persistent chemicals are released, dispersed and taken up in ecosystems are legion and were reviewed in Chapter 2. So numerous and varied are these pathways that it is not unreasonable to doubt the importance of demonstrating precise exposure pathways as a part of the decision-making process even though such assessments are complex and time consuming. Indeed by the time such a pathway is scientifically determined most persistent substances manufactured or used in quantity may well have already found a way into the general environment and a non-occupational route to human exposures. The case of the chlorofluorocarbons and the destruction of the ozone layer with its serious implications for human health is a case in point (Gribbin, 1988). Obviously greater care in the workplace can help to lower environmental concentration levels and mercifully dispersion and dilution inevitably go hand in hand. But one cannot often say scientifically whether or not any given concentration level is 'tolerable', that is a matter which is fraught with value dimensions. Neither are we likely ever to know very precisely what is the likelihood of a chemical release into the wider environment from a particular plant.

We can conclude, however, that in a large number of plants over an extended period of time some releases quite probably will occur even in cases where it is known that significant risks are involved. Many such cases are well known. A large quantity of fire retardant containing PBBs (polybrominated biphenyls) found its way into a cattle feed which was in turn widely distributed and used in the state of Michigan in the early 1970s (Jamieson, 1977). Light fixtures containing PCBs, smoke detectors containing radioactive material, lead-based paints, and a wide variety of chemicals are all present in community solid waste landfills. Warehouses and workplaces containing hazardous chemicals burn down frequently enough for fire-fighters to have provided strong political support for community right-to-know legislation (Adams, 1983). Insurance companies have become very wary of warehouses containing agricultural chemicals largely because of the dispersion that occurs in cases of fire. The Canning incident described by Holtz in Chapter 10 (p. 246) is quite instructive in this regard. Not all present-day chemical landfills are flood-proof; deep-well injection of chemicals has been highly problematic in at least several cases. Hazardous wastes can be treated chemically and physically and residues can be vitrified, but releases may well occur within or prior to such processes. Unintentional releases of hazardous chemicals are indeed an example of what Perrow has called 'normal accidents' (Perrow, 1984).

In Perrow's language the term normal accident 'is meant to signal that given the system characteristics, multiple and unexpected interactions of failures are inevitable' (Perrow, 1984). Such accidents in Perrow's view are worrisome 'if they can produce catastrophes'. In his important work, Perrow discussed in detail the full range of potential single-incident catastrophes from transportation accidents to nuclear war, from chemical plant explosions to nuclear power meltdowns. But he did not consider the importance of his conceptualization for cumulative problems. Releases of chemicals known to be hazardous, one might say, are rare but normal. They are almost always unintended and unexpected, but at the same time virtually inevitable. The key questions are how persistent are the chemicals in question, how bioaccumulative are they, at what average rate are they likely to be released, how long and in what ways are we likely to produce and use them, and in what concentration are they harmful? The determination of average rates of release can only be made after the fact, if then. The final two questions, one must be aware, are matters which cannot be answered either with certainty or without making some ethical and even political assumptions.

Scientists can do better with questions of persistence and bioaccumulation, but even these are probably answered more reliably after the fact of manufacture and release. None the less, persistence and bioaccumulation, along with dispersion, lead us to realize that we must establish some links between occupational and environmental health policies. In the case of occupational exposures, unions and industries, governments, or even individual workers always maintain an ultimate option of eliminating future risks, if necessary at short notice. The fact that the option is only rarely exercised does not lessen its importance. Environmental exposures almost always come historically later than occupational exposures and can only rarely, and even then perhaps temporarily, be avoided. Chemical traces of the modest industrial activity of the Roman Empire have been detected in the glacier ice of Greenland millennia later. It was only a matter of years before DDT was accumulated in the livers of polar bears thousands of miles from the nearest release. Strontium–90 from nuclear tests found its way into cows' milk and PCBs to human mothers' milk. Undetectable levels of DDD in the waters of Clear Lake, California, led to tragic consequences for the grebes which fed on the larger fish of that same body of water (Carson, 1962). Chernobyl has already had distributive and bioaccumulative effects from Lapland to Afghanistan (Ahearne, 1987; Hohennemser et al., 1986). The point is that if a chemical is released and it is persistent, it will be dispersed within many, if not all, global cycles and ecosystems. It cannot be retrieved, nor can human ingestion be easily avoided.

That is not to say, of course, that a detectable proportion of present populations will necessarily be adversely affected by present levels of chemicals pervasive in the general environment. It is to say that there is a need for early detection of low-level and cumulative toxic characteristics of all widely used substances. Despite efforts in this regard, however, some harmful chemicals will probably be overlooked by whatever nets we establish to detect risks prior to production and use. More than that, it may well take decades, or longer, to effectively test the chemicals already at large in industry and the environment. Workplace epidemiological testing should thus be seen as a necessary component of any effective environmental health protection system. Recognition of this reality is decidedly not tantamount to placing industrial workers in the role of test subjects. It is simply implicit in a conclusion that toxicological testing involving animals prior to industrial use achieves a necessary, but not sufficient, level of protection for the general public.

A good deal has been written about the double standard implicit in allowable occupational, as opposed to environmental, exposures (Derr et al., 1983; Hattis et al., 1982). The discussion of comparative time-weighed average airborne exposures has been pointed and useful. See Table 7.1 for some comparisons in this regard. The rationale for higher occupational standards includes such considerations as the continuous nature of environmental exposures, the universal nature of environmental exposures (including the young, the aged and the infirm), and the partially voluntary nature of occupational exposures. Some have argued that this rationale is inadequate considering the degree of difference (Derr et al., 1981). It should also be noted that little attention has been paid in this discussion to the fact that waterborne and food-chain based exposures occur environmentally, but not in many cases occupationally. Nor, as was discussed above, can environmental exposures, especially if pervasive, be appreciably reduced

Table 7.1 Comparison of occupational and environmental studies

Material or product	EPA standard[a]	OSHA standard[b]	Natural background observed	Level at which health effects	Health effects
Carbon monoxide (CO)	9 ppm (8 hrs) 35 ppm (1 hr)	50 ppm (8 hrs)	0.1 ppm[c]	35 ppm (1 hr)	5% carbon hemoglobin in blood: heart patient symptoms, changes in oxygen intake[d]
Nitrogen oxides (NO_x)	0.05 ppm (1 yr.) (0.25 ppm [1 hr] California)	5 ppm (8 hrs) (NO_2)	0.004 ppm[c]	1.5 ppm	Increased airway resistance, respiratory irritation[f]
Ozone (O_3)	0.12 ppm (1 hr)	0.1 ppm (8 hrs)	0.05 ppm[g]	0.1 ppm	Increased airway resistance, respiratory irritation[h]
Sulphur dioxide (SO_2)	0.14 ppm (24 hr) 0.03 ppm (1 hr)	5 ppm (8 hrs)	—	1 ppm	Respiratory irritation[i]
Hydrocarbons (HC)	0.05 ppm (guideline)	500 ppm (8 hrs)	—	500 ppm	Neurological damage[j]
Lead (Pb)	1.5 μgm/m³ (30 days)	200 μgm/m³ (8 hrs)	—	30 μgm/m³	Changes in blood and urine chemistry, anemia in children[k,l]
Beryllium (Be)	0.01 μgm/m³ (30 days)	2 μgm/m³ (8 hrs)	—	2 μgm/m³	Lung disease observed[m], observed cancer induction[n]
Radon (Rn)	0.015 working levels (levels for remedial action)	0.33 WL (1/4 yr) NRC	0.003 WL[o]	25 WL	Observed cancer[p]
Noise	70 db[q] (guidelines)	90 db	50 db	85 db	Hearing loss[r]
Ionizing radiation	0.5 rem (yr) NRC	5 rem (yr) NRC	0.15 rem (yr)[s]	2.5 rem (yr)	Observed cancers[t]

a *Environment Reporter*, Bureau of National Affairs, August 29, 1980, pp. 121:0101-121:0381 and 121:0461-121:0701.
b *Environment Reporter*, Bureau of National Affairs, June 20, 1980, pp. 115-1551, 115-1600.
c NAS, *Carbon Monoxide*, Washington, 1977, p. 29.
d *Ibid.*, pp. 65, 166.
e H. S. Stoker and S. L. Seager, *Environmental Chemistry*, 1972, p. 27.
f NAS, *Nitrogen Oxides*, Washington, 1977, p. 282.
g E. Calabrese, *Methodological Approach to Deriving Environmental and Occupational Health Standards*, 1978, p. 293.
h *Ibid.*, p. 295.
i NAS, *Sulphur Oxide*, Washington, 1978, p. 166.
j NIOSH, *Occupational Exposure to Alkanes*, Washington, 1977, p. 68.

k NAS, *Lead*, Washington, 1972, p. 63.
l EPA, *Air Quality Criteria for Lead*, Washington, 1977, pp. 1-13.
m NIOSH, *Criteria for a Recommended Standard Occupational Exposure to Beryllium*, Report #22-10260, Washington, 1972.
n *Science* 190 (1975): 896-9, and 201(1977): 298-300.
o NRCP Report #45, *Natural Background Radiation in U.S.*, 1975, pp. 78-87.
p NAS, *The Effects on Population Exposed to Low Levels of Ionizing Radiation*, 1980, pp. 383-4.
q A. J. Meyer, 'Noise Control at the Federal Level', *Environmental Science and Technology* 9 (1987): 1020.
r EPA, *Toward a National Strategy on Noise Control*, 1977, p. 66.
s NAS, *Effects on Population Exposed to Low Level of Ionizing Radiation*, 1980, p. 13.
t *Ibid.*, p. 187.
Source: Derr et al. (1983).

without considerable economic and/or social costs. In the end, we humans are bound up in ecosystems more firmly than we are with almost any of our socio-economic activities.

If society has a moral obligation to minimize health risks in workplaces, there is perhaps a higher socio-political priority to maintain a healthy general environment. The two are inextricably linked. On some occasions, problems with occupational exposures are a precursor to later environmental health problems. In other cases, such as with excessive heat, cotton dust, silicon dust, and some forms of radiation, there is probably no translation from occupational hazard to environmental hazard. In still other cases, occupational health protection is achieved by venting the hazard into the surrounding community. Perhaps of even greater concern in this latter regard are the even more automated industrial workplaces of the future. With few or no workers at risk, will high-risk substances be more pervasively present in workplaces? What will be the source of pressure to restrain their use?

Several strategies are available to reduce exposure of workers to chemicals. In many instances, these strategies also reduce the exposure to the surrounding communities and the biophysical environment. They are: increased mechanization; remote control processes; continuous v. batch processes; sealing of equipment and piping; increased maintenance; safety-conscious design; and substitution of chemicals. When coupled with personal hygiene programs, periodic medical examinations and physical, chemical and biological monitoring programs, they also serve to prevent or alert workers and management to potentially serious problems.

Some specific components of a program aimed at linking the reduction of occupational and environmental hazards are:

- review of data for all material on toxicity, flammability, reactivity, corrosivity;
- audits of design of process operations, equipment and instrumentation;
- review process stability under departures from normal conditions;
- conduct of qualitative failure studies using fault-tree or event-tree analyses;
- review changes in process or chemicals;
- maintain regular quality assurance programs;
- screen workers for sensitive or susceptible groups.

Those responsible for occupational health protection should, thus, also be mindful of environmental health risks. Whatever regulatory regime is developed, it is to some extent problematic if it deals exclusively with workplace health issues. Agencies, industries and unions should not behave as if workplaces were hermetically sealed from the wider environment. Bureaucratic compartmentalization and convenience in these matters is likely to result in insufficient action as only one set of risks in one time frame are acknowledged and studied. More than that, only the political weight implicit in public concern with environmental, as distinct from occupational, risks can help to avoid the continuation of the discouraging history of delay regarding occupational hazards with which this chapter began. Some improvement in environmental protection might well be achieved through occupational health protection systems which incorporate, or at least do not exclude, consideration of environmental health impacts. I will return to this issue by way of conclusion following a brief discussion of the two principal administrative approaches to occupational health protection.

Two approaches to occupational health protection

The 1970s witnessed the establishment of new approaches to occupational health protection in many Western nations, including, for example, Great Britain, West Germany, Sweden, Canada and the United States. Most of these new approaches have included some provision for worker and management participation in the regulatory process (Noble, 1986; Wilson, 1986; Kelman, 1981). In many European countries this participation has taken place principally at a relatively centralized level. In some countries the participation has not been pursued as vigorously as it might have been (Noble, 1986). In others it has been innovative, as in Sweden where the trade union organization has engaged in independent scientific research. The system in the United States has frequently been compared with those in Western Europe (Kelman, 1988; Wilson, 1986). However, a striking contrast for our purposes can be seen in a comparison of the two systems which have evolved in North America, in Canada and the United States.

The system adopted in the United States by the Occupational Health and Safety Administration (OSHA) has been characterized as a 'command-and-control' system. OSHA, a government agency, sets substance-specific legal exposure standards and enforces those standards by means of inspectors and the courts. Its efforts in this regard have been widely, though not necessarily fairly, criticized by those on both the right and the left (Smith, 1976; Noble, 1986; Viscusi, 1983). In contrast most Canadian provinces have established what has been called an internal responsibility system. Standards and enforcement exist, but the principal institutions established by this form of legislation are mandatory worker–management health and safety committees at the workplace level. Also, legally established are a series of rights for workers, especially a right to know the substances to which one is exposed in the workplace and a limited right to refuse unsafe work.

Let us look first at OSHA and the intense debate that has ensued in the United States since the OSH Act was first introduced in Congress in the late 1960s. As Mendeloff (1979), a quite neutral observer, put it 'no one appears to be very happy with the Occupational Safety and Health Administration.' It does not follow automatically, of course, that OSHA has not done as effective a job as could have been done under the circumstances. However, aside from great enthusiasm regarding passage of the Act and some enthusiasm during the years 1976–80, labor has been critical. Business organizations and pro-business writers have been consistently hostile, often evincing outrage on a variety of grounds. Academic appraisals have also often been critical. It has been widely argued, for example, that the standard approach has been ineffective because fines have been too small, citations by inspectors too often trivial and inspections altogether too infrequent to have had a significant impact on performance. As Beaumont (1983) put it: 'The average number of inspections per enterprise ... was only 0.079 over the five year period or approximately 0.02 on an annual basis, while the average number of standards violations cited per inspection was only 3.7, and the average penalty per violation was only $25.68 from 1971–75.'

With this sort of record, it is hard at first to understand why it is that those in business or whose sympathies lie predominantly with business were so exercised by OSHA throughout the 1970s and early 1980s. The overwhelming majority of firms were never inspected and thereby were never cited. Even in the relatively pro-labor Carter/Bingham years (1976–80) the total number of inspections and citations declined significantly in response to earlier industry charges of excessive attention to trivial matters (Viscusi, 1983). In the early 1970s inexperienced inspectors, required to cite everything they saw, had sometimes pressed firms to make equipment and procedural changes which managers felt were both too costly and in some cases unnecessary. Even firms with excellent safety records were sometimes required to spend monies they believed were being spent for no good reason. The small citation fines mentioned above were far lower that the potential penalties for 'failure to abate' which would ensue when action was not taken. Business's estimates of OSHA related compliance costs (even excluding the cost of fines or even associated administrative costs) have been consistently staggering (Viscusi, 1983; Smith, 1976; Beaumont, 1983). Again, objective observers must be wary that many critics of OSHA have axes to grind and that this is an agency which, given its mandate, cannot possibly please everyone.

Indeed I cannot resolve here the issues of whether or not these cost estimates are accurate, nor whether these, or any other measure of costs are justified. Nor will I comment on whether monies expended on safety and health might have been expended more effectively. I will, however, briefly discuss three dimensions of this important debate germane to the wider themes of this chapter. First, which abatement approach generally makes more sense, engineering controls or personal protection devices? Second, has OSHA overemphasized safety relative to health? Third, given vehement business resistance to change that has ensued, should those concerned about health protection allow that a command-and-control (standard/inspection) system may be a seriously flawed approach, especially when used exclusively?

The first issue, engineering changes or personal protection devices, has sparked several major debates. Business charged that OSHA too often mandated the former. This led, in turn, to a further charge from business interests that OSHA operated in a dream world, oblivious to costs. Business argued, for example, that respirators and earplugs can achieve equivalent results at a trivial fraction of the cost. However, this issue has not to my knowledge been widely considered in relation to potential environmental health impacts. No distinction has been made in this debate between workplace noise and heat, which rarely become environmental problems and workplace chemicals, which quite often eventually do. If closed systems of process changes were required in the case of hazardous chemicals, the result might well provide environmental as well as greater workplace protection. The added costs associated with engineering changes might encourage more frequently the complete withdrawal of some chemicals from workplace processes and/or product lines. On the other hand, there has been little discussion about whether engineering changes in the form of venting, for example, are not in some cases making workplace protection itself a cause of environmental hazards.

The second issue, safety protection relative to health protection, has in a sense been less contentious. There is wide agreement that in the past OSHA has to some extent erred in this regard. One might also observe that this is as well a natural tendency within other approaches. Viscusi (1983), generally a pro-business analyst, has calculated the proportion of health (as opposed to safety) inspections conducted by OSHA at 0.05 in 1973, 0.08 in 1976 and 0.19 in 1980 and the proportion of health violations citations at roughly half this level. Generally he has argued that safety as opposed to health protection would be better achieved through economic instruments such as pay differentials and greater performance rating impacts on the worker's compensation payments made by firms. Health protection, he concedes, is less responsive to market factors and forces.

Smith (1976), quite explicitly pro-business, concluded, reluctantly, that 'it appears that an occupational health program must rely on a standards approach, despite its defects.' This in contrast to the standard setting in occupational safety which 'should be repealed and replaced by a program which sets moderate fines on each injury'. In brief, both Viscusi and Smith would seem to prefer a greater relative emphasis at OSHA on health, as against safety. Noble (1986), a strongly pro-labor analyst, concluded that OSHA has had 'little impact on safety in the aggregate and [only] a small positive effect in some situations'. It has probably, however, 'reduced the risks of occupational disease in a few industries'. Other observers on the political left have been particularly critical of the insufficiency of OSHA health, as opposed to safety, standards. This has been true especially following the Supreme Court benzene ruling in 1982. Here, of course,

OSHA attempted a tough standard but was turned back by the court.

The addition of environmental health considerations to the occupational health and safety policy debate can only provide reinforcement on this widely agreed point. The OSHA has been unable to establish very many new health standards, but has a very long list of safety standards. Market forces, it has been observed, however, work far better in the latter case than in the former. Variations in job safety are widely and immediately known within the workplace while variations in health impacts are less certain, less immediate, and less visible. Health issues as well have a relevance beyond the plant gate, in addition to that which safety carries for the families and friends of injured workers. If safety issues could be handled as well or better in other ways, freeing the safety and health agency to concentrate more of its limited resources on health, one would assume that all parties would benefit.

The third issue concerns more directly the relative merits of the various possible authority systems associated with occupational health protection. The OSHA has used a command-and-control system, involving in principle agency enforcement of a detailed system of standards, for both health and safety. Many analysts have argued, for various reasons, that such a system is inordinately and inherently inflexible. Smith (1976) has argued that in industry 'technological conditions are so diverse that centrally made rules cannot be applied to all situations without some degree of counter-productivity (including time and resources spent in justifying and deciding upon variances).' It is logic of this sort which has been part of the rationale for the so called internal responsibility system common to several Canadian provinces.

In the Canadian province of Ontario virtually all workplaces other than very small ones must, by law, establish a labor–management health and safety committee. This is true for both unionized and non-unionized firms. Workers have a general right to be informed by employers about the substances to which they may be exposed at work. In addition, the Canadian Centre for Occupational Health and Safety, a federal agency with a tripartite (government, union, management) board, makes rapidly available to both management and workers in all provinces, the findings of past and current health studies. Other sources of information are also available from other countries and international agencies such as the International Labour Office. Exposure standards are established and enforced for selected substances, but the principal responsibility for both health and safety resides within each individual workplace. Management ultimately decides, the health and safety committees advise, and workers can gain some leverage in difficult situations by means of a limited, but protected, right to refuse unsafe work. When a refusal occurs, ultimately the Ministry of Labour will rule on the justifiability of the refusal. In all, the right has been relatively infrequently exercised, even though it has been at times quite widely advertised. The overwhelming majority of refusals thus far have been found to be justified and/or appropriate. That is, even in those exceptional cases where refusing workers have not ultimately been supported, it has only rarely been found that the refusals have been taken on an unreasonable or trivial basis.

Overall this system has resulted in less controversy than has arisen within the United States. Some might say this is because in many non-unionized workplaces management simply dominates workers' representatives, especially when there is high unemployment in an industry or region. It might also be argued that the system has not worked equally well in every province. Where unionization roles and/or government commitment are high and the economy sound this system will probably work better than where those conditions do not exist. Others might point out that the larger relevant difference here is that between Canada and the United States rather than the relative merits of the two institutional/authority systems. Indeed the very willingness of several quite conservative governments in Ontario and elsewhere to establish this system suggests a recognition of a level of workplace, or even social, accommodation not present in the United States in recent decades. Neo-conservatism in Canada, one might hazard, has been less militant than in the United States, and labor more willing, or able, to come to accommodation with the workplace management. More than this, one must note that in Canada, as in Britain, there was a considerable delay in introducing enhanced occupational health protection and thereby a period within which to observe some of the difficulties associated with the pioneering approach adopted by OSHA.

These points made, I might also venture that in general, an internal responsibility system supported by standards and inspections in selected vital cases appears to provide a more effective administrative design. Some on the political left in the United States have recently stated a preference for a system of this sort. Noble (1986), though he does not specifically mention the Canadian system when he discusses others such as in Britain, West Germany, and Sweden, would clearly prefer it in principle. As he said so clearly about the US system 'by investing responsibility for changing employer practices in an executive agency instead of attempting to devolve power to workers at the plant level, Congress left the program vulnerable to changes in the short-run balance of political forces.' Noble is well aware that the political climate within workplaces also varies with the times, but he is correct that the lurches over occupational health and safety would not be so dramatic as between, for example, the Carter and Reagan administrations. Within the workplace, the persons with whom both sides must come to agreement are the persons on whom both depend for continued economic performance, for labor–management wage agreements, and even for decent day-to-day personal relationships.

The internal responsibility system is in general far less likely to strike management as intrusive. Again as Noble (1986) put it, the 'intrusive' nature of the system with which OSHA was saddled was 'all but guaranteed to alienate employers' and ultimately led to the White House reviews and court action which explicitly enfeebled the system in the 1980s (Sagoff, 1985; Tolchin and Tolchin, 1983; Noble, 1986). The British version of an internal responsibility system, as Wilson (1986) noted, differs from that of the United States principally in that it is based on persuasion rather than coercion. It also empowers a labor–management board, rather than a government agency, to establish standards. The enforcement of standards, however, rests with government inspectors. The British system, like the Canadian, blends government intervention with labor–management self-management. Unlike the Canadian, the British model involves a centralized rather than a decentralized system of authority.

Some (Kelman, 1981; Vogel, 1983) have argued that many difficulties have arisen with the OSHA system because there is no tradition in the United States of business acceptance of government power as natural. Such a tradition exists, of course, in Sweden and even Britain and Canada. Wilson (1986) has stated that the 'regulators as well as the regulated in the United States are inclined to be inflexible, dogmatic and antagonistic in their approach'. He went on to observe that the OSH Act and inspection system were deliberately designed to avoid agency capture by corporate interests. Thus, inflexibility was virtually mandated for OSHA inspectors. One must conclude, I believe, that this effort to avoid capture has been an admirably intended failure principally because it did not empower the beneficiaries of regulation and thereby increase their involvement in the regulatory process.

One must be cautious here, of course, not to simply assume that the Canadian, or any other system, has yet achieved such empowerment. Nevertheless, in several systems there is at least a potential for empowerment. Noble (1986) notes that unions in Britain and West Germany have not by any means utilized the full health and safety rights available to them. None the less, it is noted by Swinton (1982) that worker and union participation in occupational health and safety decision-making has the potential to help create increased health and safety awareness and enhanced care on the part of workers. One need not subscribe to a blame-the-victim mentality to allow that this can in itself be important. In my view, only by increasing the level of knowledge and concern within workplaces is there much prospect of achieving improvements in either occupational health or environmental protection. If the outlook within the workplace, that of either labor or management, is one of hostility and resistance to change there is probably no government that can impose its will either consistently or continuously. Both consistency and continuity are essential in these matters. In conclusion, I will offer some further reasons why it is in the interest of both labor and management to be more open and flexible regarding changes of the sort proposed here.

Conclusion: responsibility for the wider environment

It seems that a considerable restructuring of occupational health and safety policy may soon be on the agenda, particularly in the United States. Should the neo-conservative mood of the moment there wane, the changes could involve more than the further weakening of OSHA without any attempt to establish other means of enhancing worker health and safety. An integral responsibility system, such as that in place in several Canadian provinces, has much to commend it (Tucker, 1984; Swinton, 1982; Reschenthaler, 1979; Paehlke, 1981). Significantly increased use of merit (performance) ratings to determine workers' compensation charges for consistently unsafe firms is another proposal which may also make sense from all points of view.

This latter change might well allow an increased emphasis on health as opposed to safety issues both within agencies such as OSHA and within many internal workplace health and safety committees. (Some workplaces, of course, may have serious safety but few health concerns.) Under new arrangements of this sort, management in most cases would be induced to make additional safety investments on economic grounds. The combination of these two changes might also relieve some aspects of management resentment over external bureaucratic interventions. However, two cautions are in order here. First, effective protection probably cannot be achieved without some resistance and objection, regardless of which approach or combination of approaches is adopted. Second, perhaps the principal shortcoming with an internal responsibility system, especially one which emphasizes health issues, is that it is not clear that workers or management are at this time sufficiently informed or concerned about occupational, let alone environmental, health hazards. Safety impacts are more visible and immediate than health impacts. Where a government agency or ministry responsible for occupational health could simply be ordered to work in liaison with appropriate environmental health agencies, internal responsibility committees will not succeed in this task unless both employees and management are sufficiently informed and concerned.

This is perhaps obvious, but the OSHA experience has taught that even a well-intentioned and generally competent government agency cannot succeed without this condition being met. The will to achieve both occupational and environmental health protection must come from within, or come to be accepted within, the relevant workplaces. Internal committees, supported and urged on by governmental authority, could eventually become an important means to improved industry performance on environmental health protection, as well as occupational health protection. If they cannot, there is little prospect of success by any other means, certainly not by government acting alone in the face of indifference or hostility within the workplace.

Any internal responsibility system requires external supports including an agency charged with supplying scientific data and an authoritative external means of resolving intractable internal disputes. This same information agency could also be empowered to provide environmental data. Workplace committee members could be schooled in the rudiments necessary to interpret that data. Obviously such efforts would not in and of themselves be a sufficient source of overall environmental protection. However, fully effective protection from hazardous environmental chemicals can only be achieved with further advances in general public education. That public includes those who play labor, management or consumer roles, essentially everyone. Unfortunately, one means by which such education was significantly advanced, particularly in the late 1970s in the United States, the so-called Committee on Occupational Safety and Health (COSH) groups, have been hard hit by recent budgetary cutbacks. These groups, state and community-based committees on occupational safety and health, have considerable potential as one locus within which community and workplace concerns can be raised in combination. These groups have sought to establish, for example, community and state public right-to-know legislation regarding the manufacture, storage, use and transport of hazardous substances (Levenstein *et al.*, 1984; Adams, 1983; Noble, 1986). These organizations include many health professionals.

The imposition of added responsibilities on the labor movement implied by a greater dependence on internal responsibility systems is, of course, of some concern. This would be particularly true if such committees were to come to play any significant role in environmental health protection. Clearly, labor might be justified in suggesting a need for some supplementary sources of public funding for such an undertaking. More than this, one can only note that labor union members, contrary to their public image, are roughly as likely as anyone else to be concerned about the general environment (Milbrath, 1984 and sources cited therein). Also, in most North American cities air pollution exposures are correlated with income and occupation. Industrial workers and their families are on average more likely to be exposed to pollutants within their communities as well as on the job. There are thus reasons to hope that labor might be one means of carrying such concerns into the process of workplace decision-making. One might even hope that at some point responsible public interest groups could gain a hearing or a place in the workplace forum.

As was noted above, the Ministry of Labour in Ontario oversees the internal responsibility system and continues to establish exposure standards for selected hazardous substances. Only habit and bureaucratic compartmentalization prevent the incorporation of environmental consideration into their deliberations. These are not, of course, modest obstacles. However, in a few cases, one would expect, almost any attempt to combine such data would raise questions about the limits of the standards approach itself. The Environmental Protection Agency in the United States, even in the Reagan era, suggested that the best approach to asbestos regulation might be to phase out its use altogether. That solution, or a use-specific version of it, might be appropriate in the case of several substances which have been proven hazardous in both the workplace and the environment. Such an approach, used sparingly, might well be both technically and administratively more manageable than an elaborate combination of exposure, emission, and disposal standards.

But what of the larger number of less immediately visible hazards? Here we must in the end depend on some form of internal responsibility, combined with enhanced public education and 'last-resort' governmental support. That vital input of governmental support ought to aspire to be both labor-sympathetic and highly flexible, a difficult combination (Wilson, 1986; Sass, 1983; Paehlke, 1981). Flexibility is particularly important in relation to the timing of workplace improvements. Labor, for example, could be more open to the temporary use of personal protection devices, management to bringing labor into health-relevant capital investment decision-making. The flexible/pro-labor ideal may well be possible only in a general political climate different from that in present-day North America. The most ironic aspect of this observation is that it is precisely this climate, especially in its workplace manifestations, which may prevent the very productivity improvements which neo-conservatism values so highly.

In conclusion, it might be noted that one key source of Japanese and Swedish industrial success lies in the mutual respect and shared goals of labor, management and government. One need not to be a total corporatist to suggest that both enhanced management concern for worker well-being and a system of co-operative workplace decision-making could help to foster a climate conducive to productivity improvements. The goals of improved occupational health and safety and improved environmental protection could also point the way towards these other important outcomes.

References

Adams, M. L. (1983), 'Right to know: a summary', *Alternatives* 11, pp. 29–36.
Ahearne, John F. (1987), 'Implications of the Chernobyl nuclear accident', *Resources* 86, pp. 10–12.
Beaumont, P. B. (1983), *Safety at Work and the Unions* (London: Croom Helm).
Carson, R. (1962), *Silent Spring* (New York: Fawcett World Library).
Commoner, B. (1973), 'Workplace burden', *Environment* 15, pp. 15–20.
Derr, P., Goble, R., Kasperson, R. E., and Kates, R. W. (1981), 'Responding to the double standard of worker/public protection', *Environment* 25, 6–11, pp. 35–42.
Doll, R. and Peto, R. (1981), 'The cause of cancer: quantitative estimates of avoidable risks of cancer in the United States today', *J. Nat Cancer Inst.* 66, pp. 1191–1308.
Eisenbud, M. (1978), 'Environmental causes of cancer', *Environment* 20, pp. 6–16.
Epstein, S. S. (1978), *The Politics of Cancer* (San Francisco: Sierra Club Books).
Graham, J. and Shakow, D. (1981), 'Risk and reward: hazard pay for workers', *Environment* 23, 14–20, pp. 44–5.

Gribbin, J. (1988), *The Hole in the Sky: Man's Threat to the Ozone Layer* (Toronto: Bantam Books).

Hamilton, A. (1943), *Exploring the Dangerous Trades* (Boston: Little, Brown).

Hattis, D. R., Goble, R., and Ashford, N. (1982), 'Airborne lead: a clearcut case of differential protection', *Environment* 24, 14–20, pp. 33–42.

Hohennemser, C., Deicher, M., Ernst, A., Hofsass, H., Linder, G., and Recknagel, E. (1986), 'Chernobyl: an early report', *Environment* 28, pp. 6–13, 30–43.

Hunter, D. (1978), *Diseases of Occupations* (6th edn), (Boston: Little, Brown).

Jamieson, J. W. S. (1977), *Polybrominated biphenyls in the environment* (Ottawa: Environmental Protection Service).

Kazis, R. and Grossman. R. (1982), *Fear at Work* (New York: Pilgrim Press).

Kelman, S. (1981), *Regulating Sweden, Regulating America: A Comparative Study of Occupational Safety and Health Policy* (Cambridge, Mass.: MIT Press).

Levenstein, D., Boden, L., and Wegman, D. (1984), 'COSH: a grass-roots public health movement', *American Journal of Public Health* 74, pp. 964–5.

Lindblom, C. E. (1977), *Politics and Markets* (New York: Basic Books).

Mendeloff, J. (1979), *Regulatory Safety: An Economic and Political Analysis of Occupational Safety and Health Policy* (Cambridge, Mass.: MIT Press).

Milbrath, L. W. (1984), *Environmentalists: Vanguard for a New Society* (Albany: State University of New York Press).

Nelkin, D. and Brown, M. S. (1984), *Workers at Risk: Voices from the Workplace* (Chicago: University of Chicago Press).

Noble, C. (1986), *Liberalism at Work* (Philadelphia: Temple University Press).

Paehlke, R. (1979), 'Occupational Health Policy in Canada', in W. Leiss (ed.), *Ecology versus Politics in Canada* (Toronto: University of Toronto Press).

Paehlke, R. (1980), 'Guilty until proven innocent: carcinogens in the environment', *Nature Canada* 9, pp. 18–23.

Paehlke, R. (1981), 'Much to be done: exploring the interface between occupational and environmental health', *Alternatives* 10, pp. 13–20.

Perrow, C. (1984), *Normal Accidents* (New York: Basic Books).

Reschenthaler, G. B. (1979), *Occupational Health and Safety in Canada* (Montreal: Institute for Research on Public Policy).

Sagoff, M. (1985), 'Sense and sentiment in occupational safety and health programs', in D. Nelkin, *The Language of Risk* (Beverly Hills: Sage Publications).

Sass, R. (1983), 'Cancer in the work environment', *Alternatives* 11, pp. 37–43.

Schrecker, T. F. (1984), *Political economy of environmental hazards* (Ottawa: Law Reform Commission of Canada).

Selikoff, I. J., Churg, J., and Hammond, E. C. (1964), 'Asbestos exposure and neoplasia', *Journal of the American Medical Association* 188, pp. 22–6.

Smith, R. S. (1976), *The Occupational Safety and Health Act* (Washington, DC: American Enterprise Institute).

Stellman, J. M. (1977), *Women's Work, Women's Health* (New York: Pantheon Books).

Swinton, K. (1982), 'Enforcement of occupational health and safety legislation; the role of the internal responsibility system', in K. Swan and K. E. Swinton, *Studies in Labor Law* (Scarborough, Ontario: Butterworth).

Swinton, K. (1983), 'Regulating occupational health and safety; worker participation through collective bargaining and legislation', *Essays in Collective Bargaining and Industrial Democracy* (Dan Mills, Ontario: CCH Canadian Ltd).

Tataryn, L. (1979), *Dying for a Living* (Montreal: Deaneau and Greenberg).

Tolchin, S. J., and Tolchin, M. (1983), *Dismantling America* (Boston: Houghton Mifflin Company).

Tucker, E. (1984), 'The determination of occupational health and safety standards in Ontario, 1960–1982', *McGill Law Journal* 29, pp. 261–311.

Viscusi, W. K. (1983), *Risk By Choice* (Cambridge: Harvard University Press).

Vogel, D. (1983), 'Regulation, lessons from Greater Britain', *Public Interest* 72, pp. 25–50.

Wagoner, J. K. (1976), 'Occupational carcinogenesis: The two hundred years since Percival Pott', *Annals of the New York Academy of Sciences* 271, pp. 1–4.

Wallick, F. (1972), *The American Worker: An Endangered Species* (New York: Ballantine Books).

Wilson, G. K. (1986), 'Legislating on occupational health and safety: a comparison of the British and American experience', *European Journal of Political Research* 14, pp. 289–303.

8 *The economic perspective*

H. M. A. JANSEN

Introduction

After the Second World War, the world economy and especially the Western countries experienced an unprecedented economic growth, resulting in levels of luxury and per capita income which had never been reached before. It was not until the 1960s that a general awareness emerged of the costs of this economic growth (Mishan, 1968). It became apparent that the relative affluence derived from marketed goods and services was achieved at the cost of environmental services and amenities which had been considered as being free. A new scarcity of environmental goods and services was observed and it turned out that growth rates of GNP gave a wrong impression, a gross overestimation of the development of societal welfare (Hueting, 1974).

The chemical industry has been one of the driving forces of industrial growth during the last decades. Figure 8.1 depicts the growth of sixteen economic sectors in the world in the period 1965–1980. The inner circle shows the standardized 1965 situation. The plastic products and chemicals sectors are the two fastest growers.

The development of the chemical industry led to the introduction of new chemicals and materials which were used in other sectors as cost-saving, innovative inputs, and thus stimulated overall production. Statistics from the EEC (Eurostat, 1986) show that over 70 per cent of the production of the chemical industry is used as inputs in other productive sectors. The main users are: chemical industry (23 per cent); rubber and plastics (26 per cent); textiles and clothing (7 per cent); and agriculture (5 per cent). The rubber and plastics sector contributes to: rubber and plastics (7 per cent of production of rubber and plastics); motor vehicles (5 per cent); and leather, and footwear (4 per cent). Over 60 per cent of the production value of chemical industry is spent on intermediate inputs, and in the rubber and plastics sector this percentage is 56.

Through innovative chemicals and materials, through industrial inter-relations, and through its own high growth rate, the chemical industry has contributed tremendously to overall economic growth in the last decades. However, this growth has occurred at some cost. Chemical wastes have caused extensive environmental pollution and damage to human health, as is discussed in other chapters in this book.

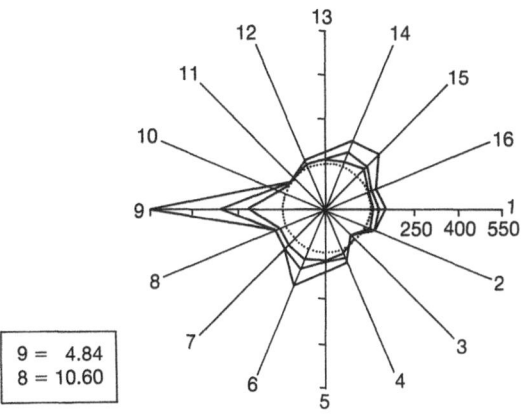

KEY:
Branches (ISIC Code):

1	Food products (311/2, 313, 314)	13	Metal products, excl. machinery (381)
2	Textiles (321, 322)	14	Non-electrical machinery (382)
3	Leather industries (323, 324)	15	Electrical machinery (383)
4	Wood and furniture (331, 332)	16	Transport equipment (384)
5	Paper and printing (341, 342)		1975–1980
6	Chemicals (351,352)		1970–1975
7	Petroleum and coal (353, 354)		1965–1970
8	Rubber products (355)	*g*	Average annual growth rate
9	Plastics products (356)		1965–1980 (percentage)
10	Non-metal mineral products (361,362,369)	*0*	Index of structural change,
11	Iron and steel (371)		1965–1980
12	Non-ferrous metals (372)		(constant prices in 1975 dollars)

Source: UNIDO data base

Figure 8.1   Growth and structural change in the world economy

Economics and externalities

The subject of economics is optimal allocation of scarce resources. In a market economy, the main mechanism for achieving optimal allocation is through the price system. It can be shown by use of mathematical models that under a series of assumptions which in reality are not all met, the price mechanism will automatically lead to an optimal allocation. Without going any deeper into the theory, it is clear that the price system has great deficiencies where environment is concerned. Environmental goods and amenities, e.g. clean air and water, are seldom marketed and as a rule there is no price for environment. In economic terms, environmental effects are *external effects* (or externalities): there is no market and consequently no price. The fact that environment is unpriced and that the consequences of pollution are not directly and completely borne by the polluter leads to a situation where environmental concerns are not automatically included in the decision-making of economic actors. In the case of external effects, the price mechanism is misleading and results in a sub-optimal allocation of

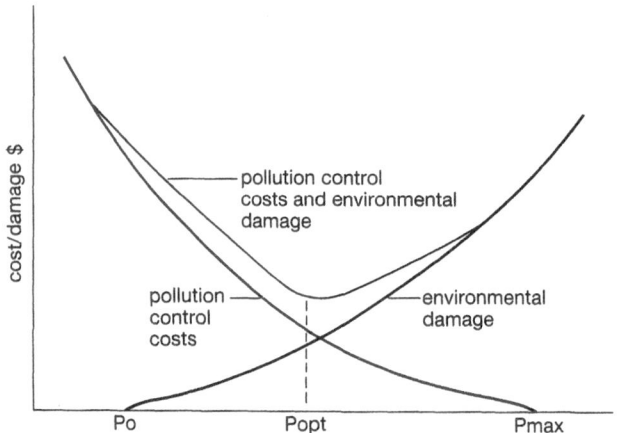

Figure 8.2   Optimal pollution control

scarce resources. The situation is illustrated in the simplified, conceptual scheme of Figure 8.2. The scheme depicts a static situation in which total social disbenefit due to pollution is measured as the sum of environmental damage and costs of environmental control.

Po is a threshold value of pollution, and at levels below Po no damage will occur. Pmax is the level of pollution without any pollution abatement. Depending on one's perspective, the optimal level of pollution may be nearer Po or Pmax. In the economist's view, the optimum level of pollution is Popt, where total social disbenefits are minimal. In the absence of any incentive for polluters to reduce pollution, Pmax will be the factual level of pollution.

If environmental damage were adequately and completely reflected in the price system, it can be shown that under various additional assumptions, the price mechanism would automatically lead to the optimal situation. This idea of including environmental damage in the prices of goods and services which cause pollution is called 'internalization' of external effects. The notion that polluters should pay for their damaging activities is also embodied in the 'polluter pays principle' (OECD, 1975). The subject of internalizing environmental goods and amenities in the price system is discussed further on pp. 203–7.

Due to practical obstacles, this solution sounds easier than the real world allows. Firstly the horizontal axis of Figure 8.2 measures 'pollution'; it is unclear which type of pollutant is meant and where or how it should be measured. There are various sources of various pollutants and concentrations vary in time and location. Some have immediate effects while others have latent effects. The overall concept of pollution used in the scheme is therefore a simplification which cannot be made operational in an unambiguous way. Secondly, relationships between pollution and bio-physical damage are manyfold and largely unknown, or at least not accurately and reliably measured. The effects may extend to the future, while the scheme is static. Thirdly, damage is measured along the same axis as costs, which means that it should be measured in the same (monetary) units. Valuation of physical damage in monetary terms is again incomplete, inaccurate and unreliable.

In Figure 8.3 the links between economics and the environment are schematized. The interrelations of both systems imply that economic activities have consequences for the environment, often measured as emissions and discharges of various pollutants. In turn, environmental control measures affect economic activities and in this chapter economic instruments of environmental protection policy are of primary interest. These topics are the subject of the next section.

When released into the environment, pollution leads to damage in complex ways which are beyond the scope of economics: dispersion, synergism and dose–effect relationships are the complex subjects of other disciplines. The task of monetary valuation of damage is again an economic one and will be discussed later, though it is acknowledged to be an incomplete and inaccurate process. Finally, techniques will be discussed for making trade-offs which should ultimately lead to decisions which bring about optimal allocation of the scarce resources of the environment. Although progress has certainly been made, the objective of guiding society to this optimal allocation of scarce resources is not met by current economic practice. For a more thorough and comprehensive treatment of environmental economics, see Kneese and Sweeny (1985), containing a treasure of information and references to the literature.

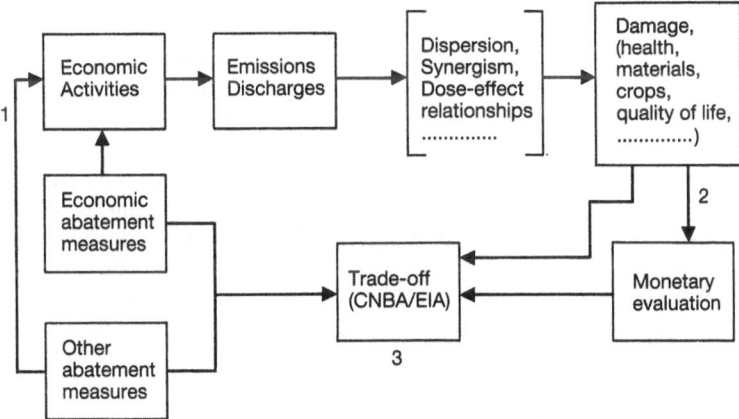

Figure 8.3 Scheme of environmental/economic interrelations

Linkages between the economic system and the environment

In this section we treat the generation of pollution by economic activities, economic instruments for environmental protection, and economic effects of environmental policy.

The relationship between economic activities and environmental pollution is too complex to be treated at a macro level. Economic activity has to be split up into various productive sectors and final demand. Figure 8.1 of the previous section provides an example of a categorization in sixteen economic sectors. By multiplying the production value of a sector by the emission coefficient of a certain pollutant one gets the emission generated by that sector. Likewise, multiplication of the final demand of a product with an emission coefficient yields the emission generated by final use. This procedure presupposes a constant emission coefficient with respect to one sector and pollutant. The more detailed the categorization of sectors is, the more chance there is that this assumption is a good approximation of reality. However, generation of very specific pollutants may be process specific instead of sector specific, and for common pollutants such as SO_2 and NO_x, different types of furnaces and other equipment may be in use within one sector. The emission coefficients approach is, therefore, a rather broad-brush approach, and the determination of emission coefficients is a complex, non-economic problem.

Input–output analysis (IOA) is an approach that was first applied in environmental economics by Leontief (1970). It analyzes intersectoral linkages in the economy. In an input–output table (see Figure 8.4) the output of economic sectors can be seen in the horizontal rows, shown as deliveries to other sectors (so-called intermediate deliveries) and to final demand (i.e. consumption, investment, export and stocks). On the vertical axis, the inputs of a sector can be seen, being deliveries from other sectors and primary inputs (i.e. wages and salaries, depreciation of capital stock, taxes, imports and profits). In IOA the assumption is made that per sector, in the vertical columns, the ratios of the inputs and total production are constant; if, for example, the production of the agricultural sector doubles, the inputs from chemical industry, used in agriculture, will also double. By making this assumption, a square matrix of IO (input–output) coefficients can be calculated per column, the ratios of intermediate inputs and total production. Primary emission coefficients of a certain pollutant can be calculated by dividing the emission in a sector by the production value. Through IO calculations, cumulative emission coefficients can be obtained, which express the emission of a pollutant, generated during all phases of production.

A recent application of IOA was carried out by James (1985), investigating consequences of Australian energy policy. Another application (Hordijk *et al.*, 1983) investigates the pollution consequences of Dutch foreign trade. It

Sector 1 2 3 ... n

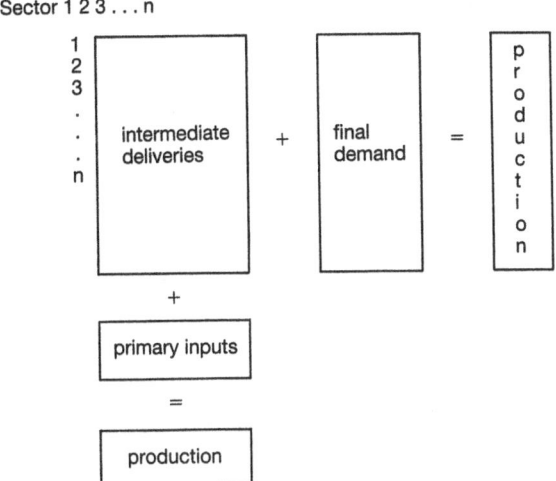

Figure 8.4 Scheme of an input–output table

compares the emission content, calculated through cumulative emission coefficients, of the Dutch package of export goods with the likewise calculated emissions of Dutch imports. The calculation is a counterpart of the balance of trade; it compares the emissions of pollutants in the Netherlands, generated in the production of export goods, with the emissions which are generated abroad in the production of Dutch import goods. Although one would not expect that the comparative advantage of the densely populated country of the Netherlands lies in its potential to absorb environmental pollution, it turned out that on average the pollution content of Dutch exports exceeded the pollution content of Dutch imports. For many pollutants, the foreign trade balance of the sector of chemical-base products accounted for a very large part of this unfavourable balance of pollution.

As discussed in the introduction, a cause of overexploitation of the environment lies in the failure of the market system to take environmental resources sufficiently into account. Direct regulations (e.g. prescriptions, licensing, zoning) are widely applied, and economic instruments can be used to introduce incentives to potential polluters to prevent or reduce environmental degradation (Baumol and Oates, 1975; James *et al.*, 1978; Brown, 1984; Bohm and Russel, 1985; Boland, 1986). Economic instruments can be regarded as corrections to the imperfect market mechanism; they include:

● taxing the polluter: if the polluter is charged for generating pollution, it can be profitable for him or her to reduce emissions. Moreover, the revenue of the tax can be used by the authorities for financing environmental improvements.

- subsidizing the polluter for the amount of prevented pollution or for financing pollution abatement investments.
- trading pollution permits or rights: if a firm is going to discharge polluting chemicals, it has to buy this right from others who will in turn reduce their discharges.
- introduction of a deposit-refund system, so as to stimulate potential polluters to bring their polluting materials to a treatment or storage facility.
- compensating the victims of pollution: although pollution itself is not reduced by this measure, the effects are compensated.
- government production: government can take over polluting activities (e.g. electricity production) so as to exert direct control over emissions, and government can install treatment or storage facilities.

Pollution charges may result in the following positive effects on pollution generating activities:

- inputs shift towards less polluting and, therefore, lower charged substitutes;
- more recovery and recycling of wastes;
- introduction of and research into new, less polluting techniques and processes;
- installation by firms of treatment facilities;
- separation of waste streams into wastes with high and low treatment costs.

Through taxation of pollution, the use which is made of environmental services during production and consumption can be influenced by means of the price system. Pollution charges can, therefore, be used to internalize environment goods and services in the price system and in the decision-making process of producers and consumers. Three types of charges can be distinguished: pigovian charges, regulatory charges and redistributional charges.

Pigovian charges are based on the theory of welfare economics (Pigou, 1962). If a charge on emission is equal to the marginal damage caused by the emission, an optimal situation will result as the polluter will reduce emissions up to the point where marginal costs of reduction equal marginal damage. In Figure 8.5, R is the curve of marginal costs to the polluter of pollution reduction. If a polluter is to reduce emission, he or she will first choose the cheapest way of reduction and further reduction will be more costly. The R curve is, therefore, declining. D is the curve of marginal damage to society, caused by the pollution and measured in monetary terms. After a certain threshold value is exceeded, measurable damage begins and in the figure it is assumed that an increment of pollution will lead to greater additional damage.

There are some practical problems for applying pigovian charges. In the first place, the concept 'level of pollution' is too abstract. Secondly, the marginal damage function, measured in monetary terms, is not easily applied to regulatory charges. Here the authorities fix a certain level of emissions L of a specified pollutant and try to find the charge T per unit of pollution, which will result in an optimal level of pollution. The situation is depicted in Figure 8.6. There is no explicit curve of marginal damage, although damage will be implicitly and intuitively included in the fixed, emission level L. Under influence of charge T, polluters will reduce emission up to L, as a reduction of emissions is cheaper than paying the charge.

The third type of charges, redistributional charges, are more frequently used in practice. They are intended to yield revenues, rather than to reduce emission. The revenue can be used by government to finance pollution abatement measures. However, the direct incentive of redistributional charges to reduce emission is often minor.

A problem of the taxing instrument is how to find an adequate assessment basis for the charge. Although societal damage should ideally be the basis, in practice it will often be units of emission of specified pollutants. This requires monitoring which may be costly, especially if pollution is caused by many minor sources. However, the charge can sometimes be coupled with inputs which cause pollution, for instance a charge of SO_2 emissions on fossil fuels with a high sulphur content.

Charges are in agreement with the 'polluter pays principle' (OECD, 1975). But the practical questions concerning this principle are, what exactly the polluter should pay and how this can be organized efficiently.

An alternative to regulatory charges is a regulatory subsidy. The case of

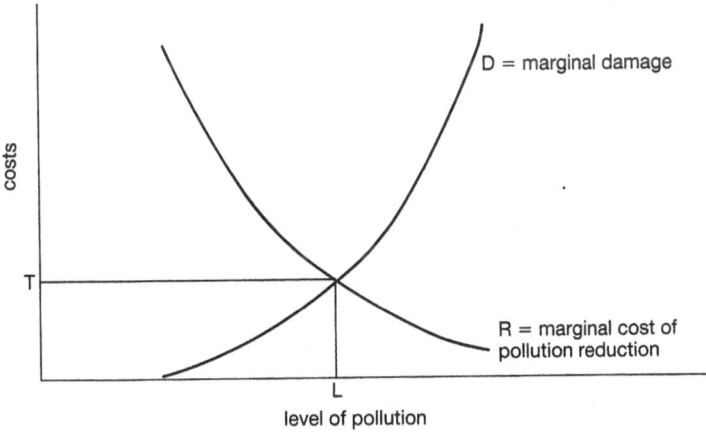

Figure 8.5 Illustration of pigovian tax

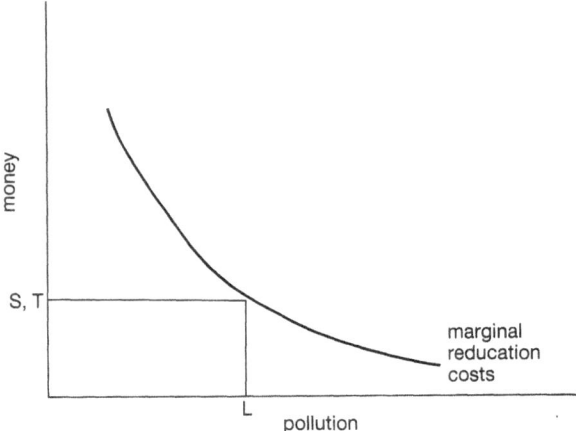

Figure 8.6   Regulatory charge/subsidy

subsidies to a polluter to encourage reduction of pollution was put forward
by Coase (1960) and critically discussed by Baumol and Oates (1975). In
Figure 8.6, a subsidy S per unit of prevented pollution (right point L) will
also induce the polluter to reduce pollution up to point L. Subsidies are not
in compliance with the 'polluter pays principle'. And although an individual polluter is equally motivated by charges and by subsidies to reduce
pollution, a subsidy may lead to more polluters, as polluters are better off
than under a system of charges. Also, subsidies will not lead to higher prices
for goods with a high pollution content, and therefore there is no internalization of environmental effects into the price system. For such reasons, a
system of subsidies is, in a situation of equilibrium, inferior to a system of
pollution charges. An advantage of subsidies is that polluters will not be
tempted to dodge the system, which they may try under a charge system.
A subsidy can be a good instrument in a situation of disequilibrium, e.g. in
the case of high costs of initial installation of abatement equipment, leading
to a high lump sum investment and subsequently low marginal costs of
reduction. Subsidies instead of charges can also be found in international
relations, where a country that is victim of transboundary pollution has no
power to levy charges.

   Trading pollution rights is also an economic instrument for internalizing
pollution into the price system. In a certain area, a certain limit to a
specified pollutant is fixed by authorities, and potential polluters can
bargain amongst each other about who is going to buy the pollution rights
at what price. If a new polluter wants to enter the area, he or she will have
to buy the right to pollute from others, who accordingly have to reduce
their emissions. A disadvantage of this system is that the market for
pollution rights can be dominated by a few powerful firms, who will

distort the system of full competition; in that case the system will not lead to an optimal internalization of pollution into the price system.

Compensating victims of pollution is not efficient if the victims are many and damage borne by each victim is low. In the case of persistent toxic wastes, it is often impossible to know who are the victims of which discharges; people die from cancer without knowing the precise underlying causes. In the case of fatalities, the cause–effect relation is clearer and compensation may be in place. The question of the amount of compensation will be settled by legal means rather than economic analysis. If a polluter expects to pay compensation, he or she will be motivated to reduce risks.

A deposit–refund system may be effective, especially in cases where many people have to be induced to bring polluting materials to a treatment or recovery installation. The system is currently used for glass and to a lesser degree for aluminum and could well be introduced for mercury and cadmium batteries. Its only disadvantage is in the organizational costs. The system is a mixture of charge (when buying the product) and subsidy (when bringing the waste product back).

An economic instrument which has very much the character of direct regulation is government production of treatment services (e.g. incineration of waste). This instrument can be coupled with a charge to users of the government facility, to finance government services. Bohm and Russel (1985) mention a number of reasons why authorities often prefer direct regulations, like bans, licences and zoning, to economic instruments for controlling pollution. The most important reason could be that economists play a minor part in the administrative work of environmental policy and that application of economic incentive instruments relies on an understanding of the market system and price effects. Another reason may be that the outcomes of a charge or subsidy depend on the reactions of the public, which are not known beforehand; the required effect on pollution may only be reached after a trial-and-error process and the financial consequences may be uncertain. Moreover, charges are strongly opposed by the polluters, whose reluctant co-operation may be needed for successful policy of pollution abatement.

Environmental pollution leads to damage and environmental control leads to costs. However, damage due to environmental pollution is difficult to measure adequately in monetary terms. As an example, let us look at the Netherlands (Opschoor, 1986). An estimate of damage has been made by the Dutch government and is given in Table 8.1. The reader should be warned that such estimates are rather unreliable and incomplete. One source of underestimation are the p.m.s in the table; they may, in fact, represent significant damage. Another source of underestimation lies in the fact that not all environmental damage is included in the table; for instance, pollution of soils is not mentioned.

The costs of environmental control in the Netherlands are slightly over

Table 8.1   Damage in monetary terms to environmental pollution, Netherlands, 1985[a].

Category	Nature of damage	Explanation[b]	Accumulated past damage	Current and future damage per year
Air				
people	health	c	p.m.[c]	700–1300
agriculture	extra lime in soil	c	p.m.	15–50
	damage to cultivated crops	b	p.m.	565–655
	incidental damage to crops	b	p.m.	0.3–3
nature management	management heather regions	c	p.m.	3–30
	other nature regions	b	200–500	p.m.
forestry	yield–reduction wood	b	p.m.	20–50
	capital loss/recovery costs	b	3,000–10,000	p.m.
cultural goods	monuments	b,c	120–200	15–30
	stained glass windows	b	20	
	carillons	c	p.m.	0.25
	textile	b	20	p.m.
	paper	b,c	620	10
use goods	steel and sheet zinc	c	p.m.	40
	value reduction residences	–	p.m.	p.m.
materials	metals	c	p.m.	113
	automobile tyres	c	p.m.	50–100
	concrete damage	b	p.m.	175–350
	cleaning facades	c	p.m.	12–25
	Subtotal air		3,980–11,360 + p.m.	1,730–2,781 + p.m.
Surface water				
recreation	swimming, sport fishing	a	p.m.	110–350
professional fishery	income loss	b	p.m.	2–4
navigation	disposal dredging mud	c	p.m.	10–40
	damage to underwater soil		p.m.	p.m.
industry	demineralization etc.	c	p.m.	11
agriculture	discharge of fertilizer	b	p.m.	20–60
greenhouse horticulture	increasing salt content	b	p.m.	25–50
households	corrosion	b	p.m.	10–30
	Subtotal water		p.m.	188–545 + p.m.
Drinking and industrial				
water	modification surface water purification systems	c	p.m.	85
	modification groundwater purification systems	c	p.m.	23–300
	damage distribution network	c	p.m.	1.5
	Subtotal drinking water		p.m.	110–387
noise	value reduction residences	b,b	1,674	80
	Subtotal noise		1,674	80
other	well-being	–	p.m.	p.m.
	Total damage		5,650–13,030 + p.m.	2,107–3,793 + p.m.

one per cent of GNP, and most of the costs are borne by government, spending about 3 per cent of its budget of consumption and investment. The costs borne by enterprises are split up in Table 8.2. Except for the basic metal sector, environmental costs of all sectors are below one per cent of production value. Environmental costs borne by enterprises are very low indeed. Macro-economic consequences of the Dutch programme of environmental management have been calculated by the (Dutch) Central Planning Bureau (1986). In the medium term they seem to be slightly positive, in the longer term slightly negative but altogether negligible.

One specific case in recent years has been that of dumping and the detection of old dumping places. Sites, sometimes used for housing, may now be heavily populated. In 1986, 7,500 contaminated sites were detected of which about 1,600 were selected for remedial operations. These were split into 30 'large cases' (on average 25,000 cubic metres of contaminated soil) and 1,600 'budget cases' (on average 3,000 square metres of contaminated soil), together circa 5.5 million cubic metres. With existing technologies not all sites can in fact be cleaned up (see Table 8.3).

The costs of cleaning up 40–70 per cent of the million cubic metres of the polluted soil will be about 1 $ Cdn (1.5 Dfl). This operation will take some thirteen years, during which period some amount of the polluted soil will have to be stored temporarily. Although public concern about the sites is high, there are also technological and budget restrictions. It is not possible to make a calculation of an economic trade-off between damage costs and clean-up costs, due to uncertainties about dose–effect relationships and problems of monetary evaluation of human health. Although it is not possible to reliably calculate what it would have cost to prevent the soil pollution in the past, experts are convinced that these prevention costs would have been many times lower than today's clean-up costs.

Economic valuation of the environment

As stated in the introductory section, the market mechanism and price system will, under certain assumptions, result in an optimal allocation of scarce resources. The assumptions are violated by externalities: scarce

---

Notes to Table 8.1
ᵃ Almost all numbers from studies since 1982 in prices of the year of publication of the study. Given the large margin of inaccuracy, it was decided not to index to 1985 prices. Health damage from air pollution was estimated at one billion in 1974. This amount is included for 1985 with an uncertainty margin.
ᵇ (a) based on willingness to pay (b) material damage (unrecovered) (c) cost of compensation measures.
ᶜ p.m. = *pro memoria*.
Source: Environmental Program of the Netherlands 1986–1990.

Table 8.2 Environmental costs of enterprises by sector, Netherlands, 1985

	Dfl mln	Dfl/1,000 Dfl production
Agriculture, forestry, fishery	129	3.4
Meat and dairy products	76	2.9
Other food	211	5.5
Beverages and tobacco	41	4.2
Textiles, clothing, footwear	34	3.6
Lumber, wood products	48	4.3
Paper, printing, publishing	60	2.8
Chemicals, rubber	447	8.8
Basic metal	206	16.7
Electrotechnical industry	65	3.0
Petroleum products	109	2.6
Water, gas, electricity	141	5.4
All other sectors	427	<2.0
Total enterprises	1,994	2.7

Source: Central Planning Bureau (1986)

Table 8.3 Classification of soils and potential for remediation

Main pollutant	%	Clean up possibility 1986	1988
Heavy metals and metalloids	25–35	–	–/o
Non-volatile chlorinated hydrocarbons	10	–/o	–/+
Polycyclic aromatic compounds	10–20	o	o/+
Cyanides	10–20	o/+	+
Volatile chlorinated hydrocarbons	5	o	o/+
Aromatic compounds and oil derivatives	15–25	+	+
Other	5	–	o

+ = good
0 = varying
– = bad
Source: Dutch Ministry for Environment (1986).

resources, such as environmental goods and services, which are not marketed and for which no prices exist. Internalization of the environment in the price system would contribute to restoring the deficiencies of the market system for achieving an optimal allocation of all scarce resources. Internalization requires an adequate measurement of environmental goods and services in monetary terms.

A good, accurate monetary valuation of environmental effects of specified plans, programmes and projects would also be a great help in decision making, as it would facilitate the comparison of benefits and disbenefits in one common measuring unit (cost–benefit analysis). A more comprehensive treatment of monetary evaluation of the environment is given in Kneese (1984) and Freeman (1979).

There are two primary methods which can be used to evaluate non-financial effects in monetary terms. These are shadow pricing and contingent valuation. In the shadow pricing approach, changes in environmental quality are linked to changes in quantities or prices of some marketable goods. If release of toxic chemicals leads to death of fish, this effect may be assessed by using fish prices. Or if water is polluted, preventing swimming, entry fees of a swimming pool can be used as an approximation of this disbenefit. Obviously, measurement of environmental effects in this way is only viable in so far as all degraded functions of the environment can be assessed in their totality by such approximations through market goods.

Another application of shadow prices is the hedonic approach. An example is measurement of environmental degradation by comparing house prices in a polluted area with prices in a clean area. To the extent that house buyers are well aware of the risks and effects of living in a polluted area, this method is logical and, if the house market is functioning without serious market imperfections, reliable. But given the limited knowledge of dose–effect relationships, one may interpret such prices as measurement of ignorance and optimism rather than as societal disbenefits. Measuring the value of human life through production or through the wage differential approach is another application of the shadow price method; (Lave and Seskin, 1977).

The contingent valuation method uses interview techniques in order to ask people directly to evaluate environmental assets in monetary terms. A distinction should be made between two different concepts as a basis for this valuation method: 'willingness to pay' and 'compensating variation'. In the first case, the respondent is asked what amount of money he or she would be willing to pay in order to prevent a certain environmental deterioration. In the other case he or she is asked what compensation would be wanted for allowing the deterioration. A difference between the two concepts lies in the budget restriction: what one is willing to pay depends partly on one's budget but compensation is independent of the budget of a respondent.

Some shrewdness is needed in preparing a good questionnaire. If the respondent does not take the possibility of environmental change seriously, the answers may reflect this. But if, on the other hand, he or she thinks that compensation may indeed be forthcoming, there is a temptation to overstate the damage. Apart from this complication, compensating variation usually results in higher values than willingness to pay, but when only minor effects are concerned, the difference between the two is expected to be small. The choice of the appropriate method to be used is of an ethical rather than an economic nature. If people are supposed to have a basic right to live in a sound, unpolluted environment, the compensating variation method should be used. But if it is taken that one has to pay for this right, willingness to pay is the appropriate concept.

Apart from technical and conceptual problems, there are questions about the validity and interpretation of the results. Again, if respondents are not well aware of the nature and effects of the environmental changes which are to be evaluated, the answer could be interpreted as an indication of ignorance or degree of optimism rather than monetary valuation of non-financial (dis)benefits. In practice, a not uncommon answer to questionnaires on monetary evaluation of significant environmental changes is 'at no price'; either the respondent evaluates the environmental change at an infinite amount of money, or he or she is not able to translate environmental effects into money units. In the first case no financial benefits could balance the environmental effects of a project. In the second, one might doubt whether money is indeed the right measuring unit for the environmental changes.

Another complication relates to the statistical population from which a sample of respondents should be drawn. If an environmental change is persistent, future generations could also be affected. Obviously, there is no possibility of including them in the samples, and estimates must be made by interviewing the existing population. Two arguments are used to justify this practice. One is assuming that future generations will have approximately the same scales of evaluation and priorities as the existing generation. However, the present generation may underestimate the disbenefits of pollution if it will not occur until many decades or centuries later; an example is the storage of radioactive waste. The other justification for not taking future generations explicitly into account is that future generations are only of concern to the extent that they are included in the preferences of the present generation.

Evaluation of external effects in monetary terms is based on the presupposition that all goods, whether they are marketable or not, are exchangeable and that loss of environmental assets can always be compensated by market goods. In societies with a relative affluence of market goods and scarcity of environmental amenities, this presupposition is doubtful, as has been seen in evaluations of air pollution in the USA (Ridker, 1967; Waddell, 1974; Crocker et al., 1979; Freeman, 1979). Even if one is prepared to accept the statistically undefined concept of a 'not too unreasonable estimate', one has to admit that the estimates are very incomplete, inaccurate and unreliable, and therefore a poor basis for practical decision making at this time.

Monetary evaluation of human life is badly needed for comparing costs and benefits of activities that may lead to the release of toxic chemicals into the environment. Although certain aspects of disease and, ultimately, death are marketed (medicines, hospital costs, doctor's bills), these are not believed to represent the essential costs of loss of human life. This more essential value of human life is not marketed and other approaches have to be used to measure it in monetary terms.

The three different approaches to valuing human life are the human capital approach, wage differential approach and cost of prevention approach. The human capital approach (Ridker, 1967; Waddell, 1974) is somewhat out of vogue. The value of human life is approximated, as a lower bound value, by calculating foregone production due to premature death. Salaries, as approximation of a person's production, are used as the statistical data on which estimates are based. This method is rather crude. It values (among others) the unemployed, retired people and housewives at nil. But the main source of criticism is that there is only a small relationship between a person's value and a person's salary.

The wage differential approach (Thaler and Rosen, 1975) is based on differences in salaries for jobs with different risks. If, on average, a job with a higher risk of death pays a certain amount more (correctly summed and discounted over the appropriate lifetime), the value of life is calculated by multiplying the amount by one thousand. This approach yields much higher estimates (approximately an order of magnitude higher) of human life; Crocker et al., (1979) uses a figure of $340,000. The wage differential approach is clearly based on the individual's own valuation of life; the value assigned by relatives and other members of society is disregarded. However, those people accepting a risky job may not constitute a random sample of the whole population. It is logical to assume that risk-inclined rather than risk-aversive people will accept such jobs and the resulting estimate may measure differences between these groups rather average value of life. A third point of criticism lies in the multiplication of the wage difference by the inverse of the additional risk. The relationship between acceptable risk and associated benefits is not necessarily proportional and has been estimated by Starr (quoted in Slovic et al., 1980) as $R \simeq B^3$. This means that the estimation used in the wage differential method is based on a linear extrapolation of a non-linear relationship. If the risk differences on which the estimates are based would have been twice as high, a value of human life of eight times higher might be expected. Finally, there is a peculiar difference between individual uncertainty and societal certainty about the consequences of a small increase of death risk. Whereas an individual may optimistically and most often realistically reason that a small risk increase will not affect him or her personally, society is very sure about the number of deaths due to a specified increased risk. For society only the names of the victims are uncertain, not the number.

The third method of valuing human life is based on the amount of money which society is effectively willing to pay to save human life. If we would take medical expenses on the most expensive patients as an indication of the value of human life, the estimate would be many times higher than a valuation based on wage differentials. Decisions on lifesaving measures are also taken regularly in policy and programme development and as they cost money they implicitly contain a lower bound estimate of the value of human life. Graham and Vaupel (1981) reviewed a number of lifesaving programmes and calculated the implicit value of life. Table 8.4

Table 8.4   Estimates of implicit valuation of human life derived from policy measures.

Problem area	Base case policy option	Alternative policy option	Net additional cost of alternative policy option	
			Per life saved	Per life-year saved
Smoke detectors	Mandatory, in sleeping rooms only	Mandatory in all rooms	1,000,000	32,000
Highway safety	Status quo prior to 55 mph speed limit	55 mph speed limit	1,200,000	29,000
Mobile source air pollution	Pre–1970 conditions	1970 Clean Air Act	1,350,000	105,000
Highway safety	Mandatory passive belts	Mandatory passive belts and air bags	1,400,000	34,000
Acrylonitrile	Status quo	2.0 pp.	3,520,000	230,000
Carcinogens in water	150 mcl rule	100 mcl rule	3,800,000	240,000
Carcinogens in water	Status quo	150 mcl rule	3,900,000	240,000
Arsenic	5 mcl rule	0.004 mcl rule	5,000,000	390,000
Carcinogens in water	100 mcl rule	50 mcl rule	6,300,000	390,000
Vinyl chloride	50 ppm	1 ppm	7,500,000	490,000
Benzene emissions	No control	97 per cent control	7,600,000	480,000
Coke ovens	Status quo	Proposed OSHA standard	12,100,000	480,000
Acrylonitrile	2.0 ppm	1.0 ppm	28,800,000	1,900,000
Benezine emissions	97 per cent control	99 per cent control	51,000,000	3,200,000
Benzene	10 ppm rule	1 ppm rule	102,000,000	6,600,000
Acrylonitrile	1.0 ppm	0.2 ppm	169,200,000	11,000,000

Source: Graham and Vaupel, (1981)

lists the higher values for some of these health and safety programmes. The authors warn that their figures 'suffer from a myriad of empirical flaws' and that 'implications should be drawn with caution'. Nevertheless, the figures suggest that the value assigned implicitly by policy makers to human life might well be of a higher order of magnitude than the value estimated by the wage differential approach.

There is a logical point in measuring all marketable goods and their loss in monetary units: if victims are compensated they can always buy the goods back. It is also logical to measure non-marketable goods in monetary units to the extent that people can sufficiently reliably indicate the amounts of money of compensation or willingness to pay: they can apparently compensate by buying other goods. In the case of human life, it is doubtful whether people can indeed assign acceptable values. Broome (1985) examines several principles for valuing changes in population and finds mostly paradoxes, such as 'the belief that it is good to save life but also good to reduce the number of babies born'. Perhaps the value of human life is beyond the realms of an anthropocentric discipline like economics. Without going into ethics or philosophy, the various ways of valuing life carry too many deficiencies and inaccuracies and too much unreliability to form a solid basis for decision making. It is true that decisions associated with loss of human life could contribute to the efficiency of decision making. The state of the art of economics does not, and perhaps never will, allow a sufficiently accurate valuation. To quote Broome (1985), 'it is one thing to say there is a manifest inefficiency, it is another to know how to do better.'

Trading off: cost–benefit analysis and environmental impact assessment

For an entrepreneur interested in financial gains and losses, cost–benefit analysis (CBA) is a simple and efficient calculation method. It tells whether the financial benefits of an activity will outweigh the costs. This appealing simplicity has led policy makers and economists to apply the method in a broader context: socio-economic CBA does not only measure financial aspects of a plan or project, but also other aspects, such as environmental ones. In order to preserve an easy comparability of the pros and cons of a project, they are all measured in one unit: money. Through application of CBA to complex, multi-dimensional decision-making problems, the complexity is reduced to the question whether the benefits of a proposed activity exceed its costs and, if so, whether alternatives yield an even better ratio or difference of benefits and costs. However, the cost of this reduction of complexity is a loss of reliability and accuracy. For an extensive treatment of the theory of CBA, see Mishan's classic work (1971).

In the theory of welfare economics, the basis of CBA is the Pareto-criterion, supplemented by Kaldor-Hicksian hypothetical compensations. According to the Pareto-criterion, a situation becomes 'better' as result of a project if at least one individual is better off and nobody's situation is worsened. But if some individuals gain at the expense of others, the Pareto-criterion by itself is of no use for determining whether a situation improves. Therefore, the theory has been extended by hypothetical compensations: if the 'gainers' of a project can or could compensate the 'losers', the project is worth carrying out, since it implies a potential Pareto improvement. This straightforward reasoning leaves two questions, especially relevant to the inclusion of environmental effects in CBA: how are non-financial effects to be evaluated? and how are flows of costs and benefits to be aggregated over time? The first question was discussed earlier. We now discuss the aggregation over time and will continue with additional problems of CBA.

Costs and benefits of a project are distributed over time. The more distant in time a cost or benefit will be realized, the less it counts in the present valuation. A dollar's value in this year is worth more than a dollar's value next year, even in a world without inflation (we disregard in this chapter depreciation of money). The reason for this phenomenon is roughly that if one gets an amount of money now, instead of next year, one has more opportunities to use it.

In order to make costs and benefits comparable over a series of years, they are discounted. Discounting is the reciprocal of compounding. Just as compounding yields a large value after sufficient time, so will discounting yield a small amount in the present valuation if the cost or benefit is sufficiently far in the future. The 'present value' of a project is calculated as the difference between the discounted benefits and discounted costs, summed over the years when the project will cause costs or benefits.

The discount rate most often chosen in practice lies between 6 and 10 per cent. This choice is based on rates of return in business and on interest rates. The reasoning is that the financial effects of a project should be compared with the alternative opportunity of depositing the money in a bank or buying shares at the stock market; if a project does not yield better results than that, it is not worthwhile. A positive present value allows the expectation that the project will at least pay back its costs with a bonus premium. One could speak of an underlying principle of conservation of money value. The higher the discount rate, the lower is the weight of costs and benefits in the more distant future. A discount rate of 10 per cent implies a 'half-life' of little over seven years: with a 10 per cent discount, the present value of one dollar now is equal to the present value of two dollars received seven years from now, and four dollars 15 years from now; costs and benefits after that period are negligible, due to the discounting procedure.

Linking the discount rates to rates of return in business or to interest rates is logical as long as costs and benefits are indeed expressed financially. In socio-economic CBA other effects are also expressed in monetary terms and discounted. However, the link between these rates of return or of interest and the valuation of future environment as compared to present environment is unclear. The discounting process reflects a rate of return on physical capital, but environment is not ordinary, reproducible physical capital. Any environmentalist would be amazed if told by an economist to use a time horizon of only twenty years, or if he were told that a specified 'amount of environment', if unused, would lose half its value in seven years.

Another problem with CBA in societal evaluation in general relates to distributional effects. In normal CBA the effects of an activity are measured in monetary terms, discounted and summed, so as to compare the costs with the benefits. However, it can be argued (Dasgupta and Pearce, 1972) that in socio-economic CBA it is not the unweighted sum of costs and benefits, but rather total 'utility' that counts. For rich people, the marginal utility of money is lower than for the poor: the gain or loss of a specified amount of money is considered less important by the wealthier groups. If pros and cons of an industry producing or using chemicals are to be measured in terms of utility instead of money, the distribution of effects over various income groups must be determined and weighed with marginal utilities before summation. Although the social justification of this procedure is clear, it introduces a certain unreliability as the marginal utility of money concept is difficult to measure reliably.

For decision making, socio-economic CBA has the attractive feature of reducing a multi-faceted problem to a one-dimensional one. But the addition of the qualification 'socio-economic' introduces a certain unreliability, as was discussed in the section on distributional effects. The unreliability is aggravated by the introduction of external effects for which no market price exists. Unfortunately, the uncertainties with respect to valuation of environment are often coupled with uncertainties with respect to dose-effect relationships, as in the following example:

In 1970, EPA made an estimation of air pollution damage in the USA (Barret and Waddell, 1970). Dose–effect relationships of air pollution effects on human life production figures were used. Nine years later, a re-estimation was made (Crocker et al., 1979) on the basis of a new epidemiological study, pointing at mortality figures roughly an order of magnitude higher. The final results of both studies were not in great contradiction. If the choice of methods were reversed, the difference would be roughly two orders of magnitude. This degree of uncertainty is underestimated, as the ranges of uncertainty within the two studies are not included. Obviously, an uncertainty of various orders of magnitude is a highly unreliable basis for decision making if such effects of a project are considered to be a significant impact of a project.

Apart from the unreliability of monetary estimation of environmental effects, there is an intrinsic tendency of bias. Some environmental effects of an activity can perhaps be captured in a monetary estimate, but not all. New environmental problems crop up regularly and some already known problems (e.g. value of nature) are so intangible that they are not included in monetary estimates. In addition, the discounting procedure results in a systematic overlooking of long-term environmental effects, even if they are included in the analysis. Thus, CBA underestimates environmental effects.

Cost-effectiveness analysis, as a variant of CBA, circumvents the problem of evaluation of the environment by fixing standards. Given those standards, cost-effectiveness analysis determines which one of a number of alternatives meets the standards at lowest cost. However, the crucial issue of which standards is not always considered in depth prior to the analysis, rendering the analysis questionable.

Although it cannot be denied that monetary measurement of environmental effects facilitates trade-offs, the state of the art is still too poor and deficiencies are too intrinsic to declare CBA as a generally acceptable method when significant environmental effects are involved.

Environmental Impact Assessment (EIA) and socio-economic CBA are complementary and, to a certain extent, even identical. Both aim at comparing social pros and cons of an intended activity. Where CBA is strongly focused on monetary evaluation, EIA primarily concentrates on measuring environmental effects in physical terms. As estimation of effects in physical terms is a prerequisite for monetary evaluation, EIA can be seen as a part of the CBA process. On the other hand, CBA is seen as one of the evaluation techniques, applicable in EIA and, therefore, CBA can be seen as a part of EIA. However, EIA does not have the one-dimensional fixation on monetary measurement which is an essential feature of CBA.

An EIA is not a single, specified method, it is rather a procedure, yielding an Environmental Impact Statement (EIS), a report on the basis of which a decision is made. Munn (1979) defines EIA as 'an activity, designed to identify and predict the impact on man's health and well-being of legislative proposals, policies, projects and operational procedures, and to communicate information on the impacts'. Evaluation of impacts is not explicitly mentioned and the trade-off of effects may indeed be left to the political process of decision making rather than to the researchers, carrying out the EIA. Identification and prediction of impacts, environmental impacts and financial impacts each in their own dimensions, can be the final information resulting from EIA. However, the communication of information is also essential and may require a comparison, in one or in more dimensions, of financial and environmental effects.

It is, of course, essential in any decision to have alternatives to choose from. Obviously, one alternative is the null-option of not carrying out the proposed activity. More alternative options of the proposed activity can be specified. It is, however, the objective of EIA to also identify possible modifications in the proposals, resulting in environmentally better options. Moreover, due to the obligation of EIA, planners will be more inclined to take environmental effects into consideration during the earlier phases of planning. An EIA can, thus, add to the integration of environmental consequences in the whole process of planning and decision making.

The assessment can take various forms. Socio-economic CBA may be part of this assessment, but many other methods have been developed for making a trade-off between the various options. An overview of the Leopold matrix, network, overlays and other techniques is given by Nichols and Hyman (1980). If an ordering is established between the various options, some form of 'multi-criteria analysis' (MCA) is used. In MCA, like in CBA, effects of an activity are compared by use of weighing procedures or other techniques. A treatment of MCA can be found in Voogd (1983). As the objective of MCA is, to a certain extent, comparing the uncomparable, it lacks an objective, internal logic which mathematical optimization techniques have. In MCA weighing procedures and standardization procedures are somewhat arbitrary. And the results of MCA tend to select the better and to eliminate the worse options, but will not identify one objectively best option.

Making a trade-off between the alternatives is not necessarily a part of the EIA procedure; an enumeration of the effects, each one estimated in its own dimensions, can also be the final result of the assessment. The final choice and decision is left to the political system in which the policy maker operates.

The assessment of impacts of toxic chemicals is often subject to uncertainty, and depends on catastrophic events that occur with a certain known or unknown probability. Other chapters of this book address the question of probability of exposure to chemicals and the assessment of risks.

References

Barret, L. B., and Waddell T. E. (1970), *The Costs of Air Pollution Damages: A Status Report*. (Washington, DC: Environmental Protection Agency).

Baumol, W. J. and Oates, W. E. (1975), *The Theory of Environmental Policy* (Englewood Cliffs, NJ: Prentice Hall).

Bohm, P., and Russel, C. S. (1985), 'Alternative policy instruments', in Kneese and Sweeney (eds), *op. cit.*, Vol. 1.

Boland, J. J. (1986), *Economic Instruments for Environmental Protection in the United States* (Paris: OECD, Environmental Committee Group of Economic Experts).

Broome, J. 1985. 'The economic value of life', *Economica* 52, pp. 281–94.

Brown, G. (1984), *Selected Economic Policies for Managing Hazardous Waste in Western Europe*, paper prepared for USEPA. University of Washington.

Central Planning Bureau (1986). *Economische Gevolgen van Voorgenomen Milieumaatregelen*. (The Hague: Central Planning Bureau).

Coase, R. H. (1960), 'The problem of social cost', *Journal of Law and Economics*, vol. 3, pp. 1–44.

Crocker, T. D., Schulze, W., Ben-David, S., Kneese, A. V. and Carlin, A., (1979), *Methods Development for Assessing Air Pollution Control Benefits*. (Washington: Environmental Protection Agency).

Dasgupta, A. K. and Pearce, D. W. (1972), *Cost Benefit Analysis: Theory and Practice*. (London: Longman).

Dutch Ministry of Environment (1985), *Environmental Programme of the Netherlands, 1986–1990*. (The Hague: Government Publishing Company).

Dutch Ministry of Environment (1986), *Notitie Verwerking Verontreinigde Grond* (draft).

Eurostat (1986), *National Accounts ESA: Input–Output Tables 1980* (Brussels: EEC).

Freeman, A. M. (1979), *The Benefits of Air and Water Pollution Control: A Review and Synthesis of Recent Estimates*. Council on Environmental Quality. (Baltimore: Johns Hopkins University Press).

Graham, J. D. and Vaupel J. W. (1981),'Value of life: what difference does it make?', *Risk Analysis*, vol. 1, no. 1, pp. 89–95.

Hordijk, L., Jansen, H. M. A., Olsthoorn, A. A., Opschoor, J. B., Reijnders, H. F. M., Stapel, J. H. A. and Vos, J. B. (1983), 'Economic structure and the environment; production, pollution and energy consumption', in: T. R. Lakshmanan and P. Jijkamp (eds), *Systems and Models for Energy and Environment Analysis* (Aldershot: Gower Publishing Company).

Hueting, R. (1974) *Nieuwe Schaarste en Economische Groei; Meer Welvaart Door Minder Productie?* (Amsterdam: Elsevier). English Edition: (1980), *New Scarcity and Economic Growth: More Welfare Through Less Production?*, (Amsterdam: North Holland Publishing Company).

James, D. (1985), *Energy Development and Environmental Quality Management.* (Canberra: Australian Government Publishing Service).

James, D., Jansen, H. M. A. and Opschoor, J. B. (1978), *Economic Approaches to Environmental Problems* (Elsevier).

Kneese, A. V. (1984), *Measuring the Benefits of Air and Water Pollution* (Washington, DC: Resources for the Future).

Kneese, A. V. and Sweeny, J. L. (eds) (1985), *Handbook of Natural Resources and Energy Economics* (Amsterdam: North Holland Publishing Company).

Lave, L. B., and Seskin, E. P. (1977), *Air Pollution and Human Health: Resources for the Future* (Baltimore: Johns Hopkins University Press).

Leontief, W. (1970), 'Environmental repercussions and the economic structure', *Review of Economics and Statistics* 52, pp. 262–71.

Mishan, E. J. (1968), *The Costs of Economic Growth* (London: Staples Press).

Mishan, E. J. (1971), *Cost Benefit Analysis* (London: George Allen and Unwin).

Munn, R. E. (ed.) (1979), *Environmental Impact Assessment. Scope 5* (New York (2nd edn): John Wiley and Sons).

Nicols, R., and Hyman, E. (1980), *A Review and Analysis of Fifteen Methodologies for Environmental Assessment*, Center for Urban and Regional Studies, University of North Carolina at Chapel Hill.

OECD, 1975. *The Polluter Pays Principle.* (Paris: OECD).

Opschoor, J. B. (1986), *A Review of Monetary Estimates of Benefits of Environmental Improvement in the Netherlands* (Amsterdam: Institute for Environmental Studies).

Pigou, A. C. (1962), *The Economics of Welfare* (London: Macmillan).

Ridker, R. G. (1967), *Economic Costs of Air Pollution: Studies in Measurement* (New York: Praeger).

Slovic, P., Fischhoff, B. and Lichtenstein, S. (1980), 'Facts and fears: understanding perceived risk', in: R. C. Schwing and W. A. Albers (eds.) *Societal Risk Assessment: How Safe is Safe Enough?* (New York: Plenum Press).

Thaler, R. and Rosen, S. (1975), 'The value of saving a life: evidence from the labor market', in L. E. Teleckyj (ed.) *Household Production and Consumption*, Nat. Bur. Econom. Res., Studies in Income and Wealth, Vol. 40, New York.

Voogd, J. H. (1983). *Multicriteria Evaluation for Urban and Regional Planning* (London: Pion).

Waddel, T. E. (1974). *The Economic Dangers of Air Pollution.* (Washington: Environmental Research Center).

9 *The legal perspective*

M. RANKIN

Introduction

> We believe the magnitude of the public risk associated with the toxic chemicals currently is increasing and will continue to do so until we are successful in controlling the introduction of these chemicals into our environment. (US Surgeon General, 1980)

As indicated in Chapter 1, there are about 100,000 chemical substances presently in commercial use in the world. The number is growing annually. Chemicals are relied upon to maintain the high standard of living in Western countries: solvents, adhesives, polishes, detergents, waxes, cleansers, pesticides, herbicides and fertilizers are just some of the chemicals in use. Many of these products find their way into the environment from the workplace and from the home, through their manufacture, use or disposal. For example, the Great Lakes Water Quality Board has identified over 450 toxic chemicals in the Great Lakes in varying concentrations.

It is said that synthetic chemicals are introduced into the market at a rate of about 1,000 per year. Many toxic chemicals currently in commercial use have been present for only a matter of years or decades, hardly enough time for ecological systems to have adapted.

In an excellent overview of the toxic substances issue, Bonine and McGarity (1984) list five characteristics of toxic pollutants which differentiate them from conventional pollutants, and which may have significance in designing a regulatory approach:

1 Much lower doses of toxic pollutants can produce adverse effects in man, animals or plants than are required for conventional pollutants.
2 Effects may not show up until decades after exposure to the substance has occurred and been discontinued.
3 Some toxic pollutants can bioaccumulate in plants and animals.
4 Some toxics may persist for long periods. Those that do break down may produce toxic by-products.
5 Most toxic pollutants do not have ubiquitous sources.

Polychlorinated biphenyls are an example of a group of chemicals which fit this description, though it is recognized that within the group there is a range of toxicities and persistence. Finkle (1983) has observed that the regulation of persistent pollutants warrants a more rigorous approach than conventionally taken in pollution control legislation: 'While there is a continuum from persistent to non-persistent pollutants, the essential differences between the two types permit a useful if somewhat imprecise policy distinction to be drawn. Persistent pollutants are less susceptible to control regimes which focus upon a permissible loading for a particular medium, because these pollutants both persist locally and also spread geographically, sometimes through quite subtle pathways.' (Doern, 1982)

As in many other legal areas, the threshold question facing those regulating toxic chemicals is one of definition. Just as regulators increasingly differentiate between non-persistent residuals of industrial activity, generally governed by pollution control legislation,[1] they must also distinguish 'toxics' from other substances for special consideration. The Canadian Environmental Protection Act provides an elaborate definition for a 'toxic substance'. First, 'substance' is defined as 'any distinguishable kind of organic or inorganic matter, whether animate or inanimate' (s. 3[1]). The definition of 'substance' goes on to address the kinds of matter, elements, uncombined radicals, and combinations of elements and molecules within a very broad framework. Second, such 'substances' are further defined as being 'toxic' if they enter into the environment in a quantity or concentration or under conditions: (1) having or that may have an immediate or long-term harmful effect on the environment; (2) constituting or that may constitute a danger to the environment on which human life depends; or (3) constituting or that may constitute a danger in Canada to human life or health (s.11).

Scientific and legal experts will differ as to the proper threshold for regulating 'toxic' substances (David, 1985). Little information exists on their toxicity at very minute levels for very long periods of time. It is unclear whether the concept of safe thresholds is applicable for some or all of these chemicals. For this reason, a 'cradle-to-grave' approach is being recommended for the management of new chemicals throughout the Western world (Environment Canada, 1986). The degree of control varies radically in most countries. Although food additives and pesticides are almost universally controlled in a relatively strict fashion,[2] other toxic chemicals may not be systematically controlled at all.[3]

This chapter will focus primarily on the regulation of the entry of toxic chemicals into the marketplace. It will not deal with regulatory controls governing chemicals in the workplace – a matter normally covered by either workers' compensation legislation or occupational health and safety legislation in the various jurisdictions (Doern, 1977; Schrecker, 1986). Neither will this chapter concentrate on the transportation of dangerous goods, nor upon the disposal of hazardous wastes, both of which may be seen as 'downstream' regulatory issues (Estrin, 1986)[4]. Instead, it will concentrate upon the 'front end' regulation of toxic chemicals, that is, prevention.

To some extent, the need for legislative intervention in this field largely mirrors intervention elsewhere in environmental law. For a variety of reasons, common law tort doctrines have proven to be largely ineffectual in deterring hazardous activities and compensating victims of such activities. In modern industrial societies, the state has had to intervene to rectify the deficiencies of the ordinary common law or civil law regimes in the face of environmental degradation. The first part of this chapter will explore the basic tort actions available throughout the common law world, the defenses to such actions and the impediments that have been encountered in their use. Next, the chapter will examine some national legislation that was passed to rectify some of these deficiencies. Emphasis will be placed on two Canadian statutes, the Environmental Contaminants Act, now repealed, and the Canadian Environmental Protection Act, assented to June 28, 1988, since the statutory reform in this field is the most contemporary and is largely representative of initiatives taken elsewhere in the Western world. Lastly, international initiatives in toxic chemical regulation will be examined and some brief conclusions drawn.

Toxic chemicals and common law

Causation

By now there is virtually universal agreement that the common law tort remedies inadequately compensate the victims of exposure to toxic substances (Ginsberg and Weiss, 1981; Trauberman, 1983). Regardless of the tort action chosen, the major barriers to recovery include (1) the need to prove causation, (2) prohibitive 'transaction costs', and (3) procedural hurdles, especially the statute of limitations.

The most significant barrier to recovery in tort is the requirement of causation, i.e the need to prove that the exposure to a toxic chemical has indeed caused a particular injury. Plaintiffs must demonstrate the way in which the toxic chemical produced the health effect: however, any one of many factors — radiation, diet, drugs and personal habits, as well as the exposure — may have caused the condition. The often limited extent of epidemiological evidence and the complex chain of causation for many non-traumatic injuries and diseases are obvious impediments to recovery. As Trauberman (1983) points out: 'Health effects from hazardous chemicals may be acute or chronic. Acute effects usually occur after high-level exposure and range in severity from temporary rashes to death. Chronic effects, often the result of long-term, low-level exposure to hazardous substances, include birth defects, cancers, and disabling lung diseases.' It is often very difficult to prove that a chronic disease was, on a balance of probabilities, caused by a particular exposure.

A central difficulty in the area of causation is the assessment of risk. A difficulty arises in the application of principles derived from scientific methodology and legal evidence. Page (1979) has contrasted two principles:

limiting false positives and limiting false negatives. A false positive is an indication in a study that some hypothesis is true when it is in fact not true; a false negative is a finding that there is inadequate evidence for some hypothesis which, in fact, is correct. To recover in civil actions, one must prove a hypothesis of causation on a 'balance of probabilities' – that is, something is more likely than not to have occurred. Therefore, the objective of a tort system, and a criminal regime as well, is to limit false positives, which is also consistent with experimental science. Before concluding that a particular toxic chemical causes a particular disease in laboratory animals, a high degree of statistical correlation is required.

As has been pointed out by Schrecker (1984) however, less attention has been paid to the probability of false negatives, where there is a finding that no significant effect exists when in fact there is an effect that, if known, would warrant recovery in tort. Bio-assay studies for potential carcinogens, for example, often produce a greater than even probability of false negatives. As Schrecker (1986) notes: 'The general principle of limiting false negatives – in contrast to the limitation of false positives which has guided most regulation of environmental hazards – could be characterized as one of erring on the side of safety. But safety in this context is defined in terms of avoiding the uncertain, but potentially disastrous adverse effects of the particular substance or activity in question.'

To establish causation requires the plaintiff to identify the specific substance injuring the plaintiff and to demonstrate that the specific named defendant was responsible for wrongly exposing the plaintiff to that substance. Moreover, the plaintiff must show that the causal link between the exposure and his or her disease was 'reasonably foreseeable'.

Identifying the substance that caused the harm may involve more uncertainty than the courts are willing to countenance.[5] With the information often unavailable due to scientific uncertainty, it may be impossible to meet the burden of proof required by common law. The scientific evidence is simply not available to enable plaintiffs to provide a critical element of their case. A plaintiff must also identify the party responsible for harm. The company at fault many years ago may no longer exist, or may lack the resources to compensate the victims fully. In areas where there are numerous sources of toxic chemical, pinning the blame on a particular defendant may prove to be an insurmountable hurdle.

The concept of 'reasonableness' is another impediment to successful toxic tort litigation. The evidence regarding reasonableness may be unavailable or the courts may analyze the reasonableness of the defendant's behavior retrospectively – by assessing it in the light of the then-current knowledge (Trauberman, 1983). As Trauberman points out, no regulatory standards may have existed for exposure at the time that the exposure allegedly occurred. Proving that the risks were 'actually known and theoretically knowable at the time of exposure as well as the costs of determining those risks' (Trauberman, 1983) may be an insurmountable barrier. The proof of fault, therefore, requires the victims to provide information that is both difficult and expensive to obtain and, if available, may only be in the hands of the defendant. Often the defendant would assert that the information requested constitutes a 'trade secret' so that a court would have to also determine whether other litigants would be able to have access to the defendant's corporate records of this kind.

The plaintiff's burden of proving causation in tort litigation may have been eased somewhat by the 1972 decision of *McGhee v. National Coal Board* [(1972) 3 All ER 1008]. The facts of the case are that the plaintiff contracted dermatitis while working in the defendant's brick kilns. The defendant admitted a breach of duty in not following the industry practice of providing washing facilities. The defendant also admitted that the disease was attributable to the work performed by the plaintiff. The defendant denied liability, however, on the basis that the plaintiff had failed to prove that the failure to provide washing facilities had caused the onset of the disease. The board argued that the cause of the disease was the susceptibility of the skin resulting from the exertion of the workman during his bicycle ride home after work, a factor over which the board had no control. The House of Lords found for the plaintiff on the grounds that the onus of proof of causation shifts to the defendant where the defendant is negligent, and where this negligence materially increases the risk of injury. The court also found that there was no difference between a finding that the defendant's breach of duty had materially increased the risk of injury to the plaintiff and a finding that the breach of duty had materially contributed to this injury.

In addition to the fundamental obstacle of causation, tort actions also generate what economists term prohibitive 'transaction costs'. For example, one important transaction cost is the difficulty plaintiffs normally encounter in acquiring access to key scientific information which, typically, is in the hands of the defendant companies producing, using or distributing the toxic chemicals (Nemetz, 1981).

Finally, there are often procedural hurdles facing potential victims injured by exposure to toxic chemicals. The most significant is the statute of limitations. If a statute of limitations begins to apply at the date of exposure, the victim's cause of action may be barred as a consequence of the latency period for many of the resulting diseases. Some jurisdictions have met this problem by adopting a 'date of discovery' rule;[6] others have failed to pass the requisite corrective legislation.

Torts

To demonstrate the shortcomings of tort recovery, the main tort actions of negligence, 'occupier's liability' (also known as Rylands and Fletcher), trespass and nuisance (both private and public) will now be assessed.

Negligence requires a legal duty owed to the plaintiff, a breach of that duty by the defendant, and damage sustained by the plaintiff as a result. In negligence suits, the plaintiff must prove that the defendant failed to meet the standard of reasonable care. The scope of the duty of care and of the damages depends upon the concept of 'foreseeability' – is there a sufficient relationship of proximity between the defendant and the plaintiff so that a reasonable person would contemplate that the defendant's carelessness would likely cause the damages suffered by the plaintiff? Often it is very difficult for the plaintiff to discharge the onus of proving that the kind of damage and extent of damage suffered by the plaintiff was reasonably foreseeable by the defendant. Defenses that may be raised include contributory negligence (that is, that the plaintiff was partly negligent as well) and voluntary assumption of risk (that the plaintiff knowingly consented to incur the potential harm).

To assist plaintiffs in situations where the defendant controls all the evidence and where the injury almost certainly resulted from someone's fault, courts have created a legal fiction of negligence law called *res ipsa loquitur* ('the thing speaks for itself'). For the doctrine of *res ipsa loquitur* to apply, the event must be one that originally does not occur in the absence of negligence, must be caused by something within the defendant's exclusive control, and must not result from any voluntary action by the plaintiff (Prosser, 1971). This doctrine has sometimes been successfully applied in the context of toxic torts. See, for example, *Reynolds Metal Company* v. *Yturbide* 258 F. 2d 321 (9th Cir. 1958), cert. denied, 358 US 840 (1958), cert. denied, 358 US 840 (1958) where the plaintiff farmers recovered from the defendant for poisoning caused by fluorides emanating from the defendant's aluminum reduction plant. Since the damage was proven, and the defendant had conceded both that it was the source of the fluorides and that it was aware of the potential danger to persons in the plaintiffs' position from exposure to 'excessive amounts' of the fluorides, the court held that the defendant's evidence as to precautions taken was insufficient to overcome the inference of negligence arising under the *res ipsa loquitur* doctrine. Meeting each of these criteria for the application of this doctrine, however, is often impossible.

The plaintiff's burden in a toxic tort suit may have been lightened somewhat by a recent US decision easing the requirement for present injury. *Ayers* v. *Jackson Township* ([1987], 525 A. 2d 287), is a case in which the defendant municipality was found to have acted, not simply negligently, but 'palpably unreasonably' in its failure to prevent its landfill from contaminating the aquifer. As a consequence, the plaintiffs' water supply was polluted with toxic pollutants. The Supreme Court of New Jersey upheld a jury award for the future cost of annual medical surveillance. In the words of the majority (with which the dissenting justice concurred in this part):

[W]e hold that the cost of medical surveillance is a compensable item of damages where the proofs demonstrate, through reliable expert testimony predicated upon the significance and extent of exposure to chemicals, the toxicity of the chemicals, the seriousness of the diseases for which individuals are at risk, the relative increase in the chance of onset of disease in those exposed, and the value of early diagnosis, that such surveillance to monitor the effects of exposure to toxic chemicals is reasonable and necessary (p. 312)

The innovative aspect of the decision is that the court granted prospective damages despite the fact that 'no claims were asserted by plaintiffs seeking recovery for specific illnesses caused by their exposure to chemicals' (p. 297) and despite the fact that no present injury was proven. The court in Ayers declined to recognize plaintiffs' 'cause of action for the unquantified enhanced risk of disease' (p. 308) but explicitly left open the possibility of a claim being upheld on the basis of the 'enhanced risk of disease that is supported by testimony demonstrating that the onset of the disease is reasonably probable' (p. 308). The court also left open the opportunity for a further easing of the requirement that a present injury be proven in this kind of negligence action.

The rule in *Rylands* v. *Fletcher* ([1868] LR 3 [HL] 330) provides that where an occupier of land brings and keeps upon it anything likely to cause damage if it escapes, he or she is strictly liable for all the direct consequences of its escape, even if the owner has not been negligent. The *Rylands* v. *Fletcher* rule involves strict liability for damages to the plaintiff, even though it may be impossible for the plaintiff to prove negligence on the part of the defendant. Only a plaintiff owning or occupying the land affected may sue under this doctrine; the defendant is normally the person in control of the dangerous thing who brought it onto the land in question. The escape must be from land, not from vehicles and the like. The defendant's use must be 'non-natural', which allows the court to determine whether a particular kind of enterprise in a given location is 'natural'. Accordingly, the courts have a great deal of flexibility in this regard.[7] The kinds of defenses available in actions based on *Rylands* v. *Fletcher* include 'acts of God' (unexpected operation of natural forces that could not reasonably be foreseen), the independent action of a third party for which it is unreasonable to held the defendant responsible, and statutory authority which will be considered below.

The tort of *trespass* involves an invasion of the plaintiff's exclusive possession of land by the intrusion of a tangible object onto his or her property. The definition of 'object' has varied from jurisdiction to jurisdiction in the common law world, with some courts dismissing actions in trespass when an intervening force like wind or water carried the object onto the plaintiff's property. (See, for example, *Esso Petroleum Co.* v. *Southport Corporation* [1956] AC 218 [HL]) No damage need be proven to recover in trespass; as long as one's right to exclusive protection is impaired, damages need not be proven. Violating property is enough. However, one major disadvantage in the action is that the trespass must be proven to be intentional.

There are two types of *nuisance* at common law: private nuisance and public nuisance. A private nuisance requires proof of either physical damage or substantial interference with the use and enjoyment of one's property. The property interest must be that of an owner, tenant or someone with a *profit à prendre*; something more than a license interest in property is required[8]. Unlike negligence, the defendant cannot discharge his or her duty by exercising reasonable care. In determining whether the defendant's interference is reasonable, courts must balance the 'gravity of harm to the plaintiff' against the 'utility of defendant's conduct'[9]. This balancing judgement involves evidence of the kind of neighborhood in question and the benefit to the public at large generated by the defendant's activities.

The tort of public nuisance involves interference or damage to the exercise of the public's rights. Generally the Attorney General of the province is the party granted standing to appear for the public in such cases. The Attorney General, in his or her absolute discretion, may consent to private citizens suing in the name of the Attorney General. Alternatively, if private citizens have suffered damages that are different in kind and degree than those suffered by the general public, then likewise an action will lie in public nuisance as well. Ironically, the more pervasive and widespread the nuisance, the more difficult it will be for any individual citizen to establish special damages. Once again, the reasonableness of the defendant's conduct may make recovery difficult.

Defence of statutory authority

A major defense to tort actions, including those actions considered above, is that of statutory authority: a court may conclude that a statute authorizes conduct that would otherwise amount to a tort. For example, in *Allen* v. *Gulf Oil Refining, Ltd* ([1981] AC 1011 ([HL]) the House of Lords held that the defendant oil company was authorized by a private statute to construct the works related to its oil refinery, so that actions for nuisance and negligence were unsuccessful, in light of the measure of statutory immunity enjoyed by the defendant. More typically, where defendants are operating within the scope of their pollution control permits or other authorizations, this defense is particularly effective.

Statutory authority is a very significant defense in practice, but in two ways courts have narrowly construed the ambit of activities authorized by a defendant's license or permit. First, the courts generally require that the defendant's conduct must not have been negligent in any way. Second, they say that the nuisance occurring must be the 'inevitable consequence' of the business that was authorized by the license or permit granted under the pertinent statute. If there are unexpected discharges beyond the terms of the license, the courts will likely consider that they are not within the realm of statutory authority. If the license grants wide discretion to the firm in question, it has the burden of proving that the way in which its discretion was exercised was the only feasible method in the circumstances.

Despite this kind of strict approach by the courts to the defence of statutory authority and despite other efforts to apply tort law to modern environmental problems, legislators in virtually all modern industrial societies have felt compelled to intervene. It seems almost self-evident that in the face of the impediments noted, common law torts as presently interpreted by the courts cannot adequately address the problems posed by toxic substances. The kinds of regulatory initiatives taken by contemporary legislators to grapple with the particular issue of toxic substances will now be considered.

National legislation

Since the 1970s, legislation has required the screening of new chemicals prior to their entry into the market. In 1969 the Swiss and in 1973 the Japanese passed the first legislation requiring pre-market notification of potentially dangerous new chemicals. The American Toxic Substances Control Act (TSCA) followed suit in 1976. Canada's Environmental Contaminants Act was passed in 1975 but was repealed in favor of the Canadian Environmental Protection Act in 1988.

The US federal agencies have been given the power to regulate almost every aspect of the production, transportation, use and disposal of toxic substances under a wide array of statutes involving several federal agencies. In 1976, US Congress passed the Resource Conservation and Recovery Act (RCRA) requiring the Environmental Protection Agency (EPA) to make regulations to govern the generation, transportation, storage and disposal of hazardous wastes. Recognizing that merely reacting to toxic problems after the fact would be inadequate, Congress enacted the TSCA in 1976. This statute gave the EPA the ability to mandate the testing of potentially dangerous chemicals before they were introduced into commerce and then into the environment. The EPA was also given the authority to regulate and even ban the production of substances posing an unreasonable risk.

The TSCA has two purposes: (1) to regulate the manufacture and use of 'chemical substances' that are dangerous to the health or the environment; and (2) to compile a comprehensive catalogue of the chemical substances produced and distributed in the United States. The chemical industry voluntarily submits data to the EPA.

In Canada, many statutes, both at the federal and provincial levels, prohibit the discharges of pollutants into air, water, or onto land, without a required permit (Rankin and Leaden, 1982). Toxic substances have been among the pollutants traditionally covered by these permits.

Environmental Contaminants Act

Until promulgation of the Canadian Environmental Protection Act, the major statute designed to regulate the life cycle of commercial chemicals was the Canadian Environmental Contaminants Act which was passed in 1975 (SC 1974–75–76, c. 72) It contained a 'schedule' (or list) of prescribed chemicals which constituted a 'significant danger' and which, by regulations, were governed as to quantities, concentrations and conditions under which their entry into the market was acceptable. The Act prohibited the 'release' of 'substances' listed in a Schedule to the Act (s.8[1]). Provision was made for inspection, and search of the premises for any products that might have been in contravention of the Act (s.10).

Under this Act, the Minister of the Environment and the Minister of National Health and Welfare were authorized to advise the Cabinet to add chemicals to the schedule (s. 7[1]), after consultation with provinces and other federal agencies (s.5[1]), to determine whether any other control action would be taken. Interested parties who disagreed with the proposed regulations or with the proposed scheduling of a substance could object in writing to the Minister of Environment (s.5[3]), who, together with the Minister of National Health and Welfare, was required to establish a Board of Review (s.5). Except in emergencies, the Minister of Environment could not proceed with regulations until hearings by the board were completed and a report was received (s.7[3]).

Notification by any person manufacturing or importing chemical compounds in excess of 500 kg per year was to be given to the Department of the Environment with respect to any danger posed by such compounds to human health or to the environment. The notification requirements for new chemicals have been found inadequate in that they do not provide the Minister with adequate information to evaluate the chemical. Under the ECA, industry needed only to identify certain chemicals and the quantities to be imported, manufactured or processed (s.4[6]). The notification was to be made only after the chemical was already in use. No systematic procedure was in place to screen chemicals for potential toxic effects before their introduction into the marketplace. Moreover, the Minister's powers to gather information applied only to 'commercial' operations; institutional, scientific and other activities that were sources of toxic chemicals were exempt.

A more salient shortcoming of the ECA was the fact that the Ministers must have had 'reason to believe' that a chemical was entering or would enter the environment in a quantity or concentration that constituted a significant danger to the human health of the environment before they notified industry of their concerns (s.4[1]). In practical terms, unless the government had already assessed test data that it has generated itself or obtained from other sources, a manufacturer or importer could not be compelled to submit the data required for an assessment. As a result, in over ten years of application, only five chemicals were assessed, added to the schedule and subsequently regulated.

The most serious problem with the ECA, however, was the requirement that government had to prove that a substance already on the market was harmful to human health and the environment. The onus was not on industry to demonstrate that new chemicals were safe, but rather on the government to prove that they were dangerous. Only if the Cabinet was satisfied that this 'significant danger' existed could the substance be listed in the schedule, then banned or severely restricted (s.7[1]). The standard of proof ('is satisfied') was quite vague and rested with the Minister. No definition was provided for 'significant danger', and with the endemic uncertainty surrounding risk assessment, it is not surprising that very few substances were regulated under the Act.

The Canadian Environmental Protection Act

The 'second generation' legislation, the Canadian Environmental Protection Act (CEPA), is designed to 'close the gap' in the Canadian regulation of toxic chemicals. As indicated, other related federal and provincial legislation is already in place; the new Act complements these other statutes while consolidating the legislation addressing toxic chemicals already within the purview of the federal Department of Environment and repealing the badly flawed ECA. Widespread disenchantment with the ECA led the Minister of the Environment to introduce the Environmental Protection Act for public discussion in late 1986. A series of public information and consultation meetings were held across the country, and first reading of a revised Canadian Environmental Protection Act was given to Parliament on June 26, 1987. Following first reading, the draft legislation was sent to a legislative committee, which heard testimony from over thirty witnesses, including labor, environment and business representatives. The CEPA passed third reading on May 5, 1988, and was promulgated on June 28, 1988. Since the Act has had the advantage of experience acquired under its predecessor statutes, as well as under other statutes such as the American TSCA, it will be examined closely as an example of modern legislation of this kind.

The CEPA proposes a comprehensive framework to control toxic chemicals from their initial development to their final disposal – i.e., 'cradle-to-grave management'. Its reach is considerably broader than the ECA. For example it:

- eases the test for ministerial intervention;
- regulates all toxic chemicals and requires the compilation and publication of various lists of chemicals for various purposes;

- permits greater public involvement through enhanced access to information and greater opportunity to 'object' to ministerial or cabinet decisions;
- shifts, to some degree, the emphasis from the protection of commercial interests to a broader conception of the public interest;
- provides more stringent penalties and greater number of remedies for breaches of the Act including a ticketing procedure; and
- includes provisions reflecting various international agreements for the protection of the environment.

Under the Act, there will no longer be a requirement that the Ministers of Environment and of National Health and Welfare have 'reason to believe' that a substance is toxic before testing by manufacturers or importers may be required. The new Act merely requires the Ministers to have 'reason to suspect' that a substance is 'toxic' before mandating that sufficient data be provided in advance of possible regulatory control (s.18[1]). An elaborate definition of the threshold notion of 'toxic' is also provided, which turns on the substance's potential 'immediate or long-term harmful effect upon the environment or human health' (s.11).

The Act covers all chemicals new to Canada, whether commercial or non-commercial. Industry is required to notify the Minister of Environment before any chemical is introduced into Canada (s.26[1]), and to submit a 'data package' on these chemicals sufficient to permit an assessment of their impact upon human life or health or the environment, including samples and information on the 'quantities, uses and composition of the substance' (s.16[2]). The basic form of these data packages has been settled through co-operation among industry, labor, provincial governments and environmental groups. With this information in hand, the government will have the power to determine whether the chemical should be banned, allowed entry subject to conditions or whether further testing is required in order to assess the substance's impact.

I   Lists kept by Minister
   (a)   Priority Substances List
   (b)   Domestic Substances List
   (c)   Non-domestic Substances List
   (d)   List of Toxic Substances Authorities
   (e)   List of Hazardous Waste Authorities
II   List in Schedules
   (a)   Schedule I – List of Toxic Substances
   (b)   Schedule II
        Part I: List of Prohibited Substances
        Part II: List of Toxic Substances Requiring Export Notification
        Part III: List of Hazardous Wastes Requiring Export or Import Notification

(c) Schedule III (re Ocean Dumping)
   Part I: List of Prohibited Substances
   Part II: List of Restricted Substances
   Part III: List of Factors

The first such list is the Priority Substances List in which the Ministers are required to note those substances in respect of which '[they] are satisfied priority should be given in assessing whether they are toxic or capable of becoming toxic' (s.12). Environment Canada has indicated that the Priority Substances List (Table 9.1) will consist of those substances which have one or more of the following properties:

1 Poisonous – it poses a severe threat to human health and/or the environment;
2 Persistent – once released it remains for an inordinate amount of time in the environment;
3 Is extensively used in the Canadian marketplace and has the potential for significant damage should release occur;
4 Is already in the environment and is causing damage.

The second list to be compiled is the Domestic Substances List which will include all chemicals presently in use in Canada (s. 25[1]). The failure of the ECA to require an inventory of existing chemicals has proven to be a major problem in practice (Castrilli, 1982). This list would name those chemicals for which industry is not required to notify the Minister or submit a data package.

The third list is the Non-domestic Substances List (s.25[2]) which will list all chemicals known to exist elsewhere in the world but which are not yet used in Canada. Anyone seeking to introduce into Canada chemicals on this list is required to notify the Minister of the Environment and submit a prescribed data package. Totally new chemicals would presumably involve the most stringent data packages.

The fourth list, which constitutes Schedule I of the Act, is the List of Toxic Substances (s.33). This Schedule will list those 'toxic' substances for which the Cabinet may make regulations for a wide variety of purposes. Ample authority is provided to control the quantity or concentration of such substances, the areas in which they may be released and the conditions under which they may be released into the environment. The Act also contemplates a 'total, partial or conditional prohibition of the manufacture, use, processing, sale, offering for sale, import or export of the substance or a product containing the substance' (s.34[1]).

The fifth list is the List of Prohibited Substances which includes those chemicals whose use is banned under the current ECA or severely restricted by other federal legislation (s.41[1]). This list forms Part I of Schedule II of the Act. Chemicals on this list may only be exported for purposes of their destruction, or for environmental or health protection (s.41[2]).

Table 9.1   Priority substances list, Canadian Environmental Protection Act

Group 1	Arsenic [7440–38–2] and its compounds
	Benzene [71–43–2]
	Effluents from pulp mills using bleaching
	Hexachlorobenzene [118–74–1]
	Methyl tertiary-butyl ether [1634–04–4]
	Polychlorinated dibenzodioxins
	Polychlorinated dibenzofurans
	Polycyclic aromatic hydrocarbons
	Waste crankcase oils
Group 2	Cadmium [7440–43–9] and its compounds
	Chlorinated wastewater effluents
	Chlorobenzene [108–90–7]
	Chromium [7440–47–3] and its compounds
	Creosote-impregnated waste materials
	Dibutyl phthalate [84–74–2]
	1.2 Dichlorobenzene [95–50–1]
	1.4 Dichlorobenzene [106–46–7]
	1.2 Dichloroethane [107–06–2]
	Dichloromethane [75–09–2]
	Di-n-octyl phthalate [117–84–0]
	bis (2-Ethylhexyl) phthalate [117–81–7]
	Inorganic fluorides
	Nickel [7440–02–0] and its compounds
	Pentachlorobenzene [608–93–5]
	Styrene [100–42–5]
	Tetrachlorobenzenes
	1.1.2.2-Tetrachloroethane [79–34–5]
	Tetrachloroethylene [127–18–4]
	Toluene [108–88–3]
	Trichlorobenzenes
	1.1.1-Trichloroethane [71–55–6]
	Trichloroethylene [79–01–6]
	Xylenes [1330–20–7]
Group 3	Analine [62–53–3]
	Benzidine [92–87–5]
	Chlorinated paraffin waxes [63449–39–8]
	bis(2-Chloroethyl) ether [111–44–4]
	bis (Chloromethyl) ether [542–88–1]
	Chloromethyl methyl ether [107–30–2]
	3.3-Dichlorobenzidine [91–94–1]
	3.5-Dimethylaniline [108–69–0]
	Methyl methacrylate [80–62–6]
	Mineral fibres
	Organotin compounds (non-pesticidal uses)

Source: Canada Gazette, Part I, February 11, 1989, pp. 543–5.

The sixth list, namely, the List of Toxic Substances Requiring Export Notification, also contemplates certain Cabinet orders (s.42 and Sched.II, Pt II). Of considerable interest in the international arena is the requirement for Canadian firms exporting chemicals included in this List to notify the authorities in countries to which the chemicals are to be exported (s.42[4]). This provision would enable Canada to fulfill its obligations as a member of the United Nations Environment Programme and of the Organization for Economic Cooperation and Development (OECD). In turn, under international treaties, Canada hopes to receive equivalent notifications from other participating countries when potentially dangerous chemicals are entering this country. The seventh list is the List of Hazardous Wastes Requiring Export or Import Notification (s.43[1] and Sched. II, Pt III). As in the case of the previous list, those importing or exporting hazardous waste must give notice to the relevant authorities (s.43[3]). The eighth and ninth lists are the List of Toxic Substance Authorities (s.42[2]) and the List of Hazardous Waste Authorities (s.43[2]), which are to specify the 'authorities, body or persons to whom notices shall be given'. The final three lists, comprising Parts I, II and III of Schedule III, relate to ocean dumping. These lists, along with other provisions of CEPA regulating ocean dumping, will be discussed below.

The CEPA permits greater public input than the ECA. The requirements of natural justice and procedural fairness represent a rapidly expanding area of administrative law in Canada and elsewhere. The CEPA has attempted to provide safeguards to meet these often amorphous legal requirements. For example, the Minister of the Environment must publish in the *Canada Gazette* a copy of every order, proposed regulatory action or emergency regulatory action contemplated under the Act. Within sixty days any person may file a notice of objection with the Minister, who, either alone or along with the Minister of National Health and Welfare, has the discretion to establish a Board of Review to inquire into the kind of danger posed by the substance to be regulated (s.89). A Board of Review would be required to 'give any person or government a reasonable opportunity, consistent with the rules of procedural fairness and natural justice, of appearing before it, presenting evidence and making representations' (s.91). Such boards have all the powers of commissioners of inquiry under Part I of the Inquiries Act (s.93), and are to submit a report to the Ministers, consisting of the recommendations and the evidence presented (s.96[1]). Generally, this report must be made public immediately upon receipt, unless it contains the sort of confidential information set out in the proposed Act (s.96[2]).

Attention is also paid to the public interest. Central to the regulatory approach found in most toxic chemical legislation is the management of information. Without adequate technical information, often found in corporate hands, the government cannot regulate effectively. Nor can public interest groups adequately participate in decisions concerning the regulation of toxic chemicals if they do not have access to the same technical information as government and the toxic chemical industry. On the other hand, health and safety studies submitted by companies may arguably contain confidential business information, which is vulnerable to 'industrial espionage' by the company's competitors through the use of freedom of information legislation. This conflict will continue to exist in

Canada under CEPA and one may safely predict controversies in this regard as the Act is implemented.

Under the Act, the Minister may publish in the *Canada Gazette* or elsewhere a notice compelling information to be submitted with a view to determining a substance's toxicity (s.16[1]). The intent of this requirement is to bring information discovered during commercial developments to the attention of the Minister. As indicated, where the Ministers have 'reason to suspect' that a substance is 'toxic', companies may be required to provide specified information or conduct certain tests to determine toxicity. As in all similar legislation, the Act protects certain confidential business information from public disclosure if a claim for confidentiality is made by its supplier (ss.19–24). Unlike previous legislative proposals, however, CEPA has shifted the emphasis from the types of information to be protected to the types of information which would normally be released (principally health and safety data) and the circumstances under which the information would normally be disclosed (including when the Minister considers that disclosure would otherwise be in 'the public interest') (ss.20[2] and [4]).

The effect of these new provisions is to grant the Minister greater discretion regarding the disclosure of information without addressing the difficulties the applicant would encounter under the Access to Information Act. The Minister acquires broader discretion primarily in two ways: first, by means of a provision permitting the disclosure of information 'as may be necessary for the purposes of the Act' (s.20 [4] [b]); and second, by a provision permitting disclosure where: (1) the disclosure is in the interest of public health, public safety or the protection of the environment; and (2) the public interest in the disclosure clearly outweighs in importance any material financial loss or prejudice to the competitive position of the person who provided the information or on whose behalf it was provided (s.20[6]). The latter provision goes well beyond its counterpart in the Access to Information Act (s.20[6]). There, the Minister is given a similar discretion to disclose environmental information, but only if the information does not constitute 'trade secrets', a term which is not defined in the Access Act. Under CEPA, the right of the Minister to disclose information in the public interest is said to apply 'notwithstanding ... the Access to Information Act' (s.20[6]). The Act requires that where practicable, the supplier is to be given twenty-four hours' notice before disclosure is made, but the Act does not appear to give the supplier any statutory right to object to or to appeal against the Minister's decision.

Public interest groups (which incidentally have not been granted explicit standing under the statute) which seek information on toxic chemicals must continue to use the cumbersome procedures set out in the Access to Information Act (CEPA, s.20[4] [e]). Nevertheless, their task is made somewhat easier by s.20[2], which explicitly permits the disclosure of health and safety data, as well as certain other specified information. However, it will still be necessary for such groups to argue that their claim to the information specified under this section of CEPA takes precedence over the supplier's claim to protection of their 'trade secrets' under the Access to Information Act (s.20).

The Minister of the Environment is given the power to order recalls of chemicals or products that are not distributed in compliance with the legislation (s.40). This provision would also include the power to require firms to publicize the recall order and even to compel offenders to re-purchase or replace the chemical or product in question. Unlike in the case of the ECA, under the new Act the Minister may compel polluters to clean up controlled chemicals released into the environment. In the event that the polluter fails to comply with such orders, the government has the authority to take remedial action and recover costs from the polluter directly (s.36). The legislation creates new offenses, as well as new civil actions, such as for failure to provide the Minister with pertinent information, failure to perform tests, fraudulent testing and fraudulent reporting on testing (ss. 112–14).[10]

Under CEPA, courts are also empowered to impose a fine related to the profits realized through any action taken in contravention of the Act (s.129). In such cases, no maximum fine is set. Courts may also impose community service orders (s.130), enjoin the polluter from particular conduct, order restitution (s.131), order restoration of the damaged environment and payment of the costs of investigation, corrective measures and prosecution. The objective is to 'internalize' the harmful activity as much as possible. Perhaps the most significant provisions, at least from a theoretical point of view,[10] are provisions for fines of up to $1,000,000 and five years' imprisonment for knowingly providing the Minister with false information (s.114), and for unlimited fines and five years' imprisonment for: (1) intentionally or recklessly caus(ing) a disaster that results in a loss of the use of the environment, or (2) show(ing) wanton or reckless disregard for the lives or safety of other persons and thereby caus(ing) a risk of death or harm to another person (s.115), with each day that a prohibited activity continues considered a separate offense (s.118). Obviously it remains to be seen how the legislation is enforced and, in turn, how the courts will react to the proposed penalties in practice.

The CEPA also includes international aspects: provisions for regulating international air pollution and ocean dumping. In legislating on these matters, the Government of Canada is fulfilling obligations it voluntarily assumed as a signatory to various international treaties, the general nature of which will be outlined below.

The provisions relating to international air pollution in Part V, while not extensive, permit the Governor-in-Council to make regulations, require the publication of proposed regulations and agreements between Canada and any province relating to air pollution, allow the Minister to require the submission of information and samples from (potential) polluters, and permit the public to make objections to proposed regulations. The most serious defect of the provisions, and consequently a serious defect of the Act, is the lack of a provision covering interprovincial air pollution.

The provisions relating to ocean dumping are contained in Part VI of the Act. They are far more extensive than those addressing international air pollution. This reflects the fact that this part of the Act is meant to give legislative force to the 1972 London Convention on the Prevention of Marine Pollution by Dumping of Wastes and Other Matter (s.66[1] and s.86[1] [a]), of which Canada was a signatory, and also to repeal and replace the Ocean Dumping Control Act (sc 1974–75–76, c.55), (s.148).

This Part contains the usual provisions relating to regulations, inspections, publication in the *Canada Gazette* and notice of objection. There are also sections addressing the detention, seizure, and ultimate forfeiture of ships, aircraft, platforms or other man-made structures that are involved or suspected of being involved in a contravention of the Act (ss.78–85). As noted above, Schedule III includes three relevant lists: a List of Prohibited Substances (Part I), a List of Restricted Substances (Part II), and a List of Factors (Part III) that *shall* be considered by the Minister in determining whether to grant permits for ocean dumping or to suspend, revoke or vary a permit. These lists are taken almost verbatim from the Convention.

The legal effect of the lists contained in Parts I and II of Schedule III is unclear. The only section making any mention of the two lists is s.86 in which the Cabinet is empowered to make regulations to implement the Convention, and to amend the lists by adding other things to be prescribed in the lists. The Act itself contains a broader prohibition than that contained in the Convention (s.67[1]). Under the Act, the dumping of any 'substance' is prohibited. The term 'substance' in this Part includes the same definition as employed elsewhere in the Act (s.3[1]), except that it is more elastic in that it includes 'any other matter' (s.66[1]).

International initiatives

There have been a number of important hurdles in achieving international co-operation in this area. Castrilli (1982) has listed some of them as follows:

1   The poor quality of existing test data available.
2   The lack of internationally accepted testing protocols.
3   Differences in national requirements regarding the timing of new chemicals' notification to government agencies.
4   Industry and some governments' concern about the release of 'confidential' information due to fears of competitive market positions being harmed.
5   The lack of public participation requirements in many national chemical control laws.

6   The establishment of unnecessary trade barriers due to differing control requirements in each country.
7   The export or 'dumping' of hazardous chemicals from countries with stringent controls to those with less stringent controls.

The foreign aid programs of many industrialized countries support the use of chemicals that are banned in the developed countries. Ethical issues abound: for example, should DDT be used in malarial control programs in less developed nations; or should developed countries refuse to dump hazardous chemicals anywhere?

There are about 50,000 chemicals traded internationally each year. It was reported in 1981 that the estimated value of this trade ranged from 11 billion to 14 billion dollars (US), (Halter, 1987). Pesticides are initially emphasized in the international arena over other toxic chemicals due to their widespread distribution and to comparative importance in the agricultural economies of developing countries.

In 1984, the Governing Council of the United Nations Environment Program (UNEP) adopted a 'provisional notification scheme for banned and severely restricted chemicals'. Under this system, exporting countries were urged to provide information to assist importing countries with respect to potentially harmful chemicals. A 'banned or severely restricted chemical' was defined to include:

any chemical that is the subject of a control action taken by a competent authority in the country of export: (a) to ban or severely restrict the use or handling of the chemical in order to protect human health or the environment domestically; or, (b) to refuse a required authorization for a proposed first-time use of the chemical based upon a decision in the country of export that such would endanger human health or the environment. (UNEP, 1984)

The kind of information to be provided would include the chemical identification/specification of the chemical in question, as well as a summary of any control action taken and the reasons for such action. The notification is to be addressed not only to the national authority in the importing country, but also to the International Register of Potentially Toxic Chemicals (IRPTC).

It must be stressed that these guidelines are only voluntary in nature. The IRPTC commenced its operations in Geneva in 1976, with the objective of increasing the capability of the UN to provide an early warning system for the deleterious effects of synthetic pollutants. However, the program has suffered from a chronic lack of financial support. Problems such as language barriers, differences in the developmental stages of countries and disagreement on methodology for testing chemicals persist. As Smith (1984) has noted:

The mere gathering and assessment of environmental issues and indicators are of little value unless there is a scientific understanding of the
environmental system of which they are a part. The real question
concerning IRPTC, then, is the basic value and use of its various
informational systems by importing countries to stop unwanted shipments of chemicals. Unless adequate regulatory schemes exist in the
importing country, the IRPTC information is of little value.

In the United States, notification requirements are found in the Federal
Insecticide, Fungicide and Rodenticide Act (FIFRA) – see 7 USC, ss. 136
(a)–(y) (1982); and in the Toxic Substances Control Act (TSCA) (15 USC, s.
2601 [1982]). Whereas FIFRA requires notification solely for pesticides, TSCA
applies also to chemicals subject to proposed regulations, enforcement
proceedings and testing requirements. See s. 12 (b) of the Act, as amended
by 42 USC 6928. Lastly, the Resource Conservation and Recovery Act
(RCRA)[11] requires that countries receiving the hazardous wastes must
provide their prior consent to shipments, under pain of criminal penalty. It
is hoped that these statutory requirements for export notification found in
American legislation, as well as in the Canadian Environmental Protection
Act, will have some positive impact in this regard.

The Organization for Economic Cooperation and Development (OECD)
has an Environment Directorate which established a chemical controls
program in 1978. Efforts have been made to improve the exchange of
testing data concerning chemicals and to co-ordinate controls among
member countries. In May, 1980, OECD members agreed to accept each
other's data if specified guidelines for testing and laboratory standards were
followed. In addition, a model agreement is being developed by OECD for
the exchange of confidential business information among member
governments.

The OECD has also been active in trying to harmonize the laws and
administrative practices with respect to chemical regulation in its member
countries. Noting that information about potential hazards of chemicals 'is
a pre-requisite for the effective protection of human health and the
environment', a group of experts led by France has reviewed the legal
provisions respecting confidentiality of data on chemicals in the various
member countries and attempted to identify where international co-
operation might enhance chemical trade (OECD, 1982).

In April, 1984, the OECD adopted 'Guiding Principles on Information
Exchange Related to Export of Banned or Severely Restricted Chemicals'
(OECD, Doc. No. C [84] 37 [1984]). The purpose of these guidelines was to
encourage member countries to comply with the UNEP program. The
Environment Committee of the OECD was also instructed to develop
guidelines for the export of hazardous wastes by December 31, 1987 (OECD,
1985).

The Council of the European Communities has taken action to control toxic chemical substances as well. In September 1979, the 'Sixth Amendment' to a 1967 EEC Directive was adopted. It contemplated standardization of domestic laws concerning the classification, packaging and labelling of dangerous substances. Pre-market notification procedure for new chemical substances is required, and manufacturers or importers must submit a technical package with information sufficient to evaluate foreseeable risks to human health or the environment. Confidential business information is protected and member countries are allowed to require additional tests or studies under certain stipulated conditions. It has been argued that government authorities cannot ban marketing or restrict usage if the appropriate notification is received; all that can be done is to ask for more information. Only if a member state can show that a substance constitutes a hazard may it temporarily prohibit or restrict its use (Kelly, 1985).

Member countries were required to incorporate provisions of the Directive in national legislation by 1981; however, most members were unable to meet this deadline. It should be noted that the Sixth Amendment is merely a guideline, with the method of implementation left to individual member nations. Accordingly, a considerable variety of notification systems have been developed in Europe. The EEC is drafting a regulation that would be binding on its member countries requiring them to enact legislation to provide export notification for certain banned and severely restricted chemicals (Halter, 1987).

In March 1978, the EEC also issued a Directive on Toxic and Dangerous Waste (EEC, 1978). The Directive broadly defines toxic or dangerous wastes, lists candidates and encourages member states to prevent the accumulation of toxic waste and, where it does exist, to process and recycle it. Toxic waste is to be kept separate from other waste, appropriately labelled, recorded and identified as to the sites where it is or has been deposited. Installations dealing with the storage, treatment and/or deposit of toxic waste are to be licensed by the relevant national authorities. Transportation is to be closely supervised and all establishments producing, holding or disposing of it are subject to inspection and supervision.

There are numerous international conventions and agreements affecting the use and disposal of toxic chemicals. The various agreements fall into four major types: (1) those focusing on land-based pollution; (2) those concerned with ocean dumping; (3) those relating to vessel-source pollution; and, finally, (4) those governing harmful air pollutants (Kindt, 1986).

Within these broad categories of international agreements are several strictly regional treaties and bilateral initiatives. Examples of treaties in the first category include the Convention for the Prevention of Marine Pollution from Land-Based Sources of 1974 and the Convention for the Protection of the Mediterranean Sea Against Pollution of 1986. The most significant treaty of the second category is the London Convention on the Prevention of Marine Pollution by Dumping Wastes and Other Matter of 1972, which is considered above (p. 238). It lists wastes that are absolutely prohibited from being dumped and those for which a special permit is required. Lastly, the 1985 Vienna Convention for the Protection of the Ozone Layer and the Montreal Protocol on Ozone of 1987 are examples of the fourth type of international initiative, by which strict regulation of the toxic substance known as chlorofluorocarbon is contemplated.

Conclusions

Increasing harmonization at the international level must be accompanied by improvements at the domestic level. In Canada, as elsewhere, there is a great need to reform the civil law to enhance the compensation of victims of toxic waste pollution. The requirement of causation must be altered so that a probabilistic causation approach is enacted (Strand, 1983). The knowledge base of toxic substances must be enhanced. Inevitably, statutory reform for compensating victims of toxic substances pollution will be required (Trauberman, 1983). More legislation like that of the so-called 'Spills Bill'[12] of the Province of Ontario should be seen as a model for other jurisdictions. This Bill provides absolute liability for the cost of cleaning up spills and strict liability for the damage caused by spills. Negligence is no longer a factor in recovery under the Act. The concept of 'loss or damage' is very broadly defined and any person has the right to recover for such a loss. There is strict liability for such a loss and damage and absolute liability for the cost of clean-up imposed on the owner or person in control of the spill (Makuch, 1986). As indicated, this major departure from common law would do a great deal to overcome the many impediments to successful lawsuits under the various common law torts.

But aside from the importance of compensating victims of toxic torts, the existing regulatory framework must be further improved to address the influx of new toxic chemicals. The Canadian Environmental Protection Act represents an attempt at more comprehensive legislation. In an increasingly interdependent world, however, increasing harmonization will be required to protect the global environment from risks that only could be imagined a few short decades ago.

Notes

1 In British Columbia, for example, the Waste Management Act, RSBC 1982, c. 41 differentiates between 'waste' defined as air contaminants, litter, effluent and refuse and 'special waste' (elsewhere usually termed 'hazardous waste') which is prescribed by regulation for intensive 'cradle to grave' management.

2 See, for example, Food and Drugs Act, RSC 1970, ch. F–27; Pest Products Control Act, RSC 1970, ch. P–10; and Pesticide Control Act, RSBC 1979, ch. 322.

3 See interview with D. F. Chant, reproduced in Lee, (n.d.).

4 In Canada, for example, there is overlapping legislation passed by the federal

and provincial governments: the Transportation of Dangerous Goods Act, sc 1980, c.36, as amended, along with Transportation of Dangerous Goods Regulations, sor/85–77 and Transport of Dangerous Goods Act, sbc. 1985, c.17.

5 This is explored in greater detail in judicial attitudes towards legal and scientific proof of cancer causation. *Columbia Journal of Environmental Law*, 3, 344. 1977.

6 For example, the bc Limitation Act, rsbc, 1979, c.236, s.6(3). The majority of American states have also adopted rules of this kind. By way of contrast, ground-breaking developments have occurred in Japanese legislation where no limitation period is applicable. The Pollution–Related Health Damage Compensation Law adopts a statistical approach to victim compensation. The Act lists those regions in which diseases related to specific toxic substances are compensated with the government establishing the causal relation between the specific disease and specific toxic substances. Levies on particular polluters in the relevant regions are paid to a joint government–industry association which distributes funds to compensate victims and to provide for rehabilitation programs. The deterrent effect upon polluters remains unclear in the Japanese experience to date. Law No. 111 of 1973 (Japan). This statute is described in Gresser, *et al.* (1981).

7 See, for example *North York v. Kert Chemical Industries, Inc.* (1985) 33 cclt 184 (Ont. hc). Toxic wastes were being discharged into a municipal sewer system by two corporate defendants. Krever, J. noted that 'The use of land in an industrial subdivision for the manufacture and processing of cleansers requiring the use of chemicals may not be a non-natural use of land.' (*Ibid.*, at p. 200).

8 See *Bolton et al. v. Forest Pest Management Institute et al.* (1985) 66 bclr 126 (bcca). On an appeal of an interlocutory injunction restraining the defendants from spraying a herbicide, the plaintiff, a native with a registered trapline under the Wildlife Act was held to have a *profit a prendre* and not a mere license. Therefore, he had standing to sue in private nuisance.

9 *Restatement (Second) of Torts* s.826 (1979). See the similar approach of McIntyre, J.A. (now of the Supreme Court of Canada) in *Royal Anne Hotel Co. Ltd. v. Ashcroft*, [1979] 2 wwr 462 (bcca).

10 Emphasis was placed on the word theoretical since a recurring theme in environmental law reform proposals is the need to enforce existing legislation. See, for example, Rankin and Finkle (1983).

11 usc, ss.6921–34, requiring notices for exports of hazardous wastes (except materials intended for recycling or beneficial re-use).

12 Part IX of the Environmental Protection Act, rso, 1980, c.141, proclaimed in force on November 29, 1985.

References

Bonine, J. and McGarity, T. (1984), *The Law of Environmental Protection* pp. 657–8. (St Paul, Minnesota: West Publishing Co).

Castrilli, J. F. (1982), 'Control of toxic chemicals in Canada: an analysis of law and policy', *Osgoode Hall Law Journal* 20, p. 322.

David, D. (1985), 'The "shot-gun wedding" of science and law: risk assessment and judicial review', *Columbia Journal of Environmental Law* 10, p. 67.

Doern, G. B. (1977), *Regulatory Processes and Jurisdictional Issues in the Regulation of Hazardous Products in Canada* (Background Study No. 41) (Ottawa: Science Council of Canada).

Doern, G. B. (1982), *The Politics of Risk: The Identification of Toxic and Other Hazardous Substances in Canada* (A study for the Royal Commission on Matters of Health and Safety Arising from the use of Asbestos in Ontario). (Toronto).

EEC (1978), *Council Directives of 20 March, 1978, on Toxic and Dangerous Waste*, Official Journal of the European Communities, No. L–84/43, 31.3.78.

Environment Canada (1986), *From Cradle to Grave: A Management Approach to Chemicals.* (Ottawa: Supply and Services).

Estrin, David (1986), *Handle with Caution: Liability in the Production, Transportation, and Disposal of Dangerous Substances* (Toronto: Carswell).

Finkle, P. (1983), 'Canadian environmental law in the 80s: problems and perspectives', *Dalhousie Law Journal* 7, pp. 257 and 259–60.

Ginsberg, W. and Weiss, L. (1981), 'Common law liability for toxic torts: a phantom remedy', *Hofstra Law Review* 9, 859.

Gresser, J., Fujikura, K. and Morishama, A. (1981), *Environmental Law in Japan* (Cambridge, Mass.: MIT Press), Ch. 6.

Halter, F. (1987), 'Regulating information exchange and international trade in pesticides and other toxic substances to meet the needs of developing countries', *Columbia Journal of Environmental Law* 12, pp. 1 and 7, note 30.

Kelly, M. (1985), 'International regulation of transfrontier hazardous waste shipments: a new EEC Environmental Directive', *Texas International Law Journal* 21, p. 85.

Kindt, J. W. (1986), 'International environmental law and policy: an overview of transboundary pollution', *San Diego Law Review* 23, p. 583.

Lee, B. L. (n.d.), *Coping with Chemicals* (Ottawa, Ont.: Environment Canada). Reprinted from a series in the *Hamilton Spectator*.

Makuch, S. (1986), *The Spills Bill: Duties, Rights and Compensation* (Toronto: Butterworth).

Nemetz, P. (1981), *Regulation of Toxic Chemicals in the Environment*, Working Paper No. 20 (Ottawa: Economic Council of Canada).

OECD (1982), *Confidentiality of Data and Chemicals Control* (Paris: OECD).

OECD (1985), Resolution of the Council on International Cooperation Concerning Transfrontier Movements of Hazardous Wastes, OECD Doc. C(85) 100.

Page, T. (1979), 'A generic view of toxic chemicals and similar risks', *Ecology Law Quarterly* 7, p. 207.

Prosser, W. (1971), *Handbook of the Law of Torts* (4th edn), (St Paul, Minnesota: West Publishing Co).

Rankin, M. and Finkle, P. (1983), 'The enforcement of environmental law: taking the environment seriously', *University of British Columbia Law Review* 17, p. 34.

Rankin, M. and Leadem, T. (1982), 'The Fisheries Act and water pollution', *Advocate* 40, p. 519.

Schrecker, T. (1984), *Political Economy of Environmental Hazards*, a study paper for the Law Reform Commission of Canada (Ottawa: Supply and Services Canada).

Schrecker, T. (1986), *Workplace Pollution*. Working Paper No. 53. (Ottawa: Law Reform Commission of Canada).

Smith, G. P. (1984), 'The United Nations and the environment: sometimes a great notion?', *Texas International Law Journal* 19, 335 at 349, n.64.

Strand, P. (1983), 'The inapplicability of traditional tort analysis to environmental risks: the example of toxic waste pollution victim compensation', *Stanford Law Review*, 43, p. 575.

Trauberman, J. (1983), 'Statutory reform of "toxic torts": relieving legal scientific and economic burdens on the chemical victim', *Harvard Environmental Law Review* 7, p. 177.

UNEP (1984), *Environmental Law Guidelines and Principles: No. 6: Banned and Severely Restricted Chemicals*. (New York: United Nations Environment Programme).

US Surgeon General (1980), *Health Effects of Toxic Pollution*. A report from the Surgeon General and a brief review of selected environmental contamination incidents with a potential for health effects. (US Government Publishers Office).

10 The social and political perspective: an environmentalist's viewpoint

S. HOLTZ

A case study: the story of Canning, Nova Scotia

In the early hours of the morning on May 31, 1986, residents of the village of Canning, in the heart of Nova Scotia's fruit-growing Annapolis Valley, woke to the roaring flames and acrid smoke of a major fire. A warehouse belonging to one of the village's oldest firms, Maple Leaf Farm Supplies, was ablaze in the center of the town.

Volunteer firefighters called to the scene from the surrounding area did not at first realize they were dealing with anything more than a burning building. Fire departments need to know about any local situations that will require special firefighting measures, and the usual contents of the warehouse were known: the agricultural supply company used the building to store machinery. Farm chemicals were stored in the main warehouse.

But it was almost June, with the fruit trees coming to full bloom throughout the Valley, and the company's inventory of pesticides and fertilizers was at its peak. Extra storage space was needed. Unknown to the fire crew, the fiercely blazing warehouse held a huge overflow of farm chemicals from the main building: forty-eight different agricultural products, some 12 tonnes of them.

In a fairly short time, the fumes had alerted firefighters to the fact that they were battling a chemical blaze. Very fortunately, their volunteer fire brigade had recently received training in chemical fires, and they, therefore, knew some of the protective measures to take. Officials hastily decided on an evacuation of the immediate area, which encompassed much of the residential part of the town, including a senior citizen's home. But within hours, a number of the firefighters who had been first on the scene began to experience nausea, headaches, throat irritations and dizziness – classic symptoms of acute pesticide poisoning.

Had the volunteer firefighters realized in time what the building contained, they would probably have let it burn and merely tried to contain the fire. Instead, they used normal procedure for an ordinary burning building – this compounded the problem. More than 1.3 million liters of water were poured onto the inferno of burning chemicals, sending toxic run-off in a stream down the main street, into the storm sewers, down to a farm pond and the nearby Habitant River. The slug of toxic run-off water poured across lawns, fields, and into the river, killing all the living things in its path. Within a day, everything in the water column of the Habitant River was dead. Dozens of earthworms lay dead along each meter of the path taken by the contaminated water.

By the day after the fire, the evacuees – who had left abruptly in the middle of the night and included nearly the entire community – naturally wanted to get back to their homes. Local business people worried about the financial effects of suspending business in the basically marginal rural village economy. Pressure to allow people to go home quickly grew. At the same time, there was uncertainty about the health dangers from the unknown amount of chemical contamination. Some people who had breathed the heavy fumes were experiencing throat irritations and nausea. But with no means to determine the degree of risk themselves, people had no choice but to rely on the decisions of government officials.

Within thirty-six hours, and with no test results yet available from wells or soil, officials permitted almost all the evacuees to return, though warning them not to drink the water and to keep their children inside. If the children had to go out, parents were told to dress them in heavy clothing. Some families questioned publicly the grounds on which this decision had been made – in the absence of any new factual information, why was the risk now regarded by officials as acceptable?

Frustrations for residents about getting information continued. A promised written information sheet was never produced. Several weeks after the fire, one family who lived next to the warehouse and who had requested the test results from their well were still not able to obtain that information. The request was passed from one government office to the next and one official told them flatly that 'my superiors directed me not to give out that information'. By this time, lawns and trees were turning brown, and many residents were now complaining of irritations from the still-present fumes, especially when workers were moving materials around in the warehouse clean-up.

The clean-up also was problematic, despite prompt assistance from the chemical industry and good co-operation among the government departments responsible for such emergency situations. No provisions for safe treatment of such toxic material as the 186 barrels of 'toxic cocktail' remaining from the farm chemicals existed in Nova Scotia, nor was there a permanent transfer station to hold dangerous material for shipment elsewhere. As a result, the barrels of chemical sludge were kept in several temporary locations while arrangements were being made to send them out of the province.

The first attempt at export failed. The province of Quebec refused to accept shipment of the toxic materials across its border without prior official acceptance of it from Ontario, and the truck carrying the barrels had to return with them to Nova Scotia. The problem, ironically, was related to recent improvements in legislation dealing with hazardous materials. Canada now has a Transport of Dangerous Goods Act in place which involves a manifest system for the transport of hazardous wastes. Provinces also have matching legislation for interprovincial transport, and the province of Ontario also has stringent requirements controlling the disposal of waste within its borders. Obtaining all the permits for the difficult-to-label toxic sludge was very time consuming, and it was many months after the fire when the barrels of chemical sludge were finally shipped off for disposal at Tricil's secure landfill site in Sarnia, Ontario – perhaps not the ideal solution for such toxic material even then.

While quite a few people experienced acute health effects from the fire, these fortunately were not long lasting. It is estimated that about 200 people were directly exposed to toxic smoke, fumes or materials during the fire and clean-up; this figure includes firefighters, environmental and emergency personnel, and local residents. Blood tests were performed on many of these people on the day of the fire. Further medical follow-up was done on the few persons showing traces of pesticide exposure in their blood for some months after that.

Whether there will be any long-term health effects is less certain, although the likelihood from a one-time only exposure is not high. A number of pesticides are suspected carcinogens, cause neurological damage, and may affect the functioning of the immune system. These effects can take years to manifest themselves, and since an individual case of cancer or other health problem could have a number of possible causes it is unlikely that it could ever be definitely shown to have been caused by the fire. This is especially the case in an agricultural district like Canning, where many people routinely work with, or near, pesticides. Only in the case of birth defects could such an association more easily be traced, since there is now a program to investigate unusual numbers or types of birth defects in that part of the province, as well as in Cape Breton (since the latter has an unusually high cancer rate which many suspect may be related to industrial pollution).

The psychological aftermath of the fire probably will vary considerably. Farming is the mainstay of the village, which grew up around the agricultural supply firm that owned the warehouse. The majority in the community were thus not inclined to strain relationships by seeking damages in the courts or otherwise making too strenuous a protest; besides, it was common knowledge that Maple Leaf warehouse was no different from most others. Doubtless for many of these people, the event will recede into memory as simply another accident in life that was, after all, survived. For others, however, perhaps especially for families with young children and those people who found official responses less than helpful, a deepened sense of distrust and an increased level of background anxiety about their families' health may well be a permanent result. (This account synthesizes reports from Potter [1987], Jones [1986] and Feldman [1986].)

The context for citizens' concerns about chemicals

The Canning situation described above shows something of what it is like, suddenly and without warning, to be in the middle of a local chemical emergency. In microcosm, it bears many similarities to events of much larger magnitude and severity such as Seveso, Italy and Bhopal, India. But it is by no means the only kind of hazardous situation that arises with the near-ubiquitous use of potentially dangerous chemicals that has become a characteristic feature of modern industrial society. Let us take a brief look at some other hazards.

Transportation accidents with chemicals are fairly common; perhaps the best known in Canada was the 1979 Mississauga, Ontario, train derailment of chlorine tank cars, which required the emergency evacuation of the quarter million people living in that area. In part due to planning and in part due to sheer luck no deaths could be attributed to the derailment. This accident also was the impetus for comprehensive federal legislation, the Transport of Dangerous Goods Act. Along with mirror provincial legislation, this act provides for regulation of all transportation of materials which are flammable, corrosive, reactive, or toxic. Labels with an international descriptive symbol and a product code number must be attached to dangerous cargoes, and the types and standards of containers and other safety features are also described in these regulations.

Other situations which in the last few years have frequently turned up as cause for concern about chemicals include leaking underground gasoline tanks, which can result in explosions and contamination of drinking water, and old dump sites where toxic chemicals have been unsafely buried. Love Canal in New York State along with other old dumps near the Niagara River and the Valley of the Drums in Kentucky typify such sites, which frequently cause water or soil contamination and perhaps local health problems. Every part of the country has such sites; a preliminary study mentioned in Environment Canada's 1986 State of the Environment Report for Canada identifies some 500 active and inactive waste dumps that pose high risks from chemical contamination (Bird and Rapport, 1986). In the United States, about 10,000 sites have been listed. As noted in Chapter 1, many other countries also have similar sites.

Pesticides, which include chemicals used to control insects (insecticides), weeds (herbicides), rodents (rodenticides), and various fungal organisms (fungicides) are also high on the list of citizens' environmental worries, and rightly so in that these chemicals are only useful because they are toxic to living organisms. One issue is the risk of run-off from agricultural chemicals contaminating watercourses and wells; it should be remembered that the intended fate of those 12 tonnes of chemicals in the Canning warehouse was to be freely spread around the countryside. Other pesticide issues involve large-scale insecticide spraying in forestry, and herbicide spraying in forest plantations and along utility rights of way. Risks from aerial spray operations are of particular concern to local residents, since the scale of operations is so large: spray drift and accidental contamination of wildlife, including wild berries, fish, and game; water, including drinking water supplies; and accidental direct spraying of people and property.

Another important consumer or health concern is about the hazardous use of chemicals in common products, such as lead in gasoline, asbestos in ceiling tiles, and formaldehyde fumes from urea formaldehyde foam insulation.

Finally, there are worries about the risk in the actual manufacture or industrial use of chemicals. The December 1984 accident at the Union Carbide pesticide plant in Bhopal, India, released 40 tonnes of methyl isocyanate, leaving 3,400 people dead and 60,000 with chronic ailments and killing 3,000 domestic farm animals. In Canada, although the safety record of the chemical industry has so far been quite good, a post-Bhopal government–industry review (Bissett *et al.*, 1986) recognized that a similar accident is possible here. As if to emphasize the point, in July 1987 a fire in a pesticide formulation facility in the province of British Columbia caused the evacuation of a small area and resulted in an order not to eat vegetables and fruit grown downwind. More recently, in August 1988, a warehouse storing PCBs burned in St Basile le Grand, Québec, spreading PCBs and other contaminants over a wide area. Some 3,500 people were evacuated as tests were conducted to assess the extent and potential ecological and health significance of the contamination. Although active pesticide ingredients are manufactured in Canada in only a very few locations, according to that review, a number of industry sectors routinely use large volumes of dangerous chemicals, and millions of tonnes of these chemicals are shipped by rail and truck across the country every year. The situation in many other industrialized countries where pesticides are manufactured in large quantities is potentially more serious.

Of special concern are the oil and gas production sector, the petroleum refining sector, and the chemical industry sector itself. In oil and gas production, natural gas is highly flammable with explosive potential, and there is a risk of a release of hydrogen sulphide, a deadly poisonous gas, from a 'sour gas' well or plant. Fire, explosion, or a toxic release of hydrogen sulphide is also a possibility with a petroleum refinery; and refineries are often located near urban residential areas. Chemical manufacturing plants are located in almost every province, though more than half of all the chemicals produced in Canada come from Sarnia, Ontario's 'Chemical Valley'. Some of the most commonly used dangerous industrial chemicals include hydrogen chloride, a colorless, highly corrosive gas used in many industries and shipped throughout the country; chlorine, also a highly corrosive and reactive gas, similarly widely used and shipped; and ammonia, an explosive and toxic gas, used in fertilizers, mining, and pulp and paper manufacture. Of the 150 chemicals identified in Environment Canada's large data base on chemical spills (the National Analysis of Trends in Emergencies System – NATES; listed in Bissett *et al.*, 1986) as the most frequently accidentally released, many are quite hazardous. These include the pesticides aminocarb (Matacil), carbaryl, and fenitrothion; acutely poisonous substances like sodium cyanide, phenol, and arsine; and persistent toxic substances such as mercury and PCBs.

Such facts as these merely illustrate the situation. While it is very difficult, if not impossible, to assess the overall risks from the approximately 100,000 chemicals (and the resultant wastes) that are in commercial use, it is also true that virtually every community is touched by some actual or potential chemical hazards. The reality behind public fears about chemicals is clearly one of the real dangers; the political pressure for greater safety is not a response based on irrational fears, but a response – necessarily subjective and impressionistic for most people – based on the knowledge that the Mississauga train derailment actually happened somewhere, and the potential exists from some similarly unexpected problem to occur almost anywhere. And the public is right in this perception.

What the unique public perspective on chemical dangers reflects is the sense of unanticipated, unexpected, and, therefore, poorly understood hazards intruding right into the home or immediate neighborhood. It is, thus, primarily a local perspective; a response to the situation of chemicals in our society only at the point where a problem impinges on daily life. It is, too, a response characterized by the sense of trespass – with what that term connotes about justice and liability. The immediate responsibility for a chemical problem visited upon local people does, in fact, usually and legitimately belong to someone else. This reaction, in part, explains the attitude of 'not-in-my-backyard' toward siting certain facilities, widely referred to by the media as NIMBY. But although this local response may not reflect a comprehensive overview, the anger and concern of a resident of Canning or any other community the day after an accident is entirely valid, and the gaps that such an event reveals in our society's ability to manage chemicals safely must be addressed.

The public perspective: a variety of interests

However, just as the hazards are real, so are many benefits, and thus they must both be placed in some larger perspective. It is necessary to begin to probe the concept of risk in its many dimensions in order to develop a more comprehensive view of the social and political issues involving toxic chemicals in our society.

A first step is to identify more precisely just what people and what interests are involved. There are quite a number of toxic chemicals 'stakeholders'– groups or individuals with an identifiable interest or stake in the issue at hand.

Such stakeholders, of course, include locally affected people, both individually and as local-interest groups. Broader based environmental groups also have a distinct perspective, supportive of local citizen groups but with a more policy oriented viewpoint; increasingly, their perspective reflects a global analysis, with concerns, for instance, about the export of hazardous chemicals to countries in the developing world. There, thousands of pesticide poisonings occur every year, and the lack of power and education for workers, along with the lack of resources for infrastructure, mean that it is extremely difficult to ensure the safe management of dangerous chemicals. Health and consumer groups such as the Consumer Unions of Southeast Asia frequently also have interests that overlap with these other organizations. However, some public health concerns are difficult to represent because the effects are random or widely diffused. Bystander exposure to toxic substances, especially pesticides, second-hand cigarette smoke, and contaminants in food such as antibiotics and hormones in meat and dairy products are an increasingly important interest to include in decision-making.

Industry has its own perspective, that of a business based on making profits, but at the same time often more environmentally sensitive now than environmentalists of fifteen years ago would have believed possible. Labor is also one of the most frequently affected stakeholders, with concerns about health and safety in the workplace, the adequate provision of information about hazards, and job security.

Each level of government has its own responsibilities and its characteristic concerns; municipalities often complain that they are given inadequate information about potential hazards in their own jurisdiction by senior levels of government, for instance. The provincial or state governments (in most countries where they exist) have primary jurisdiction over land use and local resources, and consequently most management decisions, such as those regarding the actual use of pesticides and toxic chemical waste disposal, are their responsibility. Federal jurisdiction relates usually to health, defense, trade and international relations. Consequently, the federal role deals with such matters as the registration of pesticides and other chemicals, national standards and guidelines for contaminants, and international treaties or agreements such as those developed by the Organization for Economic Cooperation and Development and the United Nations Environment Programme.

Other players in these issues, such as academics, health professionals, scientists, or consultants are usually affiliated (sometimes explicitly but often only in terms of their personal or institutional interest) with stakeholder interests mentioned above. It is worth noting that, with a topic that touches almost everyone, and involves basic assumptions about safety, economic benefits, and justice, there really are no objective non-participants or neutrals, even among individuals who do not themselves have a direct stake in a particular issue.

Neither is it uncommon for an individual to have conflicting interests in a situation. For some Canning residents, it would be possible to be simultaneously a homeowner, with concerns about property values; an affected resident and parent, concerned about the family's health; a worker in the warehouse, worried about whether there will still be a job after the fire; and a volunteer firefighter, exposed to toxic fumes while doing a dangerous job as a community volunteer. Some of these concerns logically must cancel each other out if the unfortunate individual in that situation is trying to develop a single emotional stance toward the accident. For instance, the property owner may want to deny that there has been any chance of soil or drinking water being contaminated, while the parent doesn't want to take any chance in remaining in a house where there is even a small increase in risk to the child's health. But how does the person decide whether to stay or to sell the house? And if the house is to be sold because of fears of health risks, what do you tell the new purchaser? These issues were, in fact, raised after the Canning fire, and on a much larger scale in such cases as Love Canal.

The moral dimension of values conflict

How individuals resolve such internal conflicts is perhaps a matter of personal psychology, but it is also a moral matter, a question of values and personal integrity.

All of those different roles − property owner, parent, employee, and resident with roots in the town − represent valid interests for a person to have. It is neither wrong nor inappropriate to be concerned about your job, your child, your home town's future, your health, or the value of your house. Where morality comes in is not in denying any of these values. It is, rather, a question first of being willing to investigate the reality of the situation as rigorously as possible even while feeling the pressure of being pulled in opposite directions by different hopes and fears. Secondly, integrity involves refusing to deny any of the values in the situation, but instead working them through so that what one comes to recognize as one's most important responsibilities are those on which final actions are based.

Although individuals don't always meet their own standards of behavior, most people would probably accept the validity of the description above of how, in a difficult situation involving conflicting internal values and requiring some research and weighing of information, one would ideally go about resolving it. We can recognize the sense of integrity that comes from trying to see reality accurately and in its entirety, and from trying to act on that information based on our deepest sense of what is right.

However, this situation of conflicting values and information assessment applies, in toxic chemicals issues, not only to internal, personal situations, but also to the values and perspectives of the different stakeholders. Essentially, the internal conflicts of the Canning resident mirror the external conflicts of different stakeholder groups in toxics issues. And the same sense of integrity that can be recognized in an individual's response to such difficult conflicts can also be applied to the processes society uses to make decisions about larger issues. This assertion will be more closely examined in later sections; for now, let us conclude with the reflection that there are a variety of interests and values in toxic chemical issues, sometimes conflicting ones even within individuals or groups, and that success in dealing with these conflicts is closely related to the integrity of the process used to make decisions about them.

Risk and risk analysis

Let us here take a step back from toxic chemicals and their hazards to look at the more general and inclusive subject of technological and environmental risk. Since the mid–1970s, discussion about risk in the context of technological and environmental dangers has become increasingly frequent and increasingly sophisticated in academic and technical literature; indeed, by the early 1980s, risk was sometimes referred to as an academic 'growth industry'. Risk analysis and the perception of risk are now standard aspects of many environmental issues.

The concept of risk has become precisely defined when the work is used in a technical context. The most commonly used definition is 'Risk = Hazard (or harmful consequences) × Probability'. What this means is that the concept of risk is the product of two different factors. These are (for any given risk) the hazard, or the potential harm that could occur, multiplied by the probability, or likelihood expressed numerically, of those consequences actually happening. The probability involves in part an analysis of possible exposures.

Thus, for example, a certain hazard such as death from traumatic injury per passenger-kilometer travelled can be compared for several different transportation modes. In that case, comparing the risk of death in cars, airplanes, and bicycles is actually comparing the likelihood or probability of a given hazard. However, if we muse over how we would least like to be injured in such an accident, we are considering a situation in which the probability of the hazard is not at issue, but the different qualitative aspects of the actual harm are of significance.

Risk in this technical sense first entered public debate about technology and the environment in the controversy about the safety and desirability of nuclear power plants in the 1970s. Several comprehensive safety studies attempted to assess the consequences and the probability of a major nuclear plant accident. In the absence of actual experience of such an accident, calculations based on 'event tree analysis' were made. This is a standard engineering tool in safety analysis, involving a calculation of the failure rate of each separate part of the plan, as well as an analysis of the cascading consequences of each failure (Farmer, 1982). In this way the results of small failures can be built up into different scenarios of where failures can, at worst, lead. At the same time, probabilities of that event can be calculated, based on reported failure rates for individual components such as valves and welds. From this accident scenario, assumptions about wind direction, exposure routes, the effectiveness of evacuation procedures, and so forth can be made, and some range of overall consequences can be calculated.

Similarly, although in reverse, a 'fault tree analysis' can be done, in which a major failure of a technology (such as a fuel meltdown in a nuclear plant, or, more innocuously, the failure of a car engine to start) can be traced back through different causes to initiating events (Slovic et al., 1980). The failure rate of different parts can be multiplied to give us a calculation of overall probability of this final event.

Using these kinds of analytical tools, analyses of different environmental and health risks from various technological sources were done in the late 1970s. Increasingly, this kind of analysis was used to compare risks from different sources directly. This technique involved reducing all of the consequences of an accident or a technology (the hazard) to a calculated common denominator. For example, if one were to make assumptions about how old an average worker is, or, more accurately, determine the age of workers who die from accidental pesticide poisoning, one could translate those premature deaths into worker-days lost from those persons' working lifetimes. Similarly, one could investigate the increased cancer and other morbidity and mortality rates from pesticides exposure in terms, again, of worker-days lost. Then one could add up these number of worker-days lost from pesticide use. One example of such comparisons comes from a short article by Reissland and Harries (1979) which calculated days of life expectancy lost due to workplace hazards for workers of different ages in different industries. Results from that analysis indicate that, for example, a 40-year-old coal miner will lose 61 days, a fisherman in commercial deep sea fishing will lose 273 days, and a construction worker 18 days. Such a study can give useful information about comparative workplace hazards in different industries, and perhaps identify or confirm suspicions that a particularly dangerous situation exists. When such calculated risks differ dramatically, as between the fisherman and the construction worker, it seems justifiable to at least investigate the situation to see whether an inherently hazardous situation is the cause, or whether safety efforts are as great as they could be. The context − that is, workplace hazards as experienced by workers of the same age – is sufficiently similar to be able to use such calculations to draw straightforward, if tentative, conclusions.

But risk calculations are sometimes compared in other contexts. Sometimes there is no comparable context, and risk calculations are made and compared for a variety of different kinds of situations, such as automobile accidents, smoking, or the risk from alternatives to pesticides. For instance, Wilson (1979) published a table of risks from daily activities, each having a calculated one in a million chance of death. These included the risk from accident when travelling 10 miles by bicycle, 50 miles by car, or 1,000 miles by jet plane, and the risk of cancer from exposure to various chemicals, including benzopyrene from eating 1,000 charcoal broiled steaks, polyvinyl chloride from living next to a polyvinyl chloride plant for 20 years, and chloroform from drinking Miami Florida's water for one year.

It is also possible to go further and to investigate the actual costs of preventing various hazards, and to compare expenditures for saving a life by other means. For example, it is possible to compare the costs and the number of lives saved from having emergency vehicles available in urban centres, by increasing emergency preparedness near a petrochemical plant, and by requiring air bags in new cars. The best option in terms of saving the greatest number of lives for each dollar spent can be directly determined, although – as will be discussed – it doesn't necessarily follow that the best public policy route can be decided on the basis of these sorts of calculations.

These statistical tools for examining technological risks have enormous analytical power because of the sweeping overview they command, and the apparent objectivity they lend to a subject on which perspectives differ widely. To some, this approach appears to offer a rational way to cut through seemingly inconsistent public attitudes and fears. However, this use of risk analysis is technically misleading without examining assumptions and data limitations. It also can be psychologically naive and ethically uninformed.

First of all, it is obvious that individuals can be quite inconsistent in the way they seem to react to different risks. People opposed to aerial spraying in forestry may routinely accept without apparent protest many risks which they themselves would acknowledge are statistically at least as probable, such as risks associated with travel, with natural contaminants such as aflatoxin in peanut butter, or the use of axes and chain saws. Medical doctors smoke and drink. People who would shun a high-risk stock market investment nevertheless buy houses on flood plains below dams, or near active earthquake faults. What are we to make of this behavior in the context of public policy debates on hazardous chemicals?

To begin, there are overriding constraints on time, energy, and knowledge for any individual. We live in a world where practical considerations mean that, for local public hazards, individuals moved to protest do not do so because they believe this is the most important threat in their lives, but perhaps out of a sense that their stake in the issue seems to be greater than anyone else's. In a complex, democratic society, there is an implicit division of labor on public issues: those with a special stake in a problem will get involved in that issue, trusting that others will take the lead when they have the most reason to do so. No single individual can possibly have the time to become a participant in every public debate that takes place.

For many, the determining factor for the personal involvement is often the feeling, which is really an intuitive political conclusion, that the responsible authorities do not have the same incentives to challenge the chemical hazard that they themselves do. Past mistakes and accidents have served to foster this view. Indeed, the reality is that politicians and other stakeholders often have an interest in seeing projects move ahead, without any open assessment of whether they are acceptably safe to those who must live with the consequences of the decision. In a project that will only have negative effects on its immediate neighbors, jobs and tax revenues can still bring benefits to some of the stakeholders. Single-issue protest campaigns are, thus, a realistic response to existing relationships of power and account-ability in society, even if they seem out of proportion to the specific hazard when it is compared with other dangers.

Finally, in this discussion of the 'rational' response to risk, it should be noted that, in fact, it is not reasonable to invest the amount of time required to evaluate accurately every imaginable risk. Instead, we humans routinely use intuitive 'rules' for evaluating most risks. Researchers into the per-ception of risk, such as Slovic et al. (1980), point to a tendency for lay persons in risk assessment (as opposed to risk experts) to build certain values into their response to risk. While for the risk experts, asked to evaluate the 'riskiness' of certain activities, the determining factor was strictly the number of deaths associated with the activity, for lay people other dimen-sions of the hazard, such as catastrophic potential, whether the risk for individuals was a fatal one, and whether the risk was known and under-stood were very important. In fact, lay people seem to attach considerable significance to what these authors refer to as the 'information content' of the risk – that is, what an accident signals about society's knowledge of that risk and ability to control it. Thus, the repeated finding of research that lay people regard a familiar risk as less serious than a new or poorly understood one, is, from this perspective, a very rational response to the need to manage technological risk, not – as it has sometimes been characterized – an irrational fear of the new and the unknown.

One of the aspects of technological risk that has had very little public discussion revolves around questions of ethics. The fundamental problem with all technological risk, including the risk from chemicals, is that the costs and benefits are not equitably distributed. In a 1982 symposium on risk, Kasperson and Kasperson identified four types of inequitable benefit–risk situations that affect whether the risk is considered tolerable. These include (1) workers v. the public; (2) present v. future generations; (3) choice of geographical location (hazardous and undesirable facilities have frequently been located in the backyards of the politically powerless); and (4) particularly severe impacts on specific social groups, such as native peoples, women, or the physically vulnerable, like asthma sufferers.

In a society in which most potentially hazardous public situations involving chemicals also provide benefits for some identifiable group in the private or public sector, it is perhaps not surprising that when legal issues arise about risk, justice is narrowly defined. The right of legal tradition in this country is unquestionably on the side of allowing individuals – which includes business enterprises – to pursue whatever activities they see fit, unless the actions involved can clearly be shown to be harmful. For example, this was the conclusion in the 1983 so-called Nova Scotia 'Herbicide Trial', *Palmer et al.* v. *Stora Kopparberg*, in which a number of Cape Breton landowners attempted to obtain a permanent injunction to prevent a forestry company from spraying land adjacent to their own with the herbicides 2,4-D and 2,4,5-T. The attempt by the landowners failed, despite the fact that these were toxic chemicals and they were to be used near drinking water sources and in area where local people hunted game and gathered berries and other wild foods. The central question in the trial for the judge (it was a civil case and there was no jury) turned on the theoretically calculated magnitude of the health risk. The explicit legal requirement was that the landowners had to prove that the herbicides would harm them. Any question of a prior right, for the landowners or the public, to be free of unwanted health risks (of any magnitude) imposed by a private company for profit simply never arose, despite the fact that those who bore the risk would receive no compensation or benefit.

In other words, the legal decision revolved around whether any probable harm would, in fact, be likely to affect the landowners; the law attempted to assess where the balance of interests lay. This legal framework assumed a moral equivalence between the interests of the landowners in wanting to be free from a risk from which they gained no benefit and the position of the pulp company, which was to continue a profit-making business which incidentally imposed a risk on its neighbors. The judge's view was that the size of the risk was the critical factor in weighing those competing interests. However, other commentators on risk, such as Kasperson and Kasperson (1982) have emphasized equity – that is, an assessment of the fairness of a situation in which there may be winners and losers, and whether there was a fair process of decision-making that provided an opportunity for informed consent and compensation – as the ethical framework for decision-making about risk.

One of the curious features of the 'Herbicide Trial' was that the judge decided that he would try the case himself, without a jury, because he believed the issues were too technical for lay people. This belief – that technical decisions should be left to the experts – characterizes one side of a watershed division in this entire field of toxic chemicals management. Opinion on the other side, toward which decision-making processes are slowly moving in Canada, sees the fact that the questions are profoundly value-laden as their most significant feature. This latter point of view recognizes that even technical assumptions are usually based on judgement, and the analytical framework – that is, the questions that get addressed – is not self-evident, but reflects only certain perspectives.

For instance, scientific research, from the seventeenth century until the first part of the twentieth century, was seen as objective, non-participant observation. Modern physics has made this perspective outdated; there is no such stance in physical terms as non-involvement. As a result of this shifting paradigm concerning the nature of scientific truth, the old assumption that scientists are inherently disinterested and without bias has come under scrutiny and is no longer accepted without question.

Some research into risk perception quite clearly indicates that experts are not superior to lay people in seeing the limitations that the analytic framework imposes on their expert opinions. Slovic *et al.* (1980) describe a study in which both lay persons and professional automobile mechanics were asked to rate the completeness of a fault tree analysis showing various causes for a car's not starting. The professional mechanics proved no more able than lay people to recognize that major branches of the fault tree had been deleted. Similarly, studies of the 'decision frame' – that is, the context and manner in which a question is presented – show both lay people and experts will give contradictory responses to identical hypothetical situations. Psychologists Tversky and Kahneman (1981) have done extensive research on this phenomenon; they report that different groups of respondents consistently will reverse their preference depending on the formulation of the problem. A typical example is that most people will favour the total elimination of a minor hazard rather than partial reduction of an equivalent number of deaths from a much more common danger. Similarly, most people will accept a negative outcome, such as a financial loss, more readily if it is described as a cost rather than a direct loss.

How public policy questions about the risks and the management of toxic chemicals are framed, therefore, becomes ethically significant. Public opinion can be manipulated merely by the way the issues are described or reported.

This new awareness of there being no 'objectively correct' perspective on issues involving hazardous chemicals had led to the development of an approach to the ethical questions through an indirect avenue, namely that of process. The kernel of the argument is that, since there is no 'right' answer, only a variety of interests and perspectives, the overriding ethical concern is that the decision-making processes must be open to all of these interests and be accepted by them as fair. Only in this context will it be possible to thoroughly canvass the thorny issues of differential costs and benefits. In fact, the discussion of ethical principles to apply to decisions about technological risks is increasingly subsumed in the discussion of what constitutes fair procedure, not in a legalistic sense but in a spirit of genuine ethical inquiry.

Developments in the decision-making processes concerning toxic chemicals

The most important development in recent years in toxic chemical decision-making is the inclusion of more stakeholders in these processes. Development of standards, regulations, and policy was for many years viewed as a technocratic process involving scientists and engineers in government agencies and, at most, scientists and engineers from the industries that were to be regulated. However, the recognition that scientific information is only one element to be considered is now fairly widely accepted. There are, though, variations in the degree and the type of involvement of other groups or stakeholders in the decision-making process. In Great Britain, the processes are often still private and informal. In the United States, public processes were opened at a reasonably early period, but all participation was structured in an extremely legalistic framework. In Canada, it was only in the mid–1980s that any regular protocols for public consultation were developed, but this work has been quite innovative.

In the early 1970s, environmental groups began calling for more public input, and at an earlier stage, into all decisions affecting the environment. Greater access to information, and publicly provided resources for participation in these processes were also considered essential if the consultation was to be meaningful. Some progress had been made toward more inclusive decision-making by the 1980s, but it was not until the federal Department of Environment in Canada initiated a major project aimed at developing consultation processes on toxic chemicals management that significant changes began to take place. The project was initiated in 1985, and it brought together working groups of different stakeholders in toxic chemicals: industry, labor, both levels of government, consumers, and environmental groups. The participants together developed protocols for consultation processes, and also set up working groups in several areas which were identified as topics on which common ground should be explored. These latter included environmental data reporting, a model framework for the management of toxic chemicals (Environment Canada, 1986), and – what proved to be the biggest project of all – a multi-stakeholder consultation process which ultimately resulted in the tabling in Parliament and promulgation of a new federal Environmental Protection Act, which replaces the old Environmental Contaminants Act (Environment Canada and Health and Welfare Canada 1986).

Other developments in the political processes relating to toxic chemicals also took place around this time. A new Pest Management Advisory Board, developed to provide an avenue for public input into the registration of pesticides (which is Agriculture Canada's responsibility) was established (Pest Management Advisory Board, 1987). In the province of Ontario, the Ontario Waste Management Corporation, after an early and disastrous failure to incorporate public input into its first attempt to site a toxic waste treatment facility, began serious efforts to incorporate citizen participation in its decisions. A similar consultative process was promoted in the United States by Ruckelshaus (1983), then administrative head of the Environmental Protection Agency, for the locally affected public in decisions made about risk levels associated with a steel making facility.

All of these efforts have been criticized for various shortcomings in the processes that were being developed; criticisms have included too-short time frames, too-restrictive an approach to policy options, and an undemocratic, *ad hoc* approach to the selection of participants. Some of these failures will be addressed in future processes through pressure brought to bear by environmental and other stakeholder groups. However, it is unquestionably true that fundamental principles have been articulated and largely accepted as a result of these developments. These principles include the need and the right for all stakeholders to participate as much as possible as equals in processes aimed at building a consensus about the course of action to be taken, whether that is new legislation, regulations, the use of a controversial chemical, or a siting decision. Resources need to be provided to some stakeholders to participate, and it is also recognized that other processes, such as formal public hearings and the legal system, will also be appropriate in some situations.

While there is still a long way to go before these principles are fully implemented in all decisions about environmental issues, Canada has come a long way in the development of what amounts to a significant extension of the democratic process. It is interesting that the impetus for this has primarily been the need to cope with toxic chemicals.

References

Bird, P. M., and Rapport, D. J. (1986), *State of the Environment Report for Canada* (Ottawa: Environment Canada and Statistics Canada).

Bhopal Aftermath Review Steering Committee (1986), *Bhopal Aftermath Review: An Assessment of the Canadian Situation* (Ottawa: Environment Canada).

Environment Canada (1986), *From Cradle to Grave: A Management Approach to Chemicals*. Report of a task force representing industry, governments, labor, environmental groups, and consumers. (Ottawa: Environment Canada).

Environment Canada and Health and Welfare Canada (1986), Final Report of the Environmental Contaminants Act Amendments Consultative Committee (Ottawa).

Farmer, F. (1982), 'Limitations and usefulness of methods for predicting risk of rare or unprecedented events', *Risk: proceedings of a symposium on the assessment and perception of risk to human health in Canada* (Ottawa: The Royal Society of Canada).

Feldman, M. (1986), 'Rude awakening for Canning', Ecology Action Centre's *Between the Issues*, July, vol. IV, no. 9.

Jones, D. (1986), 'Lessons from a chemical fire', *Atlantic Issues* September.

Kasperson, R. E., and Kasperson, J. X. (1982), 'Determining the acceptability of risk: Ethical and policy issues', *Risk: proceedings of a symposium on the assessment and perception of risk to human health in Canada* (Ottawa: The Royal Society of Canada).

Palmer et al. v. Stora Kopparberg (1983), NSR (2d), Nunn, J.

Pest Management Advisory Board (1987), *Update of Activities* (Ottawa).

Potter, F. (1987), *The Canning Fire*. Air Pollution Control Association, Atlantic Chapter, Technical Meeting, Moncton, New Brunswick, May 6 and 7.

Reissland, J., and Harries, V. (1979), 'A scale for measuring risks', *New Scientist*, September 13, vol. 83, no. 1172.

Ruckelshaus, W. D. (1983), 'Science, risk and public policy', *Science* 221, pp. 1026–8.

Slovic, P., Fischoff, B., and Lichtenstein, S. (1980), *Societal Risk Assessment: How Safe Is Safe Enough?* (New York: Plenum Press).

Tversky, A. and Kahneman, D. (1981), 'The framing of decisions and the psychology of choice', *Science* January 30, vol. 211, no. 4481.

Wilson, R. (1979), 'Analyzing the daily risks of life', *Technology Review*, February, vol. 81, no. 4.

11 *The comparative political economy of regulating chemicals*

B. DOERN

Introduction

From the outset, it is essential to appreciate that the regulation of any particular chemical or hazard is inevitably only a sub-set of a larger universe of problems and dynamics involved in regulating hundreds of substances and products where health and safety and risk–benefit issues are involved. The areas of regulation track the full cycle of production (often with separate governmental agencies) namely, the environment, the work-place, transportation systems, and export and import control regimes. There now exists an extensive literature that directly addresses key issues about regulation of this kind (Bardach and Kagan, 1982; Burton and McCullough, 1983; Lowrance, 1976; Wilson, 1980). Such analysis exists under various labels which are themselves indicative of the breadth of the concerns and variables involved. Thus, issues are examined under the general umbrella concept of social or health and safety regulation, or under such phrases as the regulation of hazardous substances or risk analysis. This literature has examined both the substance and processes involved in regulating substances or phenomena as diverse as nuclear power, acid rain, lead, asbestos, pesticides and herbicides as well as the safety of consumer products such as toys and automobiles.

One way for the reader to visualize the difficulties and scope of the regulatory terrain is to imagine that you have just been appointed the new 'minister of hazardous substances', a new super department combining environmental, occupational, consumer, transportation, and health and safety issues. Assume that you will probably be minister for only three or four years. What priority hazards would you tackle first? According to what criteria? Which stakeholders would you consult with? Which interests would you avoid or rank as less important? As the super-health and safety regulatory police officer, how tough will your compliance regime be? What kind of mixture between financial incentives and penalties will you apply? On which interests will you come down hard? For which interests will you reserve acts of gentle persuasion? To pose such questions is to enter the real political economy of regulating chemicals and hazardous substances. While each particular hazard has some unique regulatory features, the purpose of this chapter is to highlight the larger set of issues, albeit in a summary way. It focuses on eight issues which influence the development of both reasonable normative criteria for judging whether an appropriate regulatory system exists and/or can be established, and for thinking about how political and economic interests actually behave or tend to behave in regulatory settings. The eight issues are: the regulatory cycle and contending views of regulatory speed and lag; the structure of markets and the political economy of interests engaged in seeking and resisting regulation; the intergovernmental regulatory relationships within a particular level of government; the connections between elected politicians and the mass media, and the exercise of symbolic gestures and statements of concern (so-called symbolic politics); the compliance and monitoring of actual regulatory decisions including the mix of implementation activities carried out by officials on the one hand and private interests on the other; the norms and strictures governing the use of information in the regulatory process; and the role of scientific and technical controversy at all stages of the regulatory process.

This brief commentary on these issues is derived from an extensive literature, but since this chapter illustrates these issues in the context of several Western countries, some general points must be made about the current state of literature on the comparative politics of health and safety regulation. First, there are still very few studies which are genuinely comparative or researched in such a way that one can fully answer the question 'what factors explain different or converging regulatory outcomes in different countries?'. Part of the difficulty lies in the various kinds of dependent variables that might be studied. As Hobert (1986) has pointed out, the dependent variables could be: policy consequences (e.g. actual indices of air or water quality); policy content or the formal characteristics of government regulations (e.g. administrative rule, court order), and regulatory actions and instruments (e.g. allowable exposure limits, restricted use, etc.). Another part of the difficulty lies in simply obtaining good comparable data not only on these various dependent variables but also on the independent explanatory ones. Regulatory data do not show up as conveniently packaged as, for example, social welfare spending data. Nor have they been studied for as long a period of time.

A second feature of this comparative work, however, is that the few studies that exist point to a similar overall conclusion. This is that regulatory outcomes are very similar across countries but that the political processes that yield these outcomes show persistent differences. In particular there are differences between the United States on one hand, and the European countries, Canada and Japan, on the other. Hobert (1986) concludes, 'The United States has a distinctive process which is more formal, legalistic, open and adversarial. The European countries and Japan ... are less formal and open and marked by much more cooperation between business and government.' (See also Kelman, 1981; Vogel, 1986.) More specifically, the process in the United States is characterized by detailed legislation and safety standards implemented by very formal rule-making procedures and involving extensive economic analysis and scientific documentation. Its open and combative processes are starkly contrasted with European countries. There the preference, broadly speaking, is for enabling legislation that is general and abbreviated. Procedures are far less formal, participation more confined, and the mode of operation is to stress informal negotiation and compromise with key interests (Ilgen, 1985). There is, however, little consensus in these studies on what then explains the convergence on policy outcomes. These range from the imperatives of technology, the values of regulators, the leavening effect of international organizations and international co-operation in science, and the nature of international markets and corporate power.

As the eight issues below are examined it is essential, on the one hand, that they be put in this broad comparative context. On the other hand, the issues themselves, though drawn from comparative work, none the less have varying degrees of relevance to particular countries and should be looked upon as a mental checklist for anyone who seriously wishes to understand the political economy of social regulation.

The regulatory cycle and contending views of regulatory speed and lag

The full regulatory cycle is not always what it seems to be. As artificially discrete stages, the cycle can be seen to consist of: an initiating stage where, as in the case of pesticides, a company wishes to seek approval for the registration of a particular product and submits data and documentation to the regulatory body or bodies; an actual regulation-making stage during which rules of behavior, standards, or specific approvals are given after a process that may consist of varying kinds of public and private/secretive deliberations; and an implementation/compliance and monitoring stage where, through various actions, actual use of the product occurs and where one can hopefully determine whether behavior has changed in intended ways and expected results actually occur. Each of these stages is replete with situations in which contending views arise as to how swift or how just and thorough the decision processes are.

The movement from stage to stage in the ideal regulatory cycle is not a straightforward predictable path. In essence, the next seven issues elaborate on particular manifestations of why it is more a maze than a cycle. But prior to examining these issues one must emphasize two realities. The first is simply to stress the importance of time and space. The first two phases of the above cycle may involve a handful of key decisions that span, within the firm or the regulatory agency, a period of years. They occur in the capital city or in large urban settings where production occurs, often outside the country concerned. However, the latter phases of the cycle may involve hundreds of micro-level decisions over a vastly greater time period in locations far removed from the political capitals, from Cabinet ministers and from people who occupy the top rungs of public and private officialdom. By micro-level decisions, we mean discrete decisions involving particular companies, particular inspections, or particular applications of a chemical in a given location.

The second overriding reality to stress about 'the cycle' is that simultaneously there are numerous separate and related regulatory cycles underway, not only within a complex field such as nuclear power or pesticides, but also in the successively larger health and safety regulation in general (Doern, 1982; Douglas and Wildavsky, 1982; Dunlop, 1981). Many of the same firms which are subject to expanded regulation on the social side of the political equations are being simultaneously subjected to, or affected by, deregulation on the economic side. And even this extended world may be dwarfed by a government's even larger sense of where its overall economic and social priorities lie. Over any medium term period (e.g. five years), the place of a particular chemical hazard or issue (e.g. acid rain) is fundamentally effected by the pecking order of priorities. The sluggishness of the regulatory process may not be due to any structural fault but rather may be simply a product of the low rung it occupies on the greasy priority ladder.

Both of these realities increase the already high probability that contending views will occur as to what constitutes regulatory speed versus regulatory lag. One group's regulatory lag is often another group's regulatory justice. Competing interests can and do call for both speed and slowness in the name of regulatory justice, democratic fairness and even efficiency.

Markets and the political economy of interests seeking and resisting regulation

While regulation is purposeful in that it seeks to arrange a more appropriate balance of public and private costs and benefits than would otherwise pertain under free market conditions, it is also driven by the forces of political and economic self-interest (Hartle, 1979). In Canada, for example, chemical companies wanting to introduce a new herbicide for registration and use argued that they were deterred by the costs of doing so because their markets in Canada were marginal and Canadian regulators were allegedly slow. In this instance the company wanted the regulation of its products carried out expeditiously. That same company, however, when it comes to dealing with how its own production processes are regulated for broader environmental purposes may well want regulators to 'go slow' or 'be reasonable'.

Regulatory literature also suggests that it is not uncommon for regulators and the regulated to reach compromises in such a way as to produce regulations that are tough sounding as statutory statements or regulation but that are soft or at least flexible when it comes to enforcement (Hartle, 1979; Doern *et al.*, 1982). Some regulation is obviously adversarial but the self-interest of regulators also exerts pressure since to a certain extent everyone likes to lead a normal quiet life and have a sense of predictability in what they are doing. Moreover, just because regulations have been approved and duly promulgated does not necessarily mean that the government then provides the regulatory body with the necessary personnel and resources to actually enforce and monitor.

The political economy of regulation also shows that there is significant imbalance in the capacity of organized interests to participate continuously at all stages and through time. Producer groups and regulatory agencies themselves have a more permanent set of incentives that compels their involvement. Meanwhile, broader-based groups, so-called collective rights associations, such as consumers or environmentalists must confront the 'free-rider' problem. In brief, people can benefit from the activity of such associations without having to join. They obtain a free ride. Such associations, therefore, have difficulty building up a large membership. This fundamentally weakens the capacity of such groups to finance themselves and to be a continuous 'third party' presence in all phases of regulation, including compliance (Doern *et al.*, 1982; Hartle, 1979).

Each of the above points also testifies to the need to differentiate *interest groups* from *interests* if one is to understand regulation and compliance. The former are aggregations of individual members who possess diffuse powers to influence, lobby and publicize their positions. The latter are defined here to be particular entities such as individual large companies and also individual governments (particular states, provinces, or cities) which possess far more concrete powers to invest, disinvest, spend, tax or regulate. They possess actual capabilities to act and, moreover, usually have their own direct channels of access to ministers and officials.

Thus, it is insufficient simply to try to understand, let alone reform regulation, only through broad categories such as 'business-government' or 'intergovernmental relations'. In the chemical regulatory setting the following interests and interest groups have to be differentiated:

- chemical producers (domestic or foreign owned);
- large farms, forestry companies or other chemical users versus small owners, farmers, woodlot owners;
- regulatory officials in senior civil service positions versus other technical and scientific professionals in less senior positions, and in universities and companies (i.e. the relevant technical community in general);

- unionized versus non-unionized workers;
- broad environmental coalition groups versus more particular groups such as fish and game associations;
- the federal, national and individual states, provincial or local governments as separate interests.

The literature shows that one must be as analytically sophisticated about the number and nature of interests as one is about the particular chemical characteristics of individual hazards.

Leonard's analysis of industrial pollution in rapidly industrializing countries bears out this need and highlights, moreover, the international political economy of regulatory relations and decisions (Leonard, 1985). He concludes, for example, that there is no substantive evidence that the world is being divided into core countries that export industrial polluters and peripheral countries that accept these rejected industries. He also concludes that in poorer but developed countries such as Spain, pollution control standards are more stringent for foreign owned companies than for domestic firms. In countries such as Ireland, Spain and Mexico environmental policy is closely connected to regional industrial local policies. The Irish Industrial Development Authority (IDA), for example, has recognized the effect of dispersing different industries to particular areas. To disperse the chemical industry the IDA has, by incentives and persuasion, guided pharmaceutical companies with medium-sized plants and relatively little water effluent to sites near villages and inland streams. These decisions involve specific company by company negotiations, international sensitivities and knowledge of production techniques and effluent and absorptive capacities in specific locations.

Regulation and federalism

It is axiomatic to state that federalism, or more generically – intergovernmental relations – matters a great deal in health and safety regulation. In older federal countries such as Canada and the United States the division of powers between levels of government formally pre-dates the emergence of the 'health and safety regulatory state'. Jurisdictions are not always clearly demarcated in addition because hazards traverse numerous other sectors such as resources, manufacturing and transportation. For example, in Canada, mines and forests are clearly in provincial jurisdiction but the federal government has a decisive role not only as a land owner but also in all the intersecting ways in which its legislation and policies cut across these resource areas in fields such as pesticide legislation, environmental health and fishery legislation. In the agricultural field, moreover, the constitution provides for concurrent powers.

Federalism, interacting with the diversity of a country's regional economies (including the varying dependencies of each state or province on a particular industry or resource) produces a mixture of both co-operation and conflict. For example, in Canada, there is a tendency in the overall area of health and safety regulation for the provinces to rely on federal research since the latter has more resources to draw on. At the same time, in recent years, federal expenditure cuts have occurred in this very same field. A second intergovernmental feature occurs in that the federal government in some key areas essentially depends on the provinces to enforce or ensure the compliance and monitoring of regulation. But provincial compliance personnel are not numerous and hence, as discussed further below, there is a thinness to the entire compliance underbelly of regulation. In the research field, the direction of dependence is from the provinces to the federal government. In the compliance area, the direction of dependence tends to be reversed.

Conflicts occur and differences of opinion arise between jurisdictions and among junior level governments at each phase of the regulatory process. Yet, at the same time both levels of government know that they are dependent upon one another in very significant ways. Local governments also play a role since, on issues such as local aerial spraying or polluted beaches, they may have a legal responsibility conferred on them by the state or province or they may simply, in political terms, be the level of government at which political protest is focused or first arises.

For private interests, interest groups, and the general public, the knowledge that numerous state or national governments are taking care of the public interest is not always reassuring and for good reasons. Both the conflicts and even the areas of intergovernmental dependency can yield a mixture of real and perceived regulatory sins. They can produce extra decision stages that become hurdles. But, equally, they can also produce unseemly speed when a minister decides to ban the use of a product, or fails to force full compliance on a firm that is vital to the economy. It is essential to remember in this regard that speedy decision-making can be evident both through overt action and through inaction (or as decision theorists sometimes call it, 'non-decisions') (Edelman, 1971; Ham and Hill 1984).

Regulation and interdepartmental relations

The government at either level, but particularly at the national level, is not a homogeneous entity that always speaks or acts clearly in a single-minded way (Doern and Aucoin, 1979). Because several statutes impact directly or indirectly on the broad field of risk–benefit regulation, each embodying different mixtures of values and priorities, it follows that the departments and agencies and, therefore, the bureaucrats who administer laws also engage in a set of relations that produce both conflict and mutual dependence. Every country has an analogous division of statutory and bureaucratic labor to that listed below for the Canadian situation:

Department of Health and Welfare
Food and Drug Act
Radiation Protection Devices Act

Department of the Environment
Fisheries Act
Canada Water Act
Canadian Environmental Protection Act

Department of Transport
Shipping Act
Transport of Dangerous Goods Act
Motor Vehicle Act

Department of Consumer and Corporate Affairs
Hazardous Products Act

Department of Fisheries and Oceans
Fisheries Act
Fish Inspection Act

Department of Labour
Labour Code

Atomic Energy Control Board
Atomic Energy Control Act

Thus, some care must be exercised when trying to differentiate genuinely perverse bureaucratic behavior, when organizations and individuals within them run amok, from intended interagency sparring which cannot help but be part of a process in which different values and mandates are enshrined in statutes. This refers not just to such obvious agency boundary overlaps inherent in the very words, 'forestry', 'fisheries', 'environment', or 'agriculture', but also to values enshrined in such concurrently existing statutes that deal with commercial privilege, freedom of information, and secrecy laws. These too contribute to different senses of regulatory speed, lag and justice.

Ministers, symbolic politics, public opinion and the mass media

While all of the above can be seen to be a part of the political economy of regulation defined broadly, one must reserve special analytical space for the particular brand of politics that is induced by the short-term pressures that arise from the interplay of relationships among politicians, the mass media, real or perceived public opinion and the parliamentary or legislative opposition parties.

Since political careers can be made, side-tracked or broken depending on how these pressures are handled (on both big and often very small issues), assessment of the regulatory process cannot avoid this issue. At the same time there is not much that one can do about it. We have already indicated that companies may not want to promote products in marginal market situations if such promotion would only create adverse political publicity that threatens their main product markets. The media's short-term attention span and frequent 'villains and heroes' approach to coverage of such issues can easily produce enormous pressure on a responsible Cabinet minister either to take precipitous action or to simply say 'something' that precludes, or makes more difficult, future actions that would be much more sensible and acceptable.

Research on public opinion suggests that there is a strong underlying body of support for environmental and health issues among voters. This reservoir of support is easily tapped or at least triggered by particular controversies that may arise about particular hazards, real or perceived. In such situations media pressure often echoes public sentiment and the minister's own party caucus and Cabinet colleagues may add to the short-term pressure.

As in other dimensions of risk–benefit regulation, it is important again to stress that the pattern of behavior here is not at all uniform. Media and opposition political pressure can also produce inaction in the form of announcements to carry out 'further research' or, much more quietly, delays in, or a softening of enforcement.

Compliance and monitoring: the thin line of implementation

The actual implementation of regulations and approvals is composed of a range of activities which include the study and monitoring of effects as well as procedural compliance. With respect to effects, implementation involves studies to determine if nearby streams were adversely affected by pesticide use, or if emission levels were actually adhered to. Procedural implementation involves decidedly day-to-day things such as whether applicators in far-flung locations actually are properly trained and are applying the chemical in keeping with health and safety precautions.

Implementation, thus, involves a range of public or official behavior at headquarters and in the field but also extensive amounts of private behavior and actions (Doern et al., 1982; Dunsire, 1978; Ham and Hill, 1984). The literature on health and safety regulation none the less suggests the need to be aware of what can best be called 'the thin line' implementation. This is not a field of public policy where armies of inspectors patrol to see that things are being done properly. Private firms obviously have people involved and so do the main government agencies. But the overall thinness is stark. This is also where the inherent capacities of outside interest groups to be persistent monitors or 'third party presences' is important but quite unbalanced. This is due to the 'free rider' reasons already mentioned as well as to the vast number and sometimes the remoteness of sites.

Practical features of the physical situation in which one is operating loom large in the implementation equation. For example, in agriculture, pesticides of many kinds are used by literally thousands of farmers and farm workers every year in many thousands of locations. The scope of licensing, and the need to be in thousands of places simultaneously imposes practical limits on compliance. This situation contrasts with forestry. Forestry involves far fewer uses of herbicides and the applications are far less frequent, often one or two applications over the 80 or 100-year rotation period at any given site. Forest applications involve vast stretches of quite remote and inaccessible territory. These physical attributes quickly affect the inherent political attributes of different sectors subject to health and safety regulation and compliance. Farmers are more numerous, constitute a larger block of voters, and in practical terms, politicians may simply not want to subject such an important political constituency to unnecessary regulatory red-tape or heavy handedness. Forest lands, on the other hand, can easily evoke both an emotional and substantively real attachment as a recreational, aesthetic and environmental legacy. This can produce situations where herbicide application on the forests writ large are the subject of intense protest but with regard to agricultural applications or even more immediate applications in a neighbor's garden under the most lax of standards, there is nary a whimper.

Information and the regulatory process: concealing and revealing

As mentioned earlier, there are often conflicting strictures and values that govern the use, release and protection of information in the regulatory process. Some notions of regulatory justice involve the general need for openness. Statutes may or may not require open hearings or meetings. Freedom of information statutes are intended to allow greater access to documents but numerous exceptions exist such as in the conduct of international relations, commercial privilege, or conflicts with legislation on rights to privacy. Countries vary significantly on their inherent commitment to openness. The United States is regarded as the most open but also the most adversarial system.

In many Western countries, junior level governments have often protested that they cannot get full access to testing data submitted by chemical companies to the national government. Commercial privilege is the reason. Only summaries of data may be circulated – and even then only if all parties can agree on what constitutes suitable levels of aggregation in the summary of data. Environmental groups often seek access to the government's reports on such data or on other government views but with very mixed success. In the meantime, governments release official information in annual reports and speeches which artfully combines some facts with skilled obfuscation. Often this selectivity cuts to the core of essential questions. For example, a Canadian province may assert as a matter of policy that every tree that is cut is replaced by a new seedling planted. One is never told, however, if the seedling actually lived or, in short, if real reforestation is actually occurring.

Science and technology in regulation

Last but hardly least among the issues surveyed in this chapter is the role played by science and technology, particularly in the domain of risk–benefit or health and safety regulation (Doern, 1981; Grabowski and Vernon, 1983; Huber, 1983). Since this is examined in detail in other parts of this book, we simply note that several dynamics are in operation here. These include: the actual technical properties of the particular substance being used and regulated; the degree (if any) of peer group controversy, of a technical and scientific kind, as to environmental, human health, and occupational effects; the role of research and development in the actual regulatory process including the mechanisms used (if any) to explicitly air technical controversies; and the role of technical innovation such as in whether new methods of application are available (e.g. better methods of aerial or ground application).

References

Bachrach, P. and Baratz, M. S. (1963), 'Decisions and non–decisions: an analytical framework', *American Political Science Review* 57.

Bardach, E. and Kagan. R. A. (1982), *Going by the Book: The Problem of Regulatory Unreasonableness* (Philadelphia: Temple University Press).

Burton, I. and McCullough, R. (eds) (1983), *Living with Risk: Environmental Risk Management in Canada* (Toronto: University of Toronto Institute for Environmental Studies).

Doern, G. B. (1981), *The Peripheral Nature of Scientific and Technological Controversy in Federal Policy Formation* (Ottawa: Science Council of Canada).

Doern, G. B. (1982), *The Politics of Risk: The Identification of Toxic and Other Hazardous Substances in Canada*. (Toronto: Royal Commission on Asbestos, Queen's Printer).

Doern, G. B. and Aucoin, P. (eds) (1979) *Public Policy in Canada: Organization, Process and Management* (Toronto: Macmillan of Canada).

Doern, G. B., Prince, M. and McNaughton, G. (1982), *Living with Contradictions: Health and Safety Regulation and Implementation in Canada* (Toronto: Royal Commission on Asbestos, Queen's Printer).

Douglas, M. and Wildavsky, A, (1982), *Risk and Cultures: An Essay on the Selection of Technological and Environmental Dangers)* (Berkeley: University of California Press).

Dunlop, T. R. (1981), DDT: *Scientists, Citizens and Public Policy* (Princeton, NJ:

Princeton University Press).

Dunsire, A. (1978) *Implementation in a Bureaucracy* (Oxford: Martin Robertson).

Edelman, M. (1971), *Politics as Symbolic Action* (Chicago: Markham).

Grabowski, H. G. and Vernon, J. M. (1983) *The Regulation of Pharmaceuticals: Balancing the Benefits and Risks* (Washington: American Enterprise Institute.)

Ham, C. and Hill, M. (1984), *The Policy Process in the Modern Capitalist State* (Brighton: Wheatsheaf Books).

Hartle, D. (1979), *Public Policy Decision-making and Regulation* (Toronto: Institute for Research on Public Policy).

Hobert, G. Jr (1986), 'Technology, political structure and social regulation: a cross national analysis; *Comparative Politics*, April, pp. 357–75.

Huber, P. (1983), 'Exorcists vs. gatekeepers in risk regulation', *Regulation* Nov–Dec, pp. 23–32.

Ilgen, J. L. (1985), 'Between Europe and America, Ottawa and the Provinces: regulating toxic substances in Canada', *Canadian Public Policy* XI, no. 3, pp. 578–90.

Kelman, S. (1981), *Regulating America, Regulating Sweden: A Comparative Study of Occupational Safety and Health* (Cambridge, Mass.: MIT Press).

Leonard, H. J. (1985), 'Confronting industrial pollution in rapidly industrializing countries: myths, pitfalls, and opportunities', *Ecology Law Quarterly*. 12, pp. 779–816.

Lowrance, W. T. (1976), *Of Acceptable Risk: Science and the Determination of Safety* (Los Altos: William Kaufman).

Vogel, D. (1986), *National Styles of Regulation: Environmental Policy in Great Britain and the United States* (Ithaca, NY: Cornell University Press).

Wilson, J. Q. (ed.) (1980), *The Politics of Regulation* (New York: Basic Books).

12 *Managing toxic chemicals: a government viewpoint*

E. SOMERS

Introduction

Recent years have witnessed increased public concern over the actual and potential health risks that can be ascribed to chemicals. The remarkable achievements in chemical technology in the last fifty years, the proliferation of chemicals entering the market, the large-scale manufacture of chemical products, the worldwide dissemination of their production, have all significantly increased the number of people – whether workers, consumers, or the general public – exposed to health risks. The more dramatic accidents, such as those in Seveso, Mississauga, Bhopal and Basel, with their attendant direct costs to human health and the environment and indirect costs, e.g. to business confidence, research investment, technology transfer and international relations, have focused the world's attention on the problems of the production, transportation, use and disposal of chemicals.

The range of chemicals available in common use is usually estimated as between 60,000 and 70,000 of which some 3,000 account for almost 90 per cent of the total volume by weight (IRPTC, 1983). A report of the US National Academy of Sciences (NAS, 1984) considers that there are 65,725 substances to which humans may be exposed. When drugs and cosmetics are deleted, this leaves 60,500 chemicals. The rate of growth of the world's production of organic chemicals is estimated to have doubled every seven or eight years since 1950 to some hundreds of millions of tonnes per annum today. The total annual production value of chemicals is probably over $500 billion, with more than $200 billion being traded between nations. The National Academy of Sciences study concluded that for the vast majority of these chemicals there are insufficient data to make even a partial assessment of the impact on human health and natural environment. For 70 per cent of all chemicals, no toxicity information at all was available and even for drugs, the most completely tested category, available toxicity data to make a complete health hazard assessment was only available for 18 per cent. Information on the extent of human exposure was sparse for most substances.

All of this presents a daunting picture yet government and national

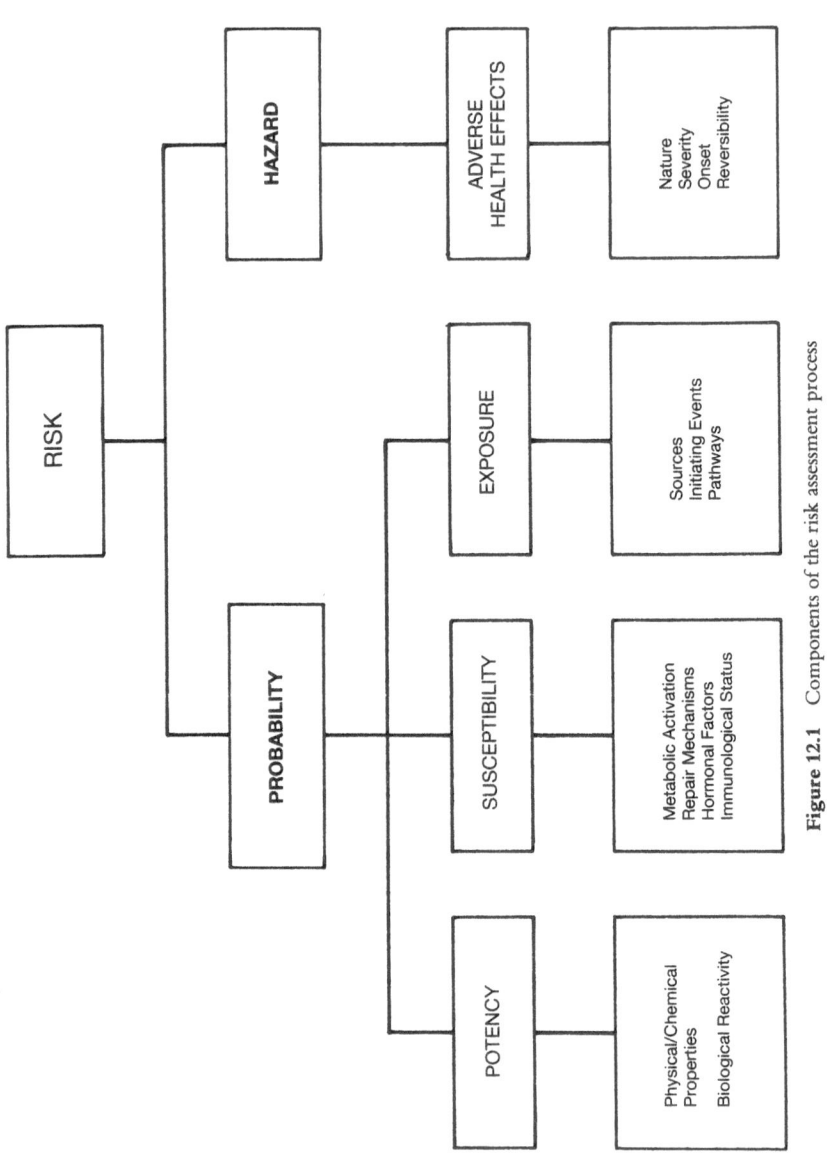

Figure 12.1 Components of the risk assessment process

agencies have been able to develop effective systems for managing the risks from many toxic chemicals in spite of what, at first glance, appears to be an insuperable problem. While the effectiveness of these assessment and decision-making systems will remain in doubt for some time, many countries such as those belonging to the OECD are at least putting more rational approaches into operation. Risk assessment and management are central to these approaches. To illustrate and develop the theme of risk management, it is proposed to consider the elements of risk and its management; comparisons of risk; risk acceptability; risk–benefit; and finally the management of strategies available to control risk. In this way, the particular hazards raised by environmental chemicals can be put in the wider context of health and environmental policy.

Elements of risk and its management

The risk presented by a chemical depends on the nature of the hazard and the probability of its occurrence, as shown in Figure 12.1. Health hazards may be characterized in terms of specific adverse effects which will be affected by such factors as their nature, the age of onset, and their reversibility. The probability of a given effect will depend on the potency of the toxicant, the level of exposure and the susceptibility of the exposed individual. ·

Models of risk assessment and management largely derive from a formulation presented by Whyte and Burton (1980) whereby three stages – risk identification, risk estimation, and risk evaluation – are recognized. The last component is concerned with comparing risks against one another, against benefits, and with a judgement on the social acceptability of the risks. This scheme clearly distinguishes the scientific determination of risk, i.e. identification and estimation, from the ultimate political judgement on the acceptability of risk and its consequences. Finally, we move to risk management. Since 1980, a number of other models for risk assessment and management have been formulated. However, although more elaborate and with some differences in terminology, they all have the basic four elements of identification, estimation, evaluation and management (Krewski and Birkwood, 1987).

In Canada, an interdepartmental group representing the federal departments with major responsibility for the regulation of toxic chemicals developed the risk assessment and risk management model shown in Figure 12.2. The first step simply involves the recognition that a hazard exists; this can be developed from epidemiology, animal experimentation, *in vitro* testing, or structure-activity analysis. However, experience has shown that for environmental chemicals, epidemiology has been used more to support public health decisions rather than to initiate them, although a number of important regulatory standards such as those for nitrogen oxides, arsenic

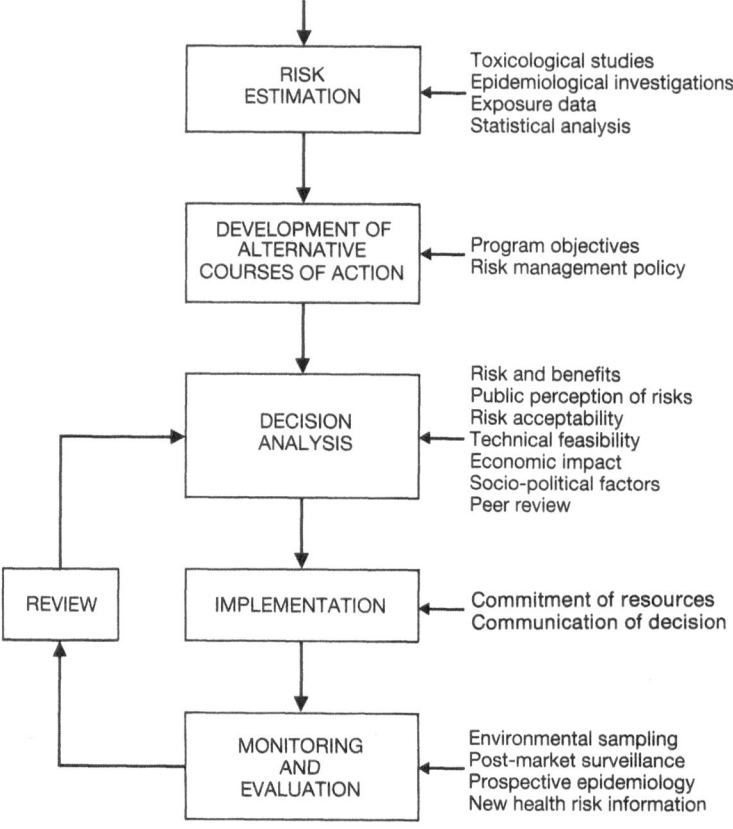

Figure 12.2 A risk assessment and management model

trioxide, and asbestos have been derived from epidemiology (Somers, 1986).

The next step is to estimate the magnitude of the risk. This may require statistical analysis of the epidemiological and toxicological data or determination of the acceptable daily intake from animal experiments. There is considerable and unresolvable uncertainty in extrapolation from high doses to low doses and in comparing animal responses to those of humans (Somers, 1984). In addition, exposure data is often inadequate or, at best, partial.

The development of alternative courses of action is required to manage the risk from a toxic chemical and, as will be discussed later, the options range from education to regulatory. Socio-economic, cultural, and political factors must come into play before a final decision is made. Implementation of a a risk management strategy will usually require a commitment of resources, and ideally requires communication to all those affected.

When the control mechanism is in place, continued monitoring is required. Exposure levels can be estimated by environmental sampling and dietary patterns can be determined by post-market surveillance programs. Prospective epidemiology may be used to estimate the effects on human health. All this information may suggest a re-evaluation or modification of the risk management strategy; in reality, risk management is an iterative process.

Comparisons of risk

When the risk from a toxic chemical has been estimated, however hedged by qualification, one needs to put that risk into a comparative context in order to develop the proper management options. The risk to health could be a 10^{-6} lifetime risk of cancer in a particular organ; it could be a 10^{-5} annual risk of death, or a 10^{-4} risk of teratogenic damage to a developing embryo. To gain an appreciation of what these risks mean in real life, it is instructive to note that annual risk of being killed by lightning in Canada is 2×10^{-7}: a one in a five million risk (Krewski *et al.*, 1988). The risk to males is three times that for females, presumably related to the former's greater involvement in such sports as golf!

As long as the risk can be quantified in terms of exposure to the chemical, it can be compared to some other risk to aid the decision-making. By doing so we achieve a sense of perspective. Table 12.1 gives some examples of this exercise. A useful separation in comparing risks is to consider the risks in three ways: elevated risk, in terms of natural background; alternative risks; and other risks (Whyte and Burton, 1980). Comparison to the natural background can suggest that we have tolerated low-level exposure to certain chemicals without apparent ill effect, and as such it can also put the exposure in context. Mercury in fish and fluoride in drinking water, for example, have always been with us but obviously the potential for increased exposure does exist. Mercury has been used in industrial processes and fungicides. Fluoride is emitted from industries and is used in consumer products. The introduction of lead as a gasoline additive increased the exposure of urban populations to the chemical, but human populations have always been exposed to lead. The best example of an elevated comparison to a natural background is that of ionizing radiation. Table 12.2 provides a useful comparison of radiation doses from a Canadian viewpoint. The importance of natural sources, including radon, compared with the effects from the nuclear industry is clearly shown.

The choice of alternatives is at the heart of the risk management process. The knowledge that other fire retardants for textiles were available influenced the decision to ban TRIS (2,3-dibromopropyl phosphate) in Canada in 1977 after it had been shown to be both a potent mutagen and carcinogenic to rodents (Somers, 1984). Similarly, the availability of less hazardous and more effective insulation materials contributed to the deci-

Table 12.1 Some commonplace risks (mean values with uncertainty)

Action	Annual risk	Uncertainty
Motor vehicle accident (total)	2.4×10^{-4}	10%
Motor vehicle accident (pedestrian only)	4.2×10^{-5}	10%
Home accidents	1.1×10^{-4}	5%
Electrocution	5.3×10^{-6}	5%
Air pollution, eastern United States	2×10^{-4}	Factor of 20 downward only
Cigarette smoking, one pack per day	3.6×10^{-3}	Factor of 3
Sea-level background radiation (except radon)	2×10^{-5}	Factor of 3
All cancers	2.8×10^{-3}	10%
Four tablespoons peanut butter per day	8×10^{-6}	Factor of 3
Drinking water with EPA limit of chloroform	6×10^{-7}	Factor of 10
Drinking water with EPA limit of trichloroethylene	2×10^{-9}	Factor of 10
Alcohol, light drinker	2×10^{-5}	Factor of 10
Police killed in line of duty (total)	2.2×10^{-4}	20%
Police killed in line of duty (by felons)	1.3×10^{-4}	10%
Frequent flying professor	5×10^{-5}	50%
Mountaineering (mountaineers)	6×10^{-4}	50%

Source: Wilson and Crouch (1987).

Table 12.2 Average radiation doses (mSv/year)

Natural background (excluding radon)	0.80
Medical diagnosis	1.00
Radon	0.80
Fallout – weapons testing	0.04
Consumer products (food, water products, dentures, eyeglasses, tobacco products)	0.01
Air flight (Halifax–Vancouver return)	0.04
Nuclear power	0.05[a]
Average total dose	2.74

[a] This dose would be the maximum under normal operating conditions by a hypothetical individual living at the nuclear plant boundary.

Source: Johnson and Tutiah (1985).

sion to ban urea formaldehyde foam insulation (UFFI) (Somers, 1984). If the disinfection of drinking water by chlorine, with its attendant risks from trihalomethanes, is to be replaced by ozonation, we need to know the toxicity of the reaction products as well as the microbial quality of the treated water. Asbestos is an acknowledged human carcinogen but if it is to be replaced as an insulation material by man–made material fibers, their potential toxicity should first be known. The toxic properties of agricultural pesticides have led to the development of more degradable compounds that do not possess the undesirable bioaccumulation properties of the organochlorines.

The different energy systems used to generate electricity have different health and safety risks. Inhaber (1982), in comparing the total risk for eleven different energy systems, concluded that nuclear power and natural gas were the least hazardous. Hamilton (1984), in his comparative analyses of the health impacts from coal and nuclear fuel cycles, has estimated the health risk of acidic air pollution to be responsible form some 50,000 premature deaths in the United States and Canada.

The commonest comparison of risk is to other risks, usually in terms of mortality. Thus, the annual death rates for workers exposed to a range of chemical or physical hazards can be compared to each other or to the fatalities from motor vehicle accidents, disasters at sea, football, or horse racing (Pochin, 1974). Wilson and Crouch (1987) have compared the carcinogenic risk from such disparate activities as flying, eating peanut butter, or smoking cigarettes whilst Pochin's (1974) classic comparison shows that a one in a million risk of death can be given by 1.5 cigarettes, 1.5 minutes of rock climbing, 250 miles of air travel, or 20 minutes of being a 60-year-old man! The *reductio ad absurdum* is provided by Cohen (1978), who applied this technique to the risk of death from bladder cancer incurred from diet drinks containing saccharin: failure to install a smoke alarm is as dangerous as ingesting three diet drinks per day, and an average street crossing is about as dangerous as a diet drink. It should be noted that mortality is probably not the best measure of the impact of environmental chemicals on humans. Morbidity and economic loss warrant more consideration.

Wide ranging comparisons of this sort are more suited to the making of debating points than in formulating public policy. When the risk comparisons are more focused, more useful decisions can be made. Ames et al. (1987) have derived potency indices for dietary carcinogens that show the importance of natural food toxins in our total risk from cancer. In order to put the potential hazard from very low levels of asbestos in context, the Ontario Royal Commission (Dupré et al., 1984), who estimated that exposure of one million population to 1.001 fiber/cm^3 of asbestos for 10 years could cause 20 deaths, noted that this is of the order of one-fiftieth as dangerous as driving to and from work over this time period. The Commission concluded that removal of asbestos from schools which could significantly increase the degree of exposure over short periods of time could well cause more harm than good. Essentially, the same conclusion has been drawn by a recent UK report (Anonymous, 1985). It is calculated that exposure to low asbestos levels (0.0005 fibers/cm^3 above background) inside buildings gives a lifetime cancer risk nearly two orders of magnitude lower than that experienced by nonsmokers exposed to cigarette smoke for more than seven hours a week.

When the toxic lesion is lung cancer, as from asbestos, it is often illuminating to compare the risks to those from cigarette smoking. In fact, the 'cigarette unit' is as valuable as the 'vehicle death unit' as an easily comprehended criterion. This approach was used in Canada for the recommended indoor air guidelines of 0.1 ppm formaldehyde from UFFI which was estimated to give a cancer risk of less than one-third-cigarette each day (Somers, 1984).

Risk acceptability

As risk is as old as life and will continue to the end of time there is nothing revolutionary in the concept of risk acceptance. There is, however, a distinction to be made between what is acceptable by whoever makes the assessment and what is accepted by those who have to live with the risk; more properly, what is tolerated by the recipients of the risk. The term 'acceptable risk' gained modern currency from Lowrance's (1976) thoughtful and well argued book. The literature has grown to alarming proportions since Starr (1969) in his consideration of the quantitative measurement of technological risk first suggested that there is a thousand-fold difference between our acceptance of voluntary as opposed to involuntary risks. This gave a measurement to all those well-known impulses whereby individuals subject themselves to the hazards of skiing, alcohol, or motor bikes, but react negatively to fluoridation, food additives, or video display terminals, all of which are regarded with suspicion, for they are imposed from outside. The acceptance of a risk depends not only on the ascribed level of risk, but also on our perception and degree of understanding of that risk. Public perception of risk is often markedly at variance with the estimates given by professional scientists. We are only beginning to explore the reasons for these fears and concerns (Fischhoff et al., 1979; Whyte and Burton, 1982). Sometimes the gap can be overcome by education, but more often it represents deeper, firmly held beliefs. Psychologists have shown how important in risk perception are such factors as unknown global, catastrophic, increasing, uncontrollable, and fearsome risks (Slovic, 1987). This research has been used to explain and forecast acceptance and opposition to particular technologies, particularly the nuclear industry. In the final analysis, public acceptance of a risk depends on public confidence in its effective management (Starr, 1985) while tolerance of certain risks can be linked to an expectation that alternatives will be developed, that further controls will be established or that a specific use will be discontinued.

In recent years, the use of biostatistics to provide quantitative risk assessments for carcinogenesis has led to the use of these calculated numbers to develop policy decisions on the regulatory control of chemicals. The approach has been particularly favored by the US Environmental Protection Agency (Anderson, 1983). It is usually acknowledged that a lifetime risk of 10^{-6} is so low as to be generally acceptable: nearly all regulatory decisions on chemical carcinogens involve higher levels of risk. In an analysis of 132 federal regulatory decisions, Travis et al. (1987) found that every chemical with an individual lifetime risk above 4×10^{-4} was regulated and, except for the FDA decision on dimethylnitrosamine in the nipples of baby bottles, no action was taken to reduce lifetime risks below 10^{-6}. In the development of national water quality criteria for chemical carcinogens, the EPA recommended concentrations associated with a lifetime risk in the range from 10^{-7} to 10^{-5}, although not as a judgement on acceptable risk (Anderson, 1983).

The assessment of carcinogenic risks associated with low levels of exposure to environmental toxicants requires extrapolation of results obtained at higher doses corresponding to experimentally measurable levels of risk. Although this necessarily entails certain assumptions, it is now largely held that the dose–response curve for carcinogenic agents will be linear at low doses (Krewski et al., 1987; Murdoch et al., 1987). Given this assumption, the estimation of low dose cancer risks essentially involves estimating the slope of the dose–response curve in the low dose region. Risk is then estimated by multiplying this slope by the anticipated or estimated level of exposure. Linearized estimates of the low dose slope can be obtained using the robust linear extrapolation procedure proposed by Krewski et al. (1986). This procedure yields estimates of risk similar to those obtained using other methods such as the linearized multi-stage model used by the Carcinogen Assessment Group of the US Environmental Protection Agency (Anderson, 1983).

The quantitative approach to risk assessment is being increasingly adopted in Canada. The *Guidelines for Canadian Drinking Water Quality* (Health and Welfare, Canada, 1979) contain two examples of quantitative risk assessment – for trihalomethanes and nitrilotriacetic acid (NTA). Taking the most conservative analysis of the linear one-hit statistical model, ignoring the biochemical evidence of a concentration threshold for NTA gives a maximum risk of one cancer in two million population per year for chloroform, and one cancer in five million per year for NTA at their maximum acceptable concentration in drinking water.

A scale of our attitudes, ranging from acceptable through warning to unacceptable, for different orders of magnitude of risk has been developed by Wilson (1984). Thus, annual risks of death lower than 10^{-6} are considered acceptable and those higher than 10^{-3} are judged unacceptable. Natural disasters fall at the low end and traffic accidents at the high end of this scale. It is within the middle risk zone, i.e. 10^{-6} to 10^{-3}, that worst-case and risk cost–benefit analyses are applicable.

All of this shows a commendable move towards using more rational analysis of the toxicological data to achieve an acceptability of risk. However, a note of caution must be sounded. Firstly, even with a specific disease such as cancer, the lifetime excess risk accepted by US regulatory agencies can vary from 10^{-2} to 10^{-7} (Albert, 1983; Travis et al., 1987). Secondly, the shape of the dose–response curves derived from animal experiments calculated by different statistical and stochastic models has a profound effect on the estimate of risk (Munro and Krewski, 1981).

However, as noted above, the use of the linear model for carcinogenesis has improved this situation. Thirdly, the risk estimates are based on extrapolation from animal data, with all that entails in terms of assumptions and approximations (Somers, 1984). In summary, the technique is valuable but should not be given more weight than it can bear.

The whole concept of acceptability of risk is extremely complex philosophically. In a sense, the only 'acceptable' risk is one that is judged worthwhile, in terms of costs and benefits, and is incurred by deliberate choice by those subjected to the risk (Council for Science and Society, 1977). Even if the risk is judged as 'acceptable' it may not be ultimately worthwhile to the recipient. It is also clear that however rational the framework that is applied in terms of probability estimates, costs, benefits, allowances, for voluntary as opposed to involuntary risks, and so forth, it will be the deep-rooted cultural and ethical forces that lead to the public's acceptance or rejection of a risk. For example, concern with nuclear energy is related to the concentration of power in quasi-governmental institutions as well as the fear of catastrophic nuclear disaster. Debate on contraceptive chemicals touches on profound religious convictions.

Risk and benefit

We have compared risk to risk, but decision-making requires an examination of the benefits brought by the activity or use that creates the risk. The risk of carcinogenic effects produced by the chlorination of drinking water must be weighed against the benefit of the absence of pathogenic organisms; the risk from organochlorine pesticides to the environment weighed against their benefit to agricultural productivity; the risk of cancer from x-rays weighed against their ability to detect carcinomas at an early stage. The reader should be mindful of the fact that, in some instances, there are alternatives available which do not involve the use of chemicals or at least reduce the volume or nature of chemicals used: physical exercise rather than diet drinks, integrated pest management in the case of pesticides.

Risk–benefit analysis attempts to make a direct comparison between program costs and health and other benefits; it may be regarded as a special case of cost–benefit analysis. Whilst benefits are interpreted in the net sense (the balance or ordinary benefits over the ordinary costs), the risks of loss of life and of human suffering represent extraordinary costs to be evaluated separately. The wide range of components of health program benefits including such factors as less disease, better mental health, less discomfort, and even improved physical appearance are considered by Thompson (1985) in a comprehensive review of the issue. There are also direct and indirect economic benefits due to reductions in health care costs and increased productivity as well as the aesthetic value of a cleaner environment. At the heart of these calculations lies economic quantification of cultural and social values, including the financial value of a human life, that are both elusive and an intrinsic part of our human heritage. Nevertheless, however distasteful as it may appear, society implicitly costs human lives in many of its decisions – whether it be siting an airport, engineering roads, or choosing a safety device (Fischoff et al., 1979).

In Canada, a full socio-economic impact analysis (SEIA) is now required for new government regulations relating to health, safety or fairness which are expected to result in major costs (Treasury Board of Canada, 1979). In addition to the usual economic factors, this requires consideration of non-allocative effects such as those involving market structure, international trade, or inflation. To date, SEIA's have been conducted in assessing the environmental impact of chemicals such as lead in gasoline and PCBS (Environment Canada, 1984), but have not yet been widely undertaken in assessing human risks.

In the USA the federal pesticide act (FIFRA) requires the balancing of risks and benefits. The application of this procedure has retained the registration for chlorobenzilate on citrus fruit, where no substitute existed, given a temporary registration to amitraz, and cancelled most uses of chlordane/heptachor (Anderson, 1983). For chlorobenzilate, the estimated carcinogenic risk to the general population was low, of the order of 10^{-6}, but two orders higher for applicators, so that added labelling requirements were made. For amitraz, the risk–benefit evaluation by EPA on the basis of one carcinogenesis bioassay led to a three-year temporary registration for pears but not for apples, where substitutes were available. The carcinogenicity data for chlordane/heptachlor are sufficiently strong, and the chemicals bioaccumulate, so that cancellation was in order except for termite control, where substitutes were not available.

A similar weighing led to the banning by EPA of most uses of the fumigant ethylenedibromide after it was found to be a potent animal carcinogen (Russell and Gruber, 1987). The US Supreme Court judgement on benzene emphasized that the Occupational Safety and Health Administration (OSHA) must show that there is significant risk before introducing more stringent regulations. The Federal Appeals Court held that the benefits of regulation had to be commensurate with costs (Lave, 1981).

The costs to US industry of meeting different occupational standards for acrylonitrile, a potent animal carcinogen, have been estimated by OSHA as being $24.4 million for 2 ppm, $41.6 million for 1 ppm and over $126 million for 0.2 ppm (Miller, 1983). In 1978, the 2 ppm standard was promulgated. Miller has calculated the annual cost per worker for a one in one thousand reduction in the risk of fatal cancer for different types of production processes of acrylonitrile at the three standard levels. He concluded that at from $2,000 to $8,000 per worker for the 2 ppm standard, the cost is reasonably comparable to that for OSHA's coke oven emission regulations and the Consumer Product Safety Commission's proposed safety standard for lawn mowers, all of which were judged acceptable.

These methods provide a technique for ensuring reasonable equivalence in costs for the same assurance of safety.

Management strategies

The options available to governments to manage the risks from chemicals may, for simplicity, be considered under three headings: educational, economic, and regulatory. Each one may, of course, be applied in conjunction with the others. The regulatory option, the one most immediately associated with government action, should be considered as encompassing the range from promulgated Acts to recommended guidelines and codes of practice.

The educational approach can serve to make producers, workers, and the general public aware of the risks from chemicals so that they can, voluntarily, take the requisite action to reduce or avoid the risk. Public information programs can further health promotion by advocating such sensible life-styles as avoiding smoking, taking alcohol in moderation, and exercising regularly. Governments can strengthen the impact of these programs through advertising in the media to susceptible groups, e.g. the young, and by the personal example of community leaders. Workers can be educated in the hazards of chemicals by courses, lectures, posters, films, and by explicit labelling of chemical products. Consumers can be made aware through government agencies, and via supported product testing studies, of the need to use household chemicals, cosmetics, pesticides, and fertilizers with care and attention.

Positive reinforcement, by publicity, can be given to those manufacturers and producers of chemical products that show corporate responsibility to their workers and the public in their treatment of chemicals, both within and outside their plants. Conversely, those companies which through poor management and a shoddy attitude to plant control are responsible for chemical accidents at the worksite, and pollution outside, are deservedly pilloried by the media, often with serious economic results.

Economic options can provide both positive and negative incentives to the effective control of hazardous chemicals. The 'polluter pays' is a principle espoused by the Organization for Economic Cooperation and Development (OECD, 1984) with the intention of maintaining equitable trading practices by encouraging polluters to reduce emissions. In general, the use of effluent charges will encourage polluters to reduce emissions until the marginal cost of further abatement exceeds the charge itself (Economic Council of Canada, 1981). In theory, effluent charges may be established to encompass all social costs associated with pollution (OECD, 1983). However, in practice, a full determination of societal impacts is difficult, but charges may be set to result in at least some reduction in pollution.

Insurance and other compensatory laws provide another economic approach to the control of risk in that there is a financial incentive to employers to prevent injuries to their workforce. In Canada, as in many countries, worker compensation plans provide cash benefits and reimbursement of medical costs to victims of work-related harm. In Ontario, for example, employers are required, through the Workmen's Compensation Board (Workmen's Compensation Act, 1980), to contribute a fixed sum to the fund for each employee, with compensation paid from an industry-wide pool. Employer premiums vary with the amount of compensation paid across the industry.

Other economic instruments include pollution-control delay penalties, market emission permits and subsidies (OECD, 1983). In the first case, schedules are established in which the maximum allowable emissions are decreased over time. Market emission permits may be issued specifying maximum cumulative and dispersive emission levels for particular pollutants. Subsidies such as grants for pollution abatement equipment may be used as a monetary incentive for pollution control. Tax deductions, rebates and credits all play a part in the total fiscal policy of governments' economic control of chemicals. These issues are considered in depth in a recent review by Krewski and Birkwood (in press).

Regulatory options rely on government authority to enforce compliance with specific health and safety requirements. The authority can be through the force of law and prosecution or through the more gentle route of persuasion and recommendations. In principle, the essential elements for a health protection control program follow this pathway:

Investigation/Research $\longrightarrow$ Criteria $\longrightarrow$ Standards $\longrightarrow$
Regulations $\longrightarrow$ Enforcement and Compliance $\longrightarrow$

That is to say, the investigation and research lead to an evaluation of the health hazards, formulated as criteria. From these criteria, health standards are derived which we in turn convert to a legislative reality when promulgated as regulations; the final stage is the enforcement of these regulations, so that if they are not achieving the original intention, the cycle of investigations to regulation must be repeated once more. Ideally, such measures relate achievements to needs (Somers, 1984).

Regulatory agencies, which have to control noxious agents, need a number to give reality to a regulation, standard, guidelines, or even a recommendation (Somers, 1987). Whatever the strength of the regulatory action, a number provides the reference point for enforcement and surveillance. The number can be a concentration, e.g. milligrams per liter (mg/l), micrograms per cubic meter ($\mu g/m^3$), a ratio, e.g. parts per million (ppm), an emission or discharge rate, e.g. kilograms per day (kg/day), or a deposition application rate, e.g. kilograms per hectare (kg/ha). Essentially, the number enables us to make a judgement as to what is safe – or, more properly, an acceptable risk – as opposed to what is negligent or criminal.

The number should be enforceable and be amenable to change as new knowledge develops and as social values change. At best it will be an approximation for health protection programs without giving an absolute guarantee of freedom of risk.

Guidelines, while expressed in the same way as standards, are envisaged more as recommendations that do not allow legal recourse to ensure compliance. They are particularly valuable in ensuring uniform environmental quality in countries where the responsibility for public health protection or the management and protection of natural resources is either held by or shared with other jurisdictions. Guidelines are also developed by, and appropriated to, international agencies such as those of the United Nations. In addition, guidelines are flexible, without enforcement costs, and serve as an incentive to industry. However, they have limited power – other than public persuasion – can be disregarded, and are capable of misinterpretation.

In general, environmental standards may be considered to fall into three categories: environmental quality standards, emission and effluent standards, and product standards (Somers, 1987). Environmental quality standards prescribe the maximum allowable levels of pollutants in media such as air, water, and soil. They are often designed to cover certain geographic areas, e.g. national, regional, or local. However, it may not be wise to set only a maximum allowable level for a pollutant in the environment. A lower level should be designated that provides a long-term objective for environmental quality, otherwise a license to pollute to the maximum may be inferred. This concept is incorporated into the former Canadian Clean Air Act, now Canadian Environment Protection Act, whereby national air quality objectives were established at three levels: 'tolerable', 'acceptable', and 'desirable'. The first sets the concentration which requires abatement without delay to avoid the onset of a threat to human health, the second provides adequate protection to people and the environment, while the maximum desirable level is a long-term goal for air quality and provides a basis for an anti-degradation policy for the unpolluted parts of the country and for the continued development of control technology.

Emission and effluent standards specify the maximum amounts or concentration of pollutants that may be released into the atmosphere either from stationary or mobile (e.g. automobile) sources. On occasion, process standards prescribing production processes or modes of operation may be applied to specify the methods to be used to achieve the emission standards. Products standards prescribe tolerance limits based on the product's properties, packaging, or emissions. Examples are the pollutant limits in foods for heavy metals or pesticides. Sometimes, in the case of fish, for example, the limits may serve to protect both human health and the environment in that the fish may be particularly sensitive to the chemical.

The Canadian approach to risk management derives from the appropriate legislative authority – both federal and provincial. In general, the

Canadian political system adopts a liberal democratic view of the state. It accords to market forces the primary initiatives in the introduction of new products and goods – and hence of new hazards. The state does, of course, referee the process and, on occasion, assumes direct involvement through public enterprise. Control of toxic substances can be applied from manufacture to disposal. Federal jurisdiction, particularly the criminal law, trade, and commerce powers, provides the basis for the major Acts (Table 12.1) designed to protect the health of Canadians from environmental hazards (Somers and MacDonald, 1983). In all, for toxic substances, there are some 27 federal statutes which exercise some form of control and the 10 provinces have enacted 99 pieces of legislation.

In general, the choice of measures and strategies for regulation depend on the existing legal and social constraints of the country. Some countries, such as the United Kingdom, make use of self-regulation whilst others, such as the USA, adopt a more rigorous approach with reliance on the courts (de Koning, 1987). The Canadian approach represents an intermediate position and can serve to illustrate the general principles of managing toxic chemicals.

The Canadian Environmental Protection Act controls hazards that are not effectively dealt with under other legislation and provides broad powers to collect data and conduct investigations with respect to substances entering or likely to enter the environment in concentrations which may endanger human health or the environment. Under the former Environmental Contaminants Act, only uses of PCBs, PCTs, Mirex, and chlorofluorocarbons had been prohibited. Regulations under the Food and Drugs Act prescribe standards for the quality of food and maximum levels for such toxic chemicals as lead, arsenic, and aflatoxin. In contrast, the Hazardous Products Act relies on schedules (such as the prohibition of the sale of UFFI) and regulations, rather than broad powers of prohibition. For drinking water, guidelines have been developed through federal–provincial collaboration, with the Department of National Health and Welfare fulfilling the role of the health adviser as prescribed under the Department of National Health and Welfare Act (Health and Welfare Canada, 1979). A similar approach has been taken with the recently developed guidelines for indoor air quality (Health and Welfare Canada, 1987).

It must be emphasized that regulatory strategies for risk management range from outright banning, through setting emission standards or maximum allowable concentrations, to recommendations or guidelines. In all cases, the authority derives from basic federal Acts. There have been two recent legislative developments in Canada that extend the range of regulatory strategies available to federal government. Firstly, the government has promulgated a more comprehensive Canadian Environmental Protection Act which subsumes the authority of the Clean Air, Environmental Contaminants, the Ocean Dumping Control Acts, and Part III of the Canadian Water Act. Secondly, a national system has been introduced, through amendments to the Hazardous Products Act, to protect workers handling dangerous substances. The Workplace Hazardous Material Information System (WHMIS) will provide information on chemicals used in the workplace. Some 100,000 products are expected to fall under the new regulations of which 10,000 of them contain confidential information. The WHMIS attempts to balance the workers' right to know with industry's right to protect confidential information.

Conclusions

It can be seen that risk management requires the integration of hazard identification and risk estimation with social, economic, and political issues. Although the world of chemicals looks daunting, systems have been and are being developed to provide a rational framework in which to make these decisions. The scientific method can take us so far, but we need more knowledge on the basic biochemical processes that underlie the toxicological results on which the ultimate regulatory decision is made. And we need better information on the dose of chemical, in its toxic form, to which we are exposed. After that, societal judgement comes into play; we leave science behind and move into the area of risk perception and political agendas that has little to do with numbers but is far more concerned with human values and the legitimacy of institutions (Otway and Thomas, 1982).

Examples have been given, largely from the Canadian experience, of dealing with these issues. These illustrate one approach to world problems. Solutions, on occasion, can be achieved by bilateral negotiation, as when the governments of Canada and the USA entered into the 1972 and 1978 Great Lakes Water Quality Agreements to deal with the pollution problems of the Great Lakes. International efforts to provide integration and co-ordination in the area of chemical safety culminated in the launching of the International Programme on Chemical Safety (IPCS) in 1980. The IPCS is a co-operative venture of the International Labour Organization, the World Health Organization, and the UN Environment Programme. At present, some twenty developed and developing countries participate in the joint work. The IPCS provides not just an opportunity to develop common evaluations, investigations, guidelines, and training programs. The program seeks to give international validity and cachet to the risk estimation of chemicals – essential in any international attempts to control their hazards. The main outputs of IPCS include: evaluations of the adverse effects chemicals may produce on human health (including carcinogenesis, mutagenesis, and teratogenesis); principles for establishing exposure limits for chemicals in foods, drinking water, and the working environment; guidelines for exposure assessment, toxicity testing, and epidemiology; information on chemical accidents and emergencies; and training for personnel, particularly in developing countries. In addition, IPCS provides co-ordinating links with the UNEP International Register of Potentially Toxic Chemicals and the International Agency for Research on Cancer of WHO.

As described in Chapters 1 and 9 and for a more limited group of countries, the OECD, through its Chemicals Group, provides a focus for the management of chemical safety programs. Thus, both nationally and internationally, many steps have been taken in recent years to provide a more comprehensive and rational approach to the management of risks from chemicals. With good science and sensible co-ordination, further achievements can soon be expected, improving the effectiveness of these approaches.

References

Albert, R. E. (1983), 'Discussion', *Journal of the Air Pollution Control Association* 33, pp. 836–7.

Ames, B. N., Magaw, R. and Gold, L. S. (1987), 'Ranking possible carcinogenic hazards', *Science* 236, pp. 271–80.

Anderson, E. L., and the Carcinogen Assessment Group of the US Environmental Protection Agency (1983), 'Quantitative approaches in use to assess cancer risk', *Risk Analysis* 3, pp. 277–95.

Anonymous (1985), 'Asbestos: a panic exposed', *New Scientist* 106 (1453), p. 3.

Cohen, B. L. (1978), 'Saccharin: The risks and benefits', *Nature* 271, p. 492.

Council for Science and Society (1977), *The Acceptability of Risks*, (London: Barry Rose).

de Koning, H. W. (1987), *Setting Environmental Standards*, (Geneva: WHO).

Dupré, J. S., Mustard, J. E., and Uffen, R. J. (1984), *Report of the Royal Commission on Matters of Health and Safety Arising from the Use of Asbestos in Ontario*, Vol. 2 (Toronto: Ontario Government).

Economic Council of Canada (1981), *Reforming Regulation* (Ottawa: Supply and Services Canada).

Environment Canada, Environmental Strategies Directorate, Environmental Protection Service (1984), *Socio-Economic Impact Analysis of Lead Phase-Down Control Options* (Ottawa).

Fischhoff, B., Slovic, P. and Lichtenstein, S. (1979), 'Weighing the risks', *Environment* 21 (4), pp. 17–38.

Hamilton, L. D. (1984), 'Health and environmental risks of energy systems', International Atomic Energy Agency, in *Risks and Benefits of Energy Systems*, pp. 20–57 (Vienna: IAEA).

Health and Welfare Canada (1979), *Guidelines for Canadian Drinking Water Quality (1978)* (Ottawa: Supply and Services Canada).

Health and Welfare Canada (1987), *Exposure Guidelines for Residential Indoor Air Quality* (Ottawa: Department of National Health and Welfare).

Inhaber, H. (1982), *Energy Risk Assessment* (New York: Gordon & Breach).

IRPTC (International Register for Potentially Toxic Chemicals) (1983), United Nations Environment Programme, Geneva.

Johnson, H. and Tutiah, M. (1985), *Radiation is Part of Your Life*, WNREI–501 (Pinawa, Manitoba: Atomic Energy of Canada Ltd).

Krewski, D. (1986), 'Risk and risk management: issues and approaches', in R. S. McColl (ed.) *Environmental Health Risks: Assessment and Management* (Waterloo, Ontario: University of Waterloo Press).

Krewski, D. and Birkwood, P. L. (1987), 'Risk assessment and risk management', *Risk Abstracts* 4, pp. 53–61.

Krewski, D. and Birkwood, P. L. (1988), Regulatory and non-regulatory options of risk management. In L. R. G. Martin and S. Laford (eds), *Risk Assessment and Management: Emerging Planning Perspectives*, pp. 253–71 (Waterloo: University of Waterloo Press).

Krewski, D., Murdoch, D. and Dewanji, A. (1986), 'Statistical modeling and extrapolation of carcinogenesis data', in S. H. Moolgavkar and R. L. Prentice (eds), *Modern Statistical Methods in Chronic Disease Epidemiology* (New York: Wiley-Interscience).

Krewski, D., Murdoch D., and Withey, J. (1987), reprinted in 'The application of pharmacokenetic data in carcinogenic risk assessment', in *Pharmacokinetics in Risk Assessment: Drinking Water and Health* Vol. 8, pp. 441–68 (Washington, DC: National Academic Press).

Krewski, D., Somers, E. and Birkwood, P. L. (1987), 'Risk perception in a decision-making context', *Environ. Carcinogenesis Rev. (Journal of Environmental Science and Health)* C J (2), pp. 175–209.

Lave, L. B. (1981), *The Strategy of Social Regulation* (Washington: The Brookings Institute).

Lowrance, W. W. (1976), *Of Acceptable Risk* (Los Altos, Calif.: William Kauffmann).

Miller, J. C. III. (1983), 'Occupational exposure to acrylonitrile: a benefit/cost analysis', *Toxic Substances Journal* 4, pp. 223–33.

Munro, I. D. and Krewski, D. R. (1981), 'Risk assessment and regulatory decision-making', *Food Cosmetic Toxicology* 19, pp. 549–60.

Murdoch, D., Krewski, D., and Crump, K. S. (1987), 'Mathematical models of carcinogenesis', reprinted in J. R. Thompson and B. W. Brown (eds), *Cancer Modeling* (New York: Marcell Dekker).

NAS (National Academy of Sciences), National Research Council (1984), *Toxicity Testing: Strategies to Determine Needs and Priorities* (Washington, DC: National Academy Press).

OECD. (Organization for Economic Cooperation and Development), Environmental Committee Group of Economic Experts (1983), *Economic Instruments: Review and Outlook*. Background Paper ENV/ECOOCB/–83.20, Session 9, International Conference on Environment and Economics, Paris, December 30.

OECD. (Organization for Economic Cooperation and Development), Environmental Committee Group of Economic Experts (1984), *Economic Instruments: Review and Outlook*. Background Paper ENV/ECO/CI/83.9, Session 9, International Conference of Environment and Economics, Paris, March 14.

Otway, H., and Thomas, K. (1982), 'Reflections of risk perception and policy', *Risk Analysis* 2, pp. 69–82.

Pochin, E. E. (1974), 'Occupational and other fatality rates', *Community Health* 6, pp. 2–12.

Russell, M. and Gruber, M. (1987), 'Risk assessment in environmental policy-making', *Science* 236, pp. 286–90.

Slovic, P. (1987), 'Perception of risk', *Science* 236, pp. 280–5.

Somers, E. (1984), 'Risk estimation for environmental chemicals as a basis for decision-making', *Regulatory Toxicology Pharmacology* 4, pp. 99–106.

Somers, E. (1986), 'The weight of evidence: Regulatory toxicology in Canada', *Regul. Toxicol. Pharmacol.* 6, pp. 391–8.

Somers, E. (1987), 'Making decisions from numbers', *Regul. Toxicol. Pharmacol.* 7, pp. 35–42.

Somers, E. and MacDonald, G. E. (1983), 'Regulatory control of chemicals: the Canadian experience', in F. Homberger (ed.), *Safety Evaluation and Regulation of Chemicals* (Basel: Karger).

Starr, C. (1969), 'Social benefit versus technological risk', *Science* 221, pp. 1026–8.

Starr, C. (1985), 'Risk management, assessment and acceptability', *Risk Analysis* 5, pp. 97–102.

Thompson, M. S. (1985), 'Measuring health benefits', in D. B. Clayson, D. Krewski, and I. Munro (eds), *Toxicological Risk Assessment* Vol. II (Boca Raton, Fla: CRC Press).

Travis, C. C., Richter, S. A., Crouch, E. A., Wilson, R., and Klema, E. D. (1987), 'Cancer risk management', *Environmental Science and Technology* 21, pp. 415–20.

Treasury Board of Canada (1979), 'Socio-economic impact analysis', *Treasury Board Administration Policy Manual*, Chapter 490 (Ottawa: Treasury Board of Canada).

Whyte, A. V. and Burton, I. (eds) (1980), *Environmental Risk Assessment*, SCOPE Report No. 15, (New York: Wiley).

Whyte, A. V. and Burton, I. (1982), 'Perception of risks in Canada', in I. Burton, C. D. Fowle, and R. S. McCullough (eds), *Living with Risk: Environmental Risk Management in Canada* (University of Toronto: Institute for Environmental Studies).

Wilson, R. (1984), 'Commentary: risks and their acceptability', *Science, Technology and Human Values* 9(2), pp. 11–22.

Wilson, R. and Crouch, E. A. C. (1987), 'Risk assessment and comparisons: an introduction', *Science* 236, pp. 267–70.

Workmen's Compensation Act (1980), Revised Statutes of Ontario 1980, Chapter 539, pp.1203–66.

13 *The future of toxic chemicals management*

R. P. CÔTÉ & P. G. WELLS

Introduction

The management of toxic chemicals in many countries occurs, at best, through a blend of fragmented programs and continual response to real and perceived crises. In part, this is due to the fact that countries, agencies and industries, being new to the task, are inexperienced in the complexities of the issue. In addition, they have a limited information base, limited predictive capabilities and a finite problem-solving capacity. It may also be due to a lack of recognition of the various facets of the problem at hand, which may have resulted in the choice of ineffective management strategies. Why must toxic chemicals issues be managed differently in the future? How should they be managed and by whom? What new technologies and strategies should be considered? This book has provided some answers and guidance, by exploring the toxic chemical management issue from a multi-disciplinary and multi-sectoral point of view.

The history of the development, manufacture, use and disposal of chemicals demonstrates quite conclusively that production and supply have moved faster than the capacity of institutions to ensure that the products are 'safe' in ecological and human health terms. In Chapter 1, Côté mentions a number of events which have prompted management and especially regulatory initiatives. Unfortunately, these initiatives continue to take a chemical-specific and, to a certain extent, a source-specific approach, each problem being examined and solutions sought independently. They are also viewed as reactive rather than preventive in nature. Therein lies part of the problem and the challenge of effectively managing new and existing chemicals.

Predictability of effects and exposures

The scientific and financial resources behind the development of new products and new uses (the benefits) far exceed those dedicated to assessing the potential effects (the risks) of each chemical on health of humans or the natural environment. Some progress is being made in the re-assessment of existing chemicals, though the resources available to undertake the work are very limited when compared to the number of chemicals in use today. The limitation of resources has necessitated setting priorities for chemicals subjected to the review process. While this is a step in the management process which has to be undertaken in any structured program aimed at the regulation of chemicals, it often appears that the priorities are established on the basis of emergencies, press reports and political expediencies. When this occurs, the resources available may not be deployed in the most effective manner. However, significant improvements have been made in the management of new chemicals. Somers (Chapter 12) and Rankin (Chapter 9) have described a number of initiatives taken by governments and international organizations to ensure that the situation is likely to improve in the future.

As stated in Chapter 1, there are approximately 100,000 chemicals in commercial use today and as many as 1,000 new chemicals are introduced into commerce each year. These are produced, transported, used and discarded in quantities of many millions of tonnes each year. If these chemicals were all known to be completely benign, there would be little need for structured and improved management. But many of the chemicals involved are known, and even designed to be, corrosive, explosive, flammable, oxidizers, or toxic. Some are considered poisonous, causing effects in a short period of time while others are carcinogenic and their effects may not be noted for thirty or forty years. Certain chemicals are designed to be lethal to certain organisms, notably pesticides. Others, such as chlorofluorocarbons, assumed to be quite innocuous, have had unexpected atmospheric effects.

The current state of our scientific knowledge and toxicological techniques, though rapidly evolving, is such that our predictive abilities are not advanced enough always to prevent further ecological or health problems. It may, of course, be unreasonable to expect all chemical problems will be prevented from happening. The need to improve predictive capabilities is highlighted by Wolff and Crossland in Chapter 2 in regard to the transport

Table 13.1 Some uses of polychlorinated biphenyls and mercury

Polychlorinated biphenyls	Mercury
Plastics	Fungicides
Inks	Electrical equipment
Waxes	Thermometers
Carbonless copy paper	Paints
Paints	Dentistry
Pesticides	Metal recovery
Cutting oils	Catalysts
Hydraulic fluids	
Dielectric fluids	

and fate of organic chemicals, Rand on toxicity (Chapter 3) and Sheehan on ecotoxicological effects (Chapter 4). Testing capability has been evolving rapidly since the 1970s and the available arsenal now includes a wider range of techniques than ever before. Unfortunately many of the techniques are still viewed by some regulatory agencies as experimental or preliminary. Therefore they have not been institutionalized in regulations and permitting systems to the extent that they might be.

The adoption of a more extensive suite of tests does however differ markedly for different types of chemicals. Food additives, drugs and pesticides now undergo extensive pre-market testing, though even in these cases all eventualities cannot be defined and accommodated prior to approval and use. In the case of drugs, a decision may be taken to allow a product on the market because of its benefits to sufferers of a disease or condition. However, the information on the product label must contain a list of contra-indications, adverse effects and precautions. There is an understanding among drug companies, the medical community and regulators that side effects will occur and that adverse effects may be noted as the drug is used more extensively. Once the product is released for use, the information on adverse effects is then compiled and relayed to the medical community so that better judgements can be made about the risks to individual patients. At some point, the regulatory agencies may decide that the risks outweigh the benefits and the product is taken off the market. An information system of this nature and scope does not appear to exist for any other category of chemicals, though there are some similarities in the case of pesticides.

One of the most significant shortcomings in the management of chemicals noted by various chapter authors is surveillance and monitoring. Sheehan has argued that monitoring is the weak link in ecotoxicological assessments. It is possible that major ecological and health impacts are not occurring, though the evidence in the last three decades would certainly tend to lead one to the opposite viewpoint. The real question is with the significance of those effects. Will they be serious enough to threaten natural ecosystems? Will they threaten life-supporting systems, such as the ozone layer? Will they decrease the viability of populations by interfering with immune systems? Unfortunately, the amount of interpretive monitoring undertaken by chemical producers, government agencies and others is so limited as to preclude statements with any degree of confidence. This is especially true in the case of chemicals which are widely used and dispersed in the environment, are persistent and bioaccumulate.

Clearly, improvements in predictability are important in ensuring that existing chemicals do not cause unacceptable impacts and that new chemicals present as little risk as possible. An expanded suite of tests assessing transport, transformation, fate and toxicological effects should be adopted by regulatory agencies. These tests must be supported by monitoring programs which serve in part to evaluate the validity and accuracy of predictions. International organizations, such as the OECD which represents the countries where the majority of these chemicals are produced, have already adopted a minimum pre-market data set for commercial chemicals. This data set is a step in the right direction but it represents only the beginning. The most appropriate data set remains to be identified, along with the tests which will produce those data.

Another area of investigation is that of multiple exposures. Most people are subjected to exposures to a large number of chemicals at any one time, through the air, the water and the food they ingest, in their workplaces and through the products they use. And yet, most of the chemical testing does not reflect this. The emphasis is on single chemicals. While safety factors are applied when standards are set, there is very little information available on the interaction and influence of tens if not hundreds of chemicals to which we are exposed on a daily basis. This situation also applies to animals and plants, though perhaps to a lesser degree. Paehlke has made the case that the linkages between the various exposure environments require more attention. In Chapter 7, he focused on the linkage of occupational and environmental health, stating that reduction of exposure in one area alone tends to underestimate the impact of at least some chemicals on people. As will be discussed later, it is important to recognize that in many industrialized countries, the assessment of exposures is often divided into distinct exposure categories by different agencies.

Control strategies

Understanding the influence of the above factors is necessary if the most effective control strategies are to be selected and applied. As described by Bishop in Chapter 6, a wide range of strategies has been considered by regulatory agencies. These include regulation of production and use, waste minimization, control of release, waste treatment and clean-up of contamination. The selection of the appropriate strategy or combination of strategies is influenced by the diversity of compounds and their properties, the different transport pathways, the diversity of producers and users, accidental releases, natural sources of some compounds or elements and past disposal practices. As can be seen in Chapter 6, these strategies are largely technologically based though they are accepted by the legal community in part because they can be readily defined in regulations and permits. Numbers can be assigned to the effluent or emissions from a treatment facility with some degree of confidence. In contrast, laboratory tests are viewed as being less exact, and ecotoxicological studies are even less so. This is largely due to the natural variability of biological systems. Hence, when biological test results are used in court, the end-point which is routinely accepted is death of the test organisms. From an ecological and human perspective, this end-point is inappropriate because it lacks sensitivity.

In recent years, society has been moving away from its emphasis on measuring consequences, i.e. end-points such as death, toward blocking those consequences through the application of waste treatment technologies. Bishop identified strategies such as regulation of production and use, including banning of compounds. More recently, waste minimization strategies are being viewed more favorably because of their focus on efficiency and cost effectiveness. The latter approach is still largely technologically based but emphasizes the recovery, re-use and recycling of chemicals. This approach can reduce exposures in all environmental categories.

Disciplines and interests

The emphasis on conservation of resources and sustainable development has provided the impetus for the development of a hierarchical decision-making framework for managing chemicals. The framework is based on the hazard causation model which was in turn derived from fault-tree and event-tree analysis. The purpose of the framework is to ensure that the approach taken to manage a particular type of chemical (1) defines the problem as it is understood by the different interests; (2) is comprehensive; and (3) provides an opportunity for discussion among the different interests or stakeholders. The toxic chemical issue has been perceived by many as primarily scientific and technical in nature. As such, chemists, biologists and engineers have figured prominently in the OECD countries. The legal discipline has also played an important role in the management process though its influence has varied depending on the country. For example, lawyers have been more prominent in the United States than in the United Kingdom. Scientists and lawyers are not the only other disciplines that play a role in the management of chemicals. Jansen, Holtz and Doern (Chapters 8, 10 and 11) have identified the need for economists, sociologists, perhaps psychologists, and political scientists in understanding the nature of the concerns faced by the different interests whose perspectives must be considered. The involvement of the various disciplines in the decision-making process is described in Figure 13.1.

As the previous chapters suggest, the nature of the issue can be and is often perceived differently by the scientific community, regulatory agencies, industry, unions and environmentalists. This is not to say that individuals cannot be faced with internal conflicts. As indicated by Holtz in Chapter 10, an individual can have divided loyalties as might occur if one worked for a chemical producer, was a consumer and homeowner and perhaps also a member of a local wildlife conservation organization concerned about the effects of chemicals on wildlife. But frequently, the values and concerns of individuals are subsumed into the perspective of the organizations which then appear to take on a life of their own. These become the interests which influence the management process. It is then

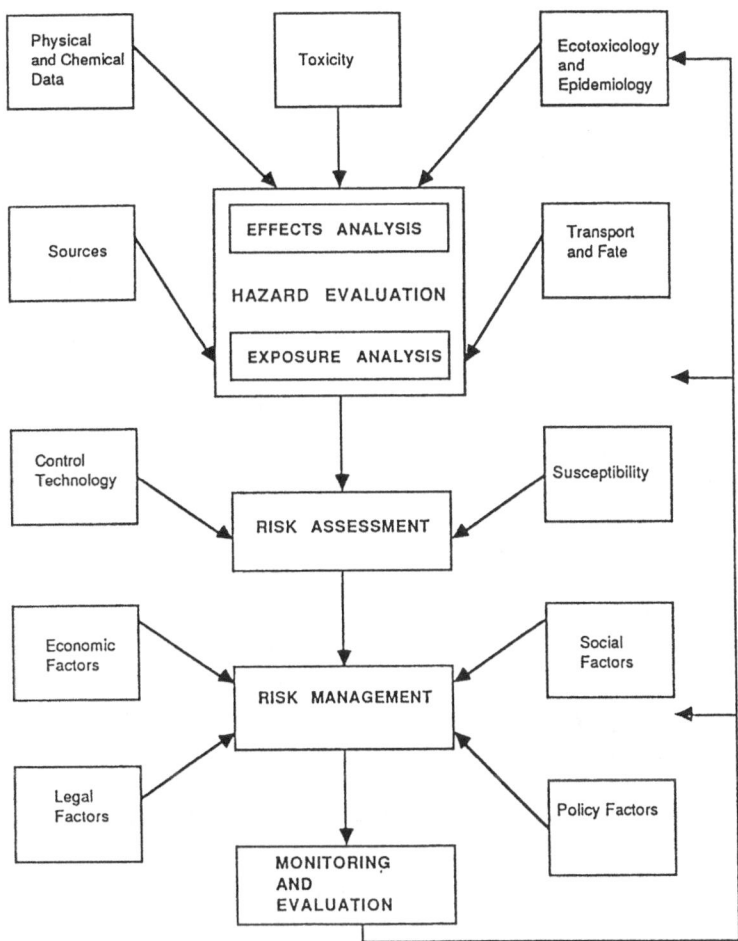

Figure 13.1 Managing toxic chemicals: a multidisciplinary effort

critical to ascertain the nature of the concerns which are driving particular interests. For many categories of chemicals, a framework can be developed to determine the level in the hierarchy at which each of the interests wishes to intervene. Figures 13.2 and 13.3 depict the application of the framework to two chemical issues. These hierarchies emphasize the fact that problem definition is central to the debate that occurs between parties. In Figure 13.2, the problem could be defined as pesticides management or pest management. In Figure 13.3, the problem could be identified as hazardous wastes management or resource management generally; in this instance, the objective could be defined as mitigation of the effects of hazardous waste or alternatively better management of residuals. In some of the debates

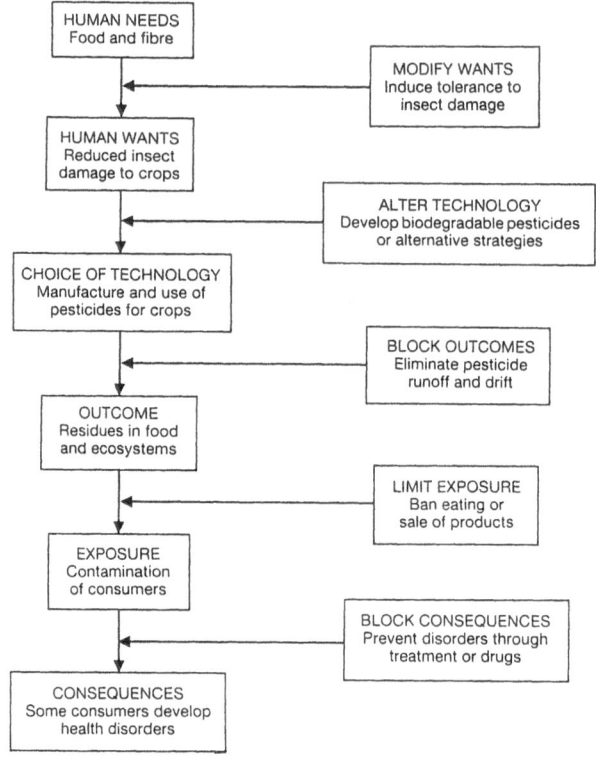

Figure 13.2 The societal decision-making framework for pest and pesticide management

between government, industry and environmentalists, the various interests may be attacking each other at different levels when their long-term objectives are the same. Once that level is identified, some common ground involving a range of alternatives might be cultivated, allowing effective management decisions to be taken.

Economic incentives and the market

Much of the debate will undoubtedly revolve around risks and costs. These are not issues to be left to a few disciplines nor to one or two interests. As indicated in Figures 13.2 and 13.3, the hierarchical level at which a decision is taken may have significant cost implications for industry, government, unions or consumers. The burden of those costs will spread as one moves up the hierarchy. Many of those costs have traditionally been externalized. As such, any burdens or risks resulting from exposure to chemicals have

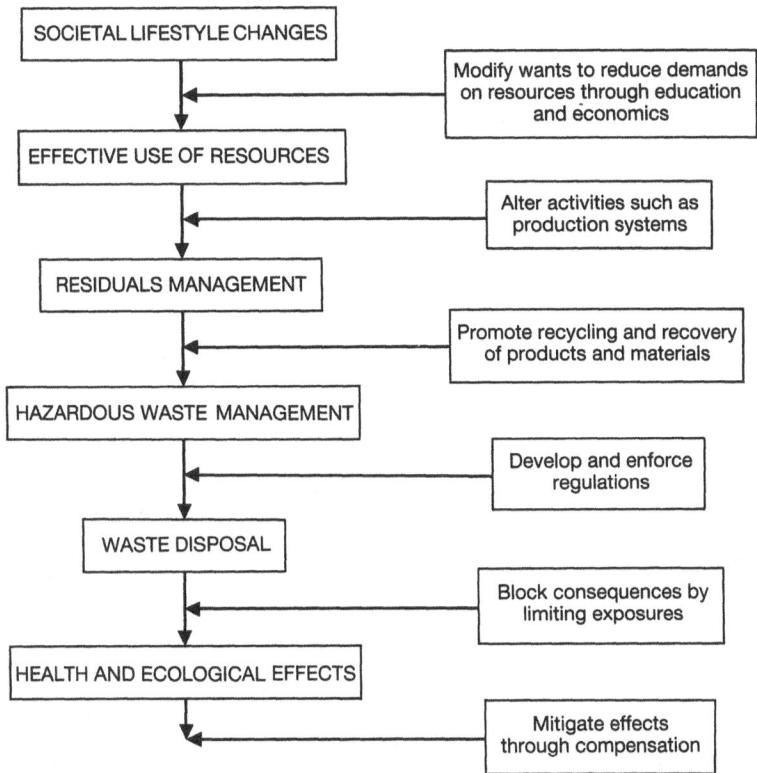

Figure 13.3 A hierarchical decision-making framework for hazardous waste management

been borne by the community at large, the taxpayer and natural eco-systems. Recently, as described by Jansen in Chapter 8, governments, pressured by environmental organizations and by taxpayers, have begun to consider ways of internalizing those costs in the prices of products, as well as reflecting them in corporate profits and dividends to shareholders. Since the use of market mechanisms has in fact been advocated by the private sector, it seems that there is now a real opportunity for testing and evaluation of economic incentives and disincentives. The argument has been put forward that this approach would foster innovation and result in less waste being produced. However, the public as represented by the environmental lobby is unlikely to accept such an approach unless there are some legal safeguards. The challenge now is to find a reasonable balance of economic incentives and legal sanctions while resolving a chemical problem.

Fragmentation

Fragmentation can be an obstacle which influences the ability of a country or organization to effectively manage the toxic chemicals issue. First, the issue itself is segmented within the bureaucracy. In some countries, Canada for example, different agencies bear the responsibility for managing specific components of the life cycle of chemicals. Thus, the Department of Transport is responsible for dangerous goods, i.e. chemicals in transport whether by ship, airplane, train or truck; the Department of Labor is responsible for protecting workers in the workplace; the Fire Marshal may be responsible for setting standards for storage of chemicals, and the Department of Environment is often responsible for waste disposal.

From a number of points of view, this situation appears quite sensible. For example, those involved in the transportation industry are able to deal with a single agency. Unfortunately, the chemicals issue is pervasive. Problems can occur at any stage in a chemical's life cycle and chemicals know no boundaries. In addition, as noted in Table 13.1, individual chemicals or compounds may have many uses, and the same chemical may be found in a large number of products. Thus co-operation and information exchange are required among all the industrial sectors and government agencies involved. Figure 13.4 highlights the interrelatedness of the exposure environments and the agencies involved.

The fragmentation of bureaucracies within government detracts from co-operation and information exchange. Each agency develops its own policies, priorities, practices, computerized information systems, etc., with the result that it has an ingrained reluctance to change, an inertia which is not easily surmounted. The consequence is delay in addressing problems involving single chemicals or classes of chemicals when the responsibilities cross bureaucratic lines. An example of this situation is that involving symbols depicting hazardous categories of chemicals in Canada. The symbols for hazardous chemicals in the workplace, in transport, and for pesticides are not consistent. This is relatively minor but it indicates the nature of the fragmentation problem.

The problem is compounded by the fragmentation within the legal regime. In Canada alone, more than twenty-seven pieces of legislation can be utilized to some degree in the management of chemicals. This situation imposes large transaction costs on all parties. Confusion is also created in the minds of industry and the public when the regulatory requirements differ for the same chemicals under different statutes. The situation is aggravated further among industry and the judiciary itself when the penalties for contravening the regulations are not consistent. Some attempts have been made to remedy this problem by the promulgation of enabling or 'umbrella' legislation such as the Toxic Substances Control Act in the United States and the Canadian Environmental Protection Act. Unfortunately, definitional problems continue to occur. By definition, a

Figure 13.4   The linkage of human health and environment as reflected by institutional divisions in government and industry

pesticide is not a toxic substance in either of these countries. This situation is related in part to the bureaucratic fragmentation described earlier wherein different agencies may be responsible for pest control products and toxic substances. Government policy makers must be aware of these situations and guard against them. Policies which apply across all agencies involved with the management of chemicals must be elaborated upon to ensure consistency, and co-ordinative mechanisms must be put into place.

The international dimension

Much of the legislation and many of the bureaucracies controlling the manufacture, transport, storage, use and disposal of chemicals in OECD countries have evolved since 1970. This action by these governments reflects a response to internal pressures, that is, pressures brought to bear by the public, by concerned individuals within government and by interested politicians. Policies were proposed and initiatives were taken in response to events which actually occurred or which were judged likely to occur because they had happened elsewhere. For example, a number of steps to exercise an increasing amount of control on PCBs were taken in Canada following specific newsworthy incidents. These included the leakage of PCBs onto the Trans Canada Highway near Kenora, Ontario in 1985, the fire in a warehouse storing waste PCBs in St Basile le Grand, Quebec, in

1988, and finally the shipment of waste PCBs to the United Kingdom in 1989. All of these events were followed by more stringent regulations than had existed before.

While international action has influenced the regulation of hazardous materials for at least forty years, beginning with international conventions dealing with the carriage of oil at sea, it is only recently that the international community has begun to exert significant influence on the management of chemicals within countries. The Basel Convention on the Transboundary Movement of Hazardous Wastes, signed in 1989, is evidence of increasing activity at the international level which will require nations to establish more stringent regulations within their own borders. In the past, international agreements have established lowest common denominators. It is possible that future conventions or treaties will set higher standards than have been the case in the past. This may place impossible burdens on less developed countries. A careful assessment of those burdens will have to be undertaken so that they are shifted to those multinational chemical companies and industrialized exporting countries which have been and are benefiting from chemicals in commerce.

Information

Several of the chapters have stressed the importance of timely and accurate information as a critical factor in the management of chemicals. There is a recognized lack of information in a number of areas but especially about the toxicological effects of chemicals. More research is clearly necessary to fill these gaps. Government and industry would also argue that there is a problem with misinformation. This view would probably be shared by unions and the environmental community, all of whom would claim that incomplete, or in some cases, biased information is presented to influence the public at large and politicians regarding the 'safety' of chemicals. This situation might be overcome by having independent, respected institutions produce technical assessments available to all parties. This type of assessment is already done in some countries, e.g. the National Academy of Sciences in the United States, the Royal Society of Canada, and by international agencies such as the World Health Organization and the United Nations Environment Program. More could be done on contentious chemicals.

Finally there is the question of access to information, which must be addressed from two perspectives. Within jurisdictions, industry argues (supported by some agencies of government), certain information should remain confidential because its release could affect an industry's competitiveness. Therefore elaborate schemes have been developed for the release of information. That information is made available in many forms to different interested parties. More effort must be made to provide information in a usable fashion to concerned individuals and groups.

There is also a need to share information between countries, especially with regard to the import and export of chemicals. Information systems are being created and expanded at a rapid pace. International agencies must ensure that different systems are connected to maximize access.

Conclusions

The management system advocated in this text should be comprehensive, addressing all stages of the life cycle of chemicals in a less fragmented manner than has been the case to date. It must ensure that the strategies being contemplated are strategically assessed. It must be multi-disciplinary in scope with a strong scientific basis. It must be more predictive as society becomes increasingly concerned with complex mixtures and multiple exposures. The system must incorporate an open decision-making process with explicit assumptions if confidence is to be re-established in producers and regulators.

The purpose of this book has been to highlight the dimensions and basic principles of the management of toxic chemicals. While the emphasis of the book is clearly on chemicals, many of the principles and concepts explored apply to the management of other environmental issues as well.

Selected readings

Conway, R. A. (ed.) (1982), *Environmental Risk Analysis for Chemicals* (New York; Van Nostrand Reinhold).

Crone, H. D. (1986), *Chemicals and Society: A Guide to the New Chemical Age* (Cambridge: Cambridge University Press).

Edelstein, M. R. (1988), *Contaminated Communities: The Social and Psychological Impacts of Residential Toxic Exposure* (Boulder, Colo: Westview Press).

Estrin, D. (1986), *Handle with Caution: Liability in the Production, Transportation and Disposal of Toxic Substances* (Toronto: Carswell).

Freedman, B. (1989), *Environmental Ecology: The Impacts of Pollution and Other Stresses on Ecosystem Structure and Function* (New York: Academic Press).

Gusman, S., von Moltke, K., Irwin, F., and Whitehead, C. (1980), *Public Policy for Chemicals: National and International Issues* (Washington: The Conservation Foundation).

Guthie, F. E., and Perry, J. J. (1980), *Introduction to Environmental Toxicology* (New York: Elsevier).

Kolaczkowski, S. T., and Crittenden, B. D. (eds) (1987), *Management of Hazardous and Toxic Wastes in the Process Industries* (Barking, UK: Elsevier Applied Science).

Lave, L. B., and Upton, A. C. (eds) (1987), *Toxic Chemicals, Health and the Environment* (Baltimore: Johns Hopkins University Press).

National Research Council (1986), *Ecological Knowledge and Environmental Problem Solving. Concepts and Case Studies* (Washington: National Academy Press).

Ramade, F. (1987), *Ecotoxicology* (Chichester: John Wiley and Sons).

Sheehan, P. J., Miller, D. R., Butler, G. C., and Bourdeau, P. (eds) (1984), *Effects of Pollutants at the Ecosystem Level*, SCOPE 22. (Chichester: John Wiley and Sons).

Index

NOTE: information contained in figures is referenced in italics; reference to information contained in Tables is prefixed with the word 'Table'. Both figures and tables have notation which includes the chapter and item number within the chapter, eg., 4.2 refers to chapter 4, second table or figure.